George Gore

The Art of Electro-Metallurgy

Including all known processes of electro-de-position. Second Edition

George Gore

The Art of Electro-Metallurgy
Including all known processes of electro-de-position. Second Edition

ISBN/EAN: 9783337218102

Printed in Europe, USA, Canada, Australia, Japan

Cover: Foto ©berggeist007 / pixelio.de

More available books at **www.hansebooks.com**

THE ART

OF

ELECTRO-METALLURGY

INCLUDING ALL KNOWN PROCESSES OF

ELECTRO-DEPOSITION

BY G. GORE, LL.D., F.R.S.

SECOND EDITION

D. APPLETON AND CO.

NEW YORK

1884

INTRODUCTION.

—◦◦—

HAVING been asked by the publishers of the Text-books of Science, to write a small volume on the subject of Electro-metallurgy, I have endeavoured to produce such a book as would be useful to scientific students, to practical workers in the art of electro-metallurgy, gilders, platers, &c., and to all persons who wish to obtain in a compact form, an explanation of the principles and facts upon which the art of electro-metallurgy is based, the circumstances under which nearly every known metal is deposited, and the special details of technical workshop manipulation in the galvano-plastic art. I have also given an historical sketch of the development of the subject, arranged in chronological order.

The book is divided essentially into four parts, viz., First, the Historical sketch, showing how from one or two isolated, and apparently unimportant facts, the great subject of Electro-chemistry arose, and by the incessant, and unremunerated labours of many eminent scientific investigators, and the exertions of practical operators, it has gradually extended, until nearly every known metal has been separated, copper has been deposited in great quantities, and the eminently useful, and beautiful products of artistic electro-deposition, have spread nearly all over the Earth, and are to be found in every civilised home. The Second part consists of the Theoretical division, being a concise statement of the chief facts and principles upon which the practical art is based, together with descriptions of the classes of phenomena usually met with in electrolytic and electro-depositing processes; the facts and principles being arranged in as systematic and logical an order as I could place them. The Third part (section A) is the first portion of the Practical division of the book ; and treats of the general methods of deposition, the selection of depositing processes, the general rules to be obeyed, and points to be observed, in actual working with all metals, followed by the special means of depositing nearly every known metal and metalloid.   The metals, &c., are arranged in their ordinary chemical classes in the following order :—Electro-negative or brittle metals, noble, base, earth and alkaline earth, alkali metals, and finally the metalloids ; and the arrangement is such, that every known instance of the electro-deposition of nearly every known metal, and metalloid, may be readily found and referred to.   It is hoped that not only students and practical workers in the art, will find this section of value to them, but that even scientific investigators may find it useful for reference. The Fourth (or concluding) section, B is of a more special and technical character, and has been composed almost entirely for the use of practical operators, including those who have not had the advantage of chemical instruction: it contains a variety of technical points of instruction necessary for the successful prosecution of the art,—information which could not be so conveniently classed or supplied in the preceding sections. This part also includes a list of all the books published on the subject, and an extensive and almost complete list of the English Patents (nearly 300 in number) relating to electro-deposition, taken out from the earliest period of the art, until the present time.

I have endeavoured not only to make the book a treatise on the practical art of Electro-metallurgy, but also to include an outline of the science of electro-chemistry, upon which that art is based ; and I have also spared no trouble in order to make it as perfect as I could ; the most complete portions are those which treat of the common methods of silvering, gilding, moulding ; the deposition of copper, nickel, brass, iron, and tin ; the special details of the art ; and the accounts of such experiments and processes with the less common metals, as scientific investigators and practical inventors, may be likely to further examine, or practically apply.

Numerous experiments of my own on the subject, (many of them, through want of previous opportunity, being now for the first time published), are scattered through the first part of the Practical section of the book ; a few of them being made to fill up missing links, whilst the book was in progress. I had hoped also to have made others in a similar way, for the purpose of settling some debateable questions still remaining ; for instance, to determine whether aluminium is, or is not, capable of being deposited from an aqueous solution ; but, owing to the deficiency of encouragement of original scientific research in this country, I have been deterred from so doing.

I beg to express my indebtedness to Mr. E. W. BALL, and to my colleague, Mr. A. BRUCE, F.C.S., for the assistance they have kindly rendered me in correcting the proof-sheets.

GEORGE GORE.

Birmingham, 1877.

CONTENTS.

HISTORICAL SKETCH.

CLASS. IV. *Base Metals.*

a

Contents. xix

SPECIAL INFORMATION RESPECTING SUBSTANCES, ETC., USED IN
THE ART.

THE

ART OF ELECTRO-METALLURGY.

————◆◆◆————

THE earliest known facts respecting the electro-deposition
of metals were those in which one metal, by being dipped
into a solution of another, became coated with the latter
metal. For instance, iron or steel, when dipped into a solution of blue vitriol, became covered with a coating of copper ;
copper, dipped into a solution of mercury, became amalgamated ; zinc, immersed in a solution of lead, formed a tree
of lead, and in one of silver produced the *arbor Dianæ*.
These and other similar facts of a chemical nature were
known long before the discovery of voltaic electricity.
Gilding and silvering on metals had been known for many
ages ; gilded statues and bronzes have been found in the
tombs of the ancient Egyptians. Both Pliny and Vitruvius
speak of processes of gilding and silvering ; but these early
processes appear to have been effected by means of amalgams of mercury, and not by electro-chemical methods.
Zosimus, however, speaks of the deposition of bright metallic copper from its solution by means of iron. Paracelsus
also and Bernard de Palissy, a thousand years later, were
acquainted with, and describe, the means of coating copper

B

and iron with silver by simple immersion in a solution of silver.

One of the earliest recorded facts in connection with voltaic electrolysis is that observed by Sulzer, who in the year 1752 remarked : 'If you join two pieces of lead and silver, so that they shall be upon the same plane, and then lay them upon the tongue, you will notice a certain taste resembling that of green vitriol, while each piece apart produces no such sensation.' ('Histoire de l'Académie des Sciences et Belles-Lettres de Berlin'). The earliest known fact of electrolysis by separate electric discharge appears to be that of Paetz and Van Troostwik, who in the year 1790 decomposed water into its two constituent gases, by passing electric sparks through it by means of very fine gold wires (De la Rive's 'Treatise on Electricity,' vol. ii., p. 443).

But all these were empirical, stagnant, and comparatively unfruitful facts ; no great progress resulted from them because they were not generalised upon, and were not recognised as instances of any great law or principle. Electrolysis did not start into active and real progress until after Volta made his great discovery of chemical electricity in the year 1799. About that time he produced his crown of cups, which was the first arrangement by means of which a current of voltaic electricity could be produced for any continued length of time.

Cruickshank soon afterwards devised his well-known trough battery, in which zinc and copper plates were fixed in vertical grooves, so as to form a more powerful and compact arrangement (Highton's 'Electric Telegraph,' pp. 13, 14, and 29).

Nicholson and Carlisle, on May 2, 1800, first decomposed water by means of a voltaic current (Highton's 'Electric Telegraph,' pp. 27 and 29).

Dr. Henry, of Manchester, also about the same year, decomposed nitric and sulphuric acids, and resolved ammonia into its constituent gases by similar means ('Encyclopedia Metropolitana,' vol. iv., pp. 221 and 611).

In the year 1801 Wollaston remarked that 'if a piece of silver, in connection with a more positive metal, be put into a solution of copper, the silver is coated over with copper, which coating will stand the operation of burnishing' ('Philosophical Transactions of the Royal Society,' 1801).

In the same year Gerboin first noticed the movements of mercury in a conducting liquid when a voltaic current was passed through the liquid metal; we now know that those movements were due to electrolysis (De la Rive's 'Treatise on Electricity,' vol. ii., p. 433). Hisinger and Berzelius also, in 1803, found by means of many experiments that, under the influence of a voltaic current, the elements of water and of neutral salts were transferred to the respective poles of the battery ('Encyclopedia Metropolitana,' vol. iv., pp. 221, 222). About the same time Cruickshank passed a voltaic current, by means of silver wires, through solutions of acetate of lead, sulphate of copper, nitrate of silver, and several other salts, and found that the metals attached. themselves to the wire connected with the zinc-end of the battery; and stated that the metals were 'revived' so completely as to suggest to him the analysis of minerals by means of the voltaic current (Wilkinson's 'Elements of Galvanism,' vol. ii., 1804, p. 54).

The first result of a decidedly practical form in electrogilding was that of Brugnatelli, who, in the year 1805, gilded two silver medals by making them the negative pole in a newly-made and well-saturated solution of ammoniuret of gold ('Philosophical Magazine,' 1805). He also electrodeposited bright metallic silver upon platinum, and observed that when the current entered the liquid by means of a pole of copper or zinc, those metals were dissolved and then deposited upon the negative pole (see his 'Annals of Chemistry'). One of the greatest discoveries in the subject, however, was that of Sir Humphry Davy, made on October 6, 1807. He passed a powerful electric current, from a battery composed of 274 cells, through a fragment of moistened potash, and deposited the metal potassium itself upon the negative platinum wire. Seebeck, of Berlin, in the year 1822 discovered thermo-electricity by observing that, when the soldered junction of two different metals (bismuth and copper) was heated, an electric current was produced. In 1824 Sir H. Davy attempted to protect the copper sheathing of ships by means of strips of zinc attached to it. Nobili, in the year 1826, discovered that when a current of voltaic electricity was passed into a solution of acetate of lead by means of a plate of platinum, and out of it by means of a platinum wire, rings of beautiful colours, caused by the formation of thin films of peroxide of lead, appeared on the platinum plate ; this effect was named ' metallo-chromy.'

Magneto-electricity was discovered by Faraday. In the year 1831 he produced a spark by pulling a keeper (covered with a coil of insulated wire) from the poles of a magnet ; he also obtained a magneto electric current by rotating a copper plate between the poles of a magnet, and by sliding a coil of insulated copper wire upon a steel-bar magnet. In 1834 he made the important discovery, that when a voltaic current was passed through different salts in solution or in a state of fusion, the amount of salt decomposed by the current was in direct proportion to the quantity of electricity ; and that the quantities of substances dissolved and set free in electrolysis were in definite proportions by weight, and that those proportions were identical with the ordinary chemical equivalents of the substances, and thus established the important law of definite electro-chemical action. He also proved that the quantity of electricity from a voltaic battery depends upon the size of the immersed portion of the plates, and that the intensity of the current depends upon the number of alternate pairs of metals ; and used a voltameter to ascertain the strength of the current.

In 1836 Mr. De la Rue devised a peculiar form of Daniell's battery, and observed that ' the copper plate is also covered with a coating of metallic copper, which is continually being deposited ; and so perfect is the sheet of copper thus formed that, being stripped off, it has the counterpart of every scratch of the plate on which it is deposited' ('Philosophical Magazine,' 1836).

In 1837 Dr. Golding Bird decomposed, by means of a voltaic current, solutions of the chlorides of sodium, potassium, and ammonium, and deposited their respective metals on a negative pole of mercury, and thus obtained their amalgams ('Philosophical Transactions of the Royal Society,' 1837, p. 37).

Several persons now made experiments upon the electro-deposition of metals at about the same time, and brought electro-metallurgy into prominent notice. Professor Jacobi, of St. Petersburg, published his galvano-plastic process, 'a method of converting any line, however fine, engraved on copper, into a relief by galvanic process, applicable to copper-plate engravings, medals, stereotype plates, ornaments, and to making calico-printing blocks and patterns for paper-hangings' ('Athenæum,' May 4, 1839). Mr. T. Spencer, of Liverpool, also on May 8, 1839, gave notice to read a paper on the 'Electrotype Process' to the Liverpool Polytechnic Society, but the paper was not read until September 13 in the same year.

Meanwhile, Mr. C. J. Jordan, on May 22, 1839, sent a letter to the 'London Mechanics' Magazine,' which was published on June 8, 1839. After stating that his experiments were made 'about the commencement of last summer, with a view of obtaining impressions from engraved copper plates,' it proceeds as follows :

'It is well known to experimentalists on the chemical action of voltaic electricity that solutions of several metallic salts are decomposed by its agency, and the metal procured in a free state. Such results are very conspicuous with copper salts, which metal may be obtained from its sulphate (blue vitriol) by simply immersing the poles of a galvanic battery in its solution, the positive wire becoming gradually coated with copper. This phenomenon of metallic reduction is an essential feature in the action of sustaining batteries, the effect, in this case, taking place on more extensive surfaces. But the form of voltaic apparatus which exhibits this result in the most interesting manner, and relates more immediately to the subject of the present communication, may be thus described : It consists of a glass tube, closed at one extremity with a plug of plaster-of-Paris, and nearly filled with a solution of sulphate of copper ; this tube and its contents are immersed in a solution of common salt. A plate of copper is placed in the first solution, and is connected by means of a wire and solder with a zinc plate, which dips into the latter. A slow electric action is thus established through the pores of the plaster, which it is not necessary to mention here ; the result of which is the precipitation of minutely crystallised copper on the plate of that metal, in a state of greater or less malleability, according to the slowness or rapidity with which it is deposited. In some experiments of this nature, on removing the copper thus formed I remarked that the surface in contact with the plate equalled the latter in smoothness and polish, and mentioned this fact to some individuals of my acquaintance. It occurred to me, therefore, that if the surface of the plate was engraved an impression might be obtained. This was found to be the case, for, on detaching the precipitated metal, the most delicate and superficial marking, from the fine particles of powder used in polishing to the deeper touches of a needle or graver, exhibited their corresponding impressions in relief with great fidelity. It is, therefore, evident that this principle will admit of improvement, and that casts and moulds may be obtained from any form of copper.

'This rendered it probable that impressions may be obtained from those other metals having an electro-negative relation to the zinc plate of the battery. With this view, a common printing type was substituted for the copper-plate, and treated in the same manner. This also was successful ; the reduced copper coated that portion of the type immersed in the solution. This, when removed, was found to be a perfect matrix, and might be employed for the purpose of casting when time is not an object.

'It appears, therefore, that this discovery may be turned to practical account. It may be taken advantage of in procuring casts from various metals, as above alluded to ; for instance, a copper die may be formed from a cast of a coin or medal, in silver, type metal, or lead, &c., which may be employed for striking impressions in soft metals. Casts may probably be obtained from a plaster surface surrounding a plate of copper ; tubes or any small vessel may also be made by precipitating the metal around a wire, or any kind of surface, to form the interior, which may be removed mechanically, by the aid of an acid solvent, or by heat.

'C. J. JORDAN.

'May 22, 1839.'

Mr. Spencer in his paper, after making some preliminary remarks, states :—' In the latter part of September 1837 I was induced to make some electro-chemical experiments with single pairs of plates, consisting of small pieces of zinc and equal-sized pieces of copper, connected together with wires of the latter metal. It was intended that the action should be slow ; the fluids in which the metallic electrodes were immersed were in consequence separated by thin discs of plaster-of-Paris. In one cell thus formed was placed sulphate of copper in solution, in the other a weak solution of common salt. I need scarcely add that the copper electrode was placed in the cupreous solution, the other being in that of the salt. I mention these experiments briefly ; not because they are *directly* connected with what I shall have to lay before the Society, but because, by a portion of their results, I was induced to come to the conclusions I have done in the following paper. I was desirous that no action should take place on the wires by which the electrodes were held together ; and to obtain this object I

varnished them with sealing-wax varnish, but in one instance
I dropped a portion on the copper electrode that was
attached. I thought nothing of this circumstance at the
moment, but put the experiment inaction.

'This operation was conducted in a glass vessel ; I had
consequently an opportunity of occasionally examining its
progress from the exterior. After the lapse of a few days,
metallic crystals had covered the copper electrode—with the
exception of *that portion* which had been spotted with the
drops of varnish. I at once saw that I had it in my power
to guide the metallic deposition in any shape or form I chose,
by a corresponding application of varnish or other non-
metallic substance.

' I had been aware of what everyone who uses a sustaining galvanic battery with sulphate of copper must know, that
the copper plates acquire a coating of copper from the action
of the battery ; but I had never thought of applying it to a
useful purpose, except to multiply the plates of a species of
battery, which I did in 1836. My present attempt was with
a piece of thin copper plate having about four inches of
superficies, with an equal-sized piece of zinc, connected as
before by a piece of copper wire. I gave the copper a coating of soft cement, consisting of bees'-wax, rosin, and a red
earth. It was compounded in the way recommended by
Dr. Faraday, in his work on *Chemical Manipulation*, but
with a larger proportion of wax. The plate received its
coating while hot. When it was cold, I scratched the initials
of my name rudely on the plate, taking special care that the
cement was quite removed from the scratches, that the
copper might be thoroughly exposed. This was put in action
in a cylindrical glass vessel, about half filled with a saturated
solution of sulphate of copper. I then took a common gas
glass, similar to that used to envelope an argand burner, and
filled one end of it with plaster-of-Paris acting as a partition
to separate the fluids, but at the same time being sufficiently

porous to allow the electro-chemical action to permeate its substance.

'I now bent the wire in such a manner that the zinc end of the arrangement should be in the saline solution, while the copper end, when in its place, should be in the cupreous solution. The gas glass, with the wire, was then placed in the vessel containing the sulphate of copper.

' It was then suffered to remain at rest, when in a few hours I perceived that action had commenced, and that the portion of the copper rendered bare by the scratches had become gradually coated with pure, bright deposited metal, whilst all the surrounding portions were not at all acted on. I now saw my former observations realised ; but whether the deposition so formed would retain its hold on the plate, and whether it would be of sufficient solidity or strength to bear the working if applied to a useful purpose, became questions which I now determined to solve by experiment. It also became a question, should I be successful in these two points, whether I should be able to produce lines sufficiently in relief to print from. This latter appeared to depend entirely on the nature of the cement or etching-ground I might use.

'This I endeavoured to solve at once ; and, I may state, it appeared at the time to be the main difficulty, as my impression then was, that little less than one-eighth of an inch would be requisite to print from.

' I now procured a piece of copper, and gave it a coating of a modification of the cement I have already mentioned, and, having covered it to about one-eighth of an inch in thickness, I took a steel print and endeavoured to draw lines in the form of network, that should entirely penetrate the cement, and leave the surface of the copper exposed. But in this I experienced much difficulty from the thickness I deemed it necessary to use, more especially when I came to draw the cross lines of the network. The cement being soft, the lines were pushed, as it were, into each other, and

when it was made of harder texture, the intervening squares
of the network chipped off the surface of the metallic plate.
However, those that remained perfect I put in action as
before.

' In the progress of this experiment I discovered that the
solidity of the metallic deposition depended entirely on the
weakness or intensity of the electro-chemical action, which I
knew I had in my power to regulate at pleasure, by the
thickness of the intervening wall of plaster-of-Paris, and by
the coarseness or fineness of the material. I made three
similar experiments, altering the texture and thickness of the
plaster each time, by which I ascertained that if the parti-
tions were *thin* and *coarse*, the metallic depositions proceeded
with great *rapidity*, but the crystals were pliable and easily
separated ; on the other hand, if I made them thicker, and
of a little finer material, the action was slower, but the metallic deposition was as solid and ductile as copper formed
by the usual methods ; indeed, when the action was exceedingly slow, I have had a metallic deposition apparently
much harder than common sheet-copper, but more brittle.

' There was one most important, and to me discouraging,
circumstance attending these experiments, which was, that
when I heated the plates to get off the covering of cement,
the meshes of copper network occasionally *came off with
it.* I at one time imagined this difficulty insuperable, as
it appeared that I had cleared the cement entirely from
the surface of the copper that I meant to have exposed; and
I concluded that there must be differences in the molecular
arrangement of copper prepared by heat and that prepared
by voltaic action which prevented their chemical combination. However, I determined, should this prove so, to
turn it to account in another manner, which I shall
relate in the second portion of the paper.

' I now occupied myself for a considerable period in
making experiments on this latter section of the subject.

' In one of them I found, on examination, that a portion

of the copper deposition, which I had been forming on the surface of a coin, adhered so strongly that I was quite unable to get it off; indeed, a chemical combination had apparently taken place. This was only on one or two spots on the prominent parts of the coin. I immediately recollected that on the day I put the experiment in action I had been using nitric acid for another purpose on the table I was operating on, and that in all probability the coin might have been laid down where a few drops of the acid had accidentally fallen. Bearing this in view, I took a piece of copper, coated it with cement, made a few scratches on its surface until the copper appeared, and immersed it for a short time in dilute nitric acid, until I perceived, by an elimination of nitrous gas, that the exposed portions were acted upon sufficiently to be slightly corroded. I washed the copper in water, and put it in action as before described. In forty-eight hours I examined it, and found the lines were entirely filled with copper. I applied heat, and then spirits of turpentine, to get off the cement, and, to my satisfaction, I found that the voltaic copper had completely combined itself with the sheet in which it was deposited.

'I then gave a plate a coating of cement to a considerable thickness, and sent it to an engraver; but when it was returned I found the lines cleared out so as to be wedge-shaped, or somewhat in the form of a V, leaving a hair-line of the copper exposed at the bottom, and a broad space near the surface; and where the turn of the letters took place the top edges of the lines were galled and rendered rugged by the action of the graver. This, of course, was an important objection, which I have since been able to remedy in some degree by an alteration in the shape of the graver, which should be made of a shape more resembling a narrow parallelogram than those in common use: some engravers have many of their tools so made. I did not put this plate in action, as I saw that the lines, when in relief, would have been broad at the top and narrow at the bottom. I took another plate, gave it a coating of the wax, and had it written on with a mere point. I deposited copper on the lines, and afterwards had it printed from.*

'I now considered part of the difficulty removed : the principal one yet remaining was to find a cement, or etching-ground, the texture of which should be capable of being cut to the required depth, without raising what is technically called a *burr*, and at the same time of sufficient toughness to adhere to the plate when reduced to a small, isolated point, which would necessarily occur in the operation which wood-engravers term cross-hatching.

'I have since learned from practical engravers that much less relief is necessary to print from than I had deemed indispensable, and that, on becoming more familiar with the cutting of the wax-cement, they would be enabled to engrave in it with facility and precision.

'I tried a number of experiments with different combinations of wax, resins, varnishes, earths, and metallic oxides, all with more or less success. One combination that exceeded all others in its texture was principally composed of bees'-wax, resin, and white-lead. This had nearly every requisite, so that I was enabled to polish the surface of the plate with it until it was nearly as smooth as a plate of glass. With this compound I had two plates, five inches by seven, coated over, and portions of maps cut on the cement, which I had intended should have been printed off. I applied the same process to these as to the others, immersing them into dilute nitric acid before putting them in action—indeed I suffered them to remain about ten minutes in the solution. I then put them into the voltaic arrangement. The action proceeded slowly and perfectly for a few days, when I removed them. I applied heat, as usual, to remove the cement, but *all* came away, as in a former instance—the voltaic copper peeling off the plate with the greatest facility. I

---

* This plate was shown to friends, and also specimens of printing from it, in 1838.

was much puzzled at this unexpected result, but, on cleaning the plate, I discovered a delicate trace of *lead*, exactly corresponding to the lines drawn on the cement previous to the immersion in the dilute acid. The cause of this failure was at once obvious : the carbonate of lead I had used to compound the etching-ground had been decomposed by the dilute nitric acid, and the metallic lead thus reduced had deposited itself on the exposed portions of the copper plates, preventing the voltaic copper from chemically combining with the sheet of copper. I was now with regret obliged to give up this compound and to adopt another, consisting of bees'-wax, common resin, and a small portion of plaster-of-Paris. This seems to answer the purpose tolerably, though I have no doubt, by an extended practice, a better may still be obtained by a person practically acquainted with the etching-grounds in use among engravers.

'I now proceed to the second, and, I believe, the most satisfactory portion of the subject. Although I have placed these experiments last, some of them were made at the same time with the others already described, and some of them before ; but to render the subject more intelligible I have placed them thus.

'The members of the Society will recollect that, on the first evening it met, I read a paper on the "Production of Metallic Veins in the Crust of the Earth," and that, among other specimens of cupreous crystallization which I produced on that occasion, I exhibited two coins,—one wholly covered with metallic crystals, the other on one side only. It was used under the following circumstances : when about to make the experiment, I had not a slip of copper at hand to form the negative end of my arrangement, and, as a good substitute, I took a penny and fastened it to one end of the wire and put it, in connection with a piece of zinc, in the apparatus already described.

Voltaic action took place, and the copper coin became covered with a deposition of copper in a crystalline form.

But when about to make another experiment, and being desirous of using the piece of wire used in the first instance, I pulled it off the coin to which it was attached. In doing this, a piece of the deposited copper came off with it ; on examining the under portion of which, I found it contained an exact mould of a part of the head and letters of the coin, as smooth and sharp in every respect as the original on which it was deposited. I was much struck with this at the time, but, on examination, the deposited metal was very brittle. This, and the fact that it would require a metallic nucleus to aggregate on, made me apprehensive that its future usefulness would be materially abridged; but it was reserved for future experiment, and in consequence laid aside for a time, until my attention was recalled to the subject in a subsequent experiment, already detailed, by the drops of varnish on a slip of copper. Finding in this instance that the deposit would take the direction of any non-conducting material, and be, as it were, guided by it, I was induced to give the previous branch of the subject a second trial, because I had, in the first instance, supposed that the deposition would only take place continuously, and not as isolated specks of a metallic surface, as I now found it would ; but the principal inducement to investigate the subject was the fact of finding that the deposited copper had much more tenacity than I at first imagined.

'Being aware of the apparent natural law which limits metallic deposition by voltaic electricity, excepting in the presence of a metallic body, I perceived that the uses of the process would in consequence be extremely limited, except in the multiplication of already-engraved plates, as, whatever ornament it might produce, it would only be done by adhering to the condition of a metallic mould.

'I accordingly determined to make an experiment on a very prominent copper medal. It was placed in a voltaic circuit, as already described, and deposited a surface of copper on one of its sides, to about the thickness of a shilling.

I then proceeded to get the deposition off. In this I experienced some difficulty, but ultimately succeeded. On examination with a lens, every line was as perfect as the coin from which it was taken. I was then induced to use the same piece again, and let it remain a much longer time in action, that I might have a thicker and more substantial mould, in order to test fairly the strength of the metal. It was accordingly put again in action, and let remain until it had acquired a much thicker coating of the metallic deposition ; but on attempting to remove it from the medal I found I was unable. It had apparently completely adhered to it.

'I had often practised with some degree of success a method of preventing the oxidation of polished steel, by slightly heating it until it would melt fine bees'-wax ; it was then wiped apparently completely off, but the pores or surface of the metal became impregnated with the wax.

'I thought of this method, and applied it to a copper coin.

'I first heated it, applied wax, and then wiped it so completely off, that the sharpness of the coin was not at all interfered with. I proceeded as before, and deposited a thick coating of copper on its surface. Being desirous to take it off, I applied the heat of a spirit lamp to the back, when a sharp crackling noise took place, and I had the satisfaction of perceiving that the coin was completely loosened. In short, I had a most complete and perfect copper mould of one side of a halfpenny.

'I have since taken some impressions from the mould thus taken, and by adopting the above method with the wax, they are separated with the greatest ease.

'By this experiment it would appear that the wax impregnates the surface of the metal to an inconsiderable depth, and prevents a chemical adhesion from taking place on the two surfaces ; and I can only account for the crackling noise on separation, by supposing it probable that the molecular arrangement of the voltaic metal is different from that subjected to percussion, and this difference causes an unequal degree of expansibility on the application of heat.

'I now became of opinion that this latter method might be applied to engraving much better than the method described in the first portion of this paper. Having found in a former experiment that copper in a voltaic circuit deposited itself on lead, with as much rapidity as on copper, I took a silver coin and put it between two pieces of clean sheet-lead, and placed them under a common screw-press. From the softness of the lead, I had a complete and sharp mould of both sides of the coin, without sustaining injury. I then took a piece of copper wire, soldered the lead to one end and a piece of zinc to the other, and put them into the voltaic arrangement I have already described. I did *not* in this instance *wax* the mould, as I felt assured that the deposited copper would easily separate from the lead by the application of heat, from the different expansibility of the two metals.

'In this result I was not disappointed. When the heat of a spirit lamp was applied for a few seconds to the lead the copper impression came easily off. So complete do I think this latter portion of the subject, that I have no hesitation in asserting that *fac-similes* of any coin or medal, no matter of what size, may be readily taken, and as sharp as the original. To test further the capabilities of this method I took a piece of lead plate, and stamped some letters on its surface to a depth sufficient to print from when in relief. I deposited the copper on it, and found it came easily off, the letters being in relief.

'Finding from this experiment that the extreme softness of the lead allowed it to be impressed on by type-metal, I caused a small portion of ornamental letter-press to be set up in type, and placing it on a planed piece of sheet-lead, it was subjected to the action of a screw-press.

'After considerable pressure, it was found that a perfectly

sharp mould of the whole had been obtained in the lead. A wire was now soldered to it, and it was placed in an apparatus similar in principle, but larger than the one already described. At the end of eight days from this time copper was deposited to one-eighth of an inch in thickness ; it was subjected to heat, when the two metals began to loosen. The separation was completed by inserting a piece of wedge-shaped wood between them.

' I had now the satisfaction of perceiving that I had obtained a most perfect specimen of stereotyping in copper, which had only to be mounted on a wooden block to be ready to print from.

' From the successful issue of this experiment, which was mainly due to the susceptibility of the lead, I was induced to attempt to copy a wood engraving by a similar method, provided the wood would bear the requisite pressure. Knowing that wood engravings are executed on the *end* of the block, I had better hopes of succeeding, the wood being less likely to sustain injury.

' I accordingly procured a small block, and placed its engraved surface in contact with a piece of sheet-lead, made very clean, and subjected it to pressure as in the former instance. I had now, as before, the gratification of perceiving that a perfect mould of the little block had been obtained, and no injury done to the original. Several wood engravings and copper plates were subjected to similar treatment, and are now in process of being deposited on in the apparatus before me.

' I now come to the third and concluding portion of the experiments on this subject, the object being to deposit a metallic surface on a model of clay, wood, or other *non*-metallic body, as otherwise I imagined the application of this principle would be extremely limited. Many experiments were made to attain this result, which I shall not detail, but content myself with describing those which were ultimately most successful.

c

'I procured two models of an ornament—one made of clay, and the other of plaster-of-Paris, soaked them for some time in linseed-oil, took them out and suffered them to dry. I gave them a thin coat of mastic varnish. When the varnish was nearly dry, *but not thoroughly so,* I sprinkled some bronze powder on that portion I wished to make a mould of. This powder is composed of mercury and sulphur, or it may be chemically termed a sulphuret of mercury. There is a sort which acts much better, in which is a portion of gold. I had, however, a complete metalliferous coating on the surface of the model, by which I was enabled to deposit a surface of copper on it by the voltaic method I have already described. I have also gilt the surface of a clay model with gold leaf, and have been successful in depositing copper on its surface. There is another, and, as I trust it will prove, a similar method of attaining this object ; but, as I have not sufficiently tested it by experiment, I shall take another op-portunity of describing it.' The reading of this paper was accompanied by an exhibition of specimens of coins, medals, and copper plates, formed by the electrotype process.

The publication of Mr. Spencer's paper excited general attention, and thousands of persons of all classes of society

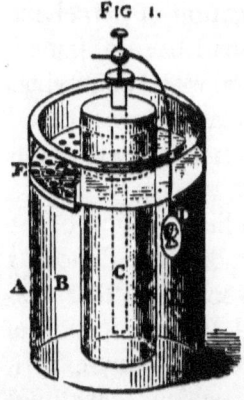

Fig 1.

at once became fascinated by the new art ; and various improvements were made which soon led to its becoming a definite manufacturing process.

The apparatus commonly employed at this period was what is termed the 'single cell,' and is shown in the annexed figure, in which A is a wide, open jar of glass or earthenware, nearly filled with a saturated solution of blue vitriol (sulphate of copper). B is a narrow and much taller jar, made of porous earthenware, containing dilute sulphuric acid, and filled to the same level as the outer vessel. C is a tall rod of cast zinc immersed in the acid, and supported in the centre of the jar. D is a coin or medal, suspended in the copper solution, and attached by means of a copper wire to the rod of zinc ; and E is a perforated copper shelf placed near the top of the liquid, upon which are placed crystals of blue vitriol to supply copper to the solution as fast as it is extracted by the depositing process. For the circular vessel A, and round porous cell B, and cast rod of zinc C, were soon afterwards substituted rectangular-shaped vessels, which are more convenient, and a plate of rolled zinc coated with mercury.

At this period several persons were attempting to make electro-plating a profitable speculation. Foremost among these were Messrs. G. R. and H. Elkington, who were engaged commercially in the year 1838 in coating military and other metal ornaments with gold and silver, by simply immersing them in solutions of those metals, particularly a boiling one of carbonate of potash containing dissolved gold. They also employed and patented, in conjunction with O. W. Barratt (July 24, 1838), a process for coating articles of copper and brass with zinc, by means of an electric current, generated by a piece of zinc attached to the articles by a wire, and immersed with them in a boiling and neutral solution of chloride of zinc. This was the first patent in which a separate metal was employed for electro-plating purposes.

During the year 1840 Mr. John Wright, a surgeon in Birmingham, and Mr. Alexander Parkes, an experimentalist in the employment of Messrs. Elkington, were engaged in electro-deposition experiments, with the object of obtaining with gold and silver, similar results to those already obtained by Jacobi, Jordan, and Spencer with copper, viz. *thick* deposits of firm, coherent metal, bright, and of good colour. As, however, there are very few solutions capable of yielding such results, their attempts were not at first completely successful.

At this juncture Mr. Wright met with a passage in Scheele's ' Chemical Essays ' (pp. 405 and 406) which soon

proved of the highest de ree of importance to the commercial success of the art, by enabling suitable deposits of silver
and gold to be obtained. Speaking of the solubility of the
oxides and cyanides of gold, silver, and copper, Scheele says
' If, after these calces' (i.e. the cyanides of gold and silver)
' have been precipitated, a sufficient quantity of the precipitating liquor be added in order to re-dissolve them, the solution remains clear in the open air, and in this state the
aërial acid ' (i.e. the carbonic acid) 'does not precipitate the
metallic calx.'

This statement suggested to Mr. Wright the probable
suitability of the cyanides of gold and silver, dissolved in solutions of the alkaline cyanides, for the purposes of electro-
plating ; and he immediately took a solution, composed of
chloride of silver dissolved in aqueous ferro-cyanide of potassium, and quickly obtained what had never been acquired
before, viz. a *thick* deposit of *firm* and *white* silver by electrolytic action. In all previous trials the coating of silver
had either been very thin, or in a state of dark-coloured,
loose powder, completely useless for the intended purpose.

The first article that received the successful coating was
a small vase, and the next was a small figure of a kid. They
were coated by Mr. Wright at his residence, and the process
adopted was as follows :—A common, porous garden-pot, containing the silver solution, was placed in dilute sulphuric acid
contained in an outer vessel ; the article to be coated was
immersed in the inner liquid, and connected by a wire with
a cylinder of zinc surrounding the porous cell, and immersed in the dilute acid. It was about a month after
this that a solution of actual cyanide (not ferro-cyanide) of
silver and potassium was first employed by Mr. Wright for
the same purpose. It is true that cyanides in several forms
had been used both for electro-coppering and silvering about
sixteen months previously ; but that was by the simple immersion process, without the use of zinc, or a single cell or battery, and by that process no thick deposits can be obtained.

Meanwhile other persons were active in other departments of the subject. On March 3, 1840, Joseph Shore patented a process by which he deposited copper from a neutral solution of its sulphate, and nickel from one of its nitrates 'kept of a neutral strength,' by connecting the articles by a copper wire with a piece of zinc in dilute sulphuric acid ; the two liquids being separated by a partition of un-glazed earthenware. 'Larger articles are covered separately, and small articles, such as iron nails, are placed in a wire basket connected by a wire to a zinc plate, and are moved from time to time to prevent any parts remaining uncovered.'

At this period Messrs. G. R. and H. Elkington were taking out another patent, dated March 25, 1840, for an electro process for coating articles with silver or gold by means of the single-cell arrangement, using solutions composed of oxide of silver dissolved in 'pure ammonia,' or prussiate of soda, or other analogous salts ; or oxide of gold dissolved in 'prussiate of potash' by boiling, and kept 'saturated with gold,' or in 'any soluble prussiate,' or 'any other analogous salt.' Mr. Wright having submitted his results to Messrs. Elkington, these were also embodied by them in their patent. At first Mr. Wright received a royalty of one shilling per ounce for every ounce of silver deposited; but on his decease, which took place soon afterwards, an annuity was paid to his widow. This patent process proved to be the basis of all successful electro-plating of gold and silver, because it included the solutions of alkaline cyanides, the only known liquids that fulfil all the necessary conditions ; and the success of those liquids was largely due to their alkaline character. From the earliest period of electro-plating, german-silver was the substance employed for forming the articles which were to be plated.

On December 8 in the same year Messrs. Elkington took out a similar patent in France; and this was quickly followed, on December 19, by a patent taken out in the same country by M. De Ruolz, a French electro-depositor, for

similar objects effected by similar means, viz. electro-gilding
by means of a solution composed either of cyanide or chloride of gold dissolved in the simple cyanide, the ferro-cyanide
or the ferrid-cyanide of potassium ; the double chloride of gold
and potassium dissolved in cyanide of potassium ; the double
chloride of gold and sodium dissolved in soda ; or the sulphide of gold dissolved in neutral sulphide of potassium ;
also electro-silvering by means of a solution of cyanide of silver
dissolved in cyanide of potassium, each liquid being acted
upon by means of the voltaic battery. His patent also included
similar solutions for the electro-deposition of platinum, copper, lead, tin, cobalt, nickel, and zinc ('Encyclopédie Roret,'
Galvanoplastie, tome ii., p. 114). A Commission of the
French Royal Academy of Sciences was appointed to report
on the new processes of Elkington and De Ruolz, and de-
cided in favour of Elkington. A dispute also arose between
the two patentees, which, after a trial at law, was finally set-
tled by a compromise.

The chief conditions of success in the process had now
been attained by the use of the cyanides, but there still remained various smaller difficulties to be overcome. In some
instances the deposited silver would rise in blisters, or portions would peel off on application of the burnishing tool.
In others the metal assumed objectionable colours, frequently
brown, yellowish, streaky, or grey ; or, instead of being even
and smooth, it was covered with asperities, nodules, lines, or
vertical grooves ; or was of unequal thickness. Great difficulty was also experienced in coating articles of iron, steel,
lead, Britannia-metal, &c., so as to secure perfect adhesion ;
and particularly in coating of an uniform colour and appearance, articles, the different parts of which were formed of
several of these metals, or which had been united by different
kinds of solder. In consequence of these various defects, particularly imperfect adhesion and blisters, multitudes of articles
were returned to the platers. Great opposition was also
experienced from the manufacturers of Sheffield wares, and

shopkeepers who sold plated goods, to the introduction of the new articles ; and in consequence of these various diffi- culties the process was not remunerative for at least seven years.

The want of adhesion between the silver and the metal beneath, arose partly from the employment of too many cells in the battery, but chiefly from want of perfect cleanliness of the receiving surfaces. It was eventually overcome by cleaning the surfaces with very great care, dipping the cleaned article in a *very weak* solution of mercury imme- diately before placing it in the depositing liquid, and lessening the number of battery cells. Non-adhesion of the silver was especially apt to occur with articles made of Britannia metal, and with this particular alloy it was not overcome for several years ; it was then remedied by first coating the surfaces with copper, by electro-process, in a liquid composed of cyanide of copper dissolved in a boiling solution of cyanide of potassium, and then silvering them. This method was patented by O. W. Barratt, September 8, 1841 ; a different and a simpler process is now employed.

One of the next important additions to the art of electro- metallurgy was that made by Mr. Murray (January 1840) ; who communicated it orally to the members of the Royal Institution, London. It consisted in coating non-conducting surfaces with plumbago, which being an electric conductor enabled deposits of metal to be formed by the electro-process upon the surfaces of non-conductors, and thus greatly ex- tended the sphere of usefulness of the art. During the same year also Professor de la Rive made known a process of electro-gilding employed by him in the year 1828 ; he gilded wires of platinum and silver 'by employing them as negative electrodes in a solution of chloride of gold.' (De la Rive's 'Treatise on Electricity,' vol. iii., p. 546.)

In the same year also another arrangement of electro- depositing apparatus, now known as the 'separate battery' apparatus, was devised by Mr. Mason. In this new arrangement a current from a single cell of a Daniell's battery, A, is caused to produce deposition in a separate vessel, B.   It is represented by the annexed sketch.

FIG. 2.

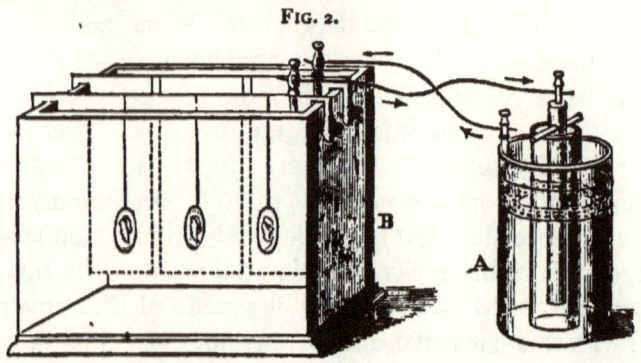

In this apparatus, as fast as the metal is removed from the solution at the negative pole, by articles being coated, it is replaced by an equal amount of metal dissolving at the positive pole, or dissolving plate, and the solution, therefore, does not require any additional supply of metallic salt.   A still further modification of Mr. Mason's arrangement was soon generally adopted, by substituting any ordinary voltaic battery for the single cell of Daniell.

Another arrangement, termed the 'compound depositing cell,' was also devised about this time, by means of which

FIG. 3.

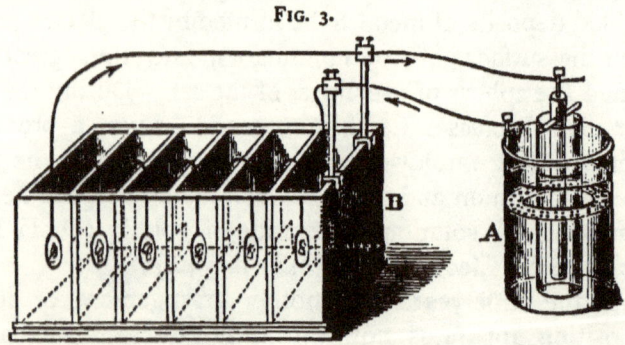

the current from a single cell, A, of a battery was caused to pass through a series of depositing cells, B (see sketch), and thus dissolve and deposit several times the amount of metal by the same amount of consumption of zinc and acid ; but it was soon abandoned on account of the slowness of action.

During the year 1841 Mr. Alfred Smee published the results of his experiments upon the electro-deposition of Antimony, Iridium, Rhodium, Palladium, Platinum, Gold, Silver, Copper, Nickel, Iron, Lead, Cadmium, Zinc, &c., in the first edition of his book on Electro-metallurgy ; he also applied the very appropriate term ' electro-metallurgy ' to the process of working in metals by means of electrolysis. In the same year also M. De Ruolz electro-deposited brass from a solution composed of the cyanides of copper and zinc dissolved in aqueous cyanides of potassium (Walker's ' Electrotype Manipulation,' last edition). In the year 1841 also Alexander Jones took out a patent (dated January 14) relating to electro-deposition, in which he renders a non-conducting surface conducting, and fit for receiving electro-deposits, by first immersing the article in a solution of nitrate of silver, then reducing a film of silver upon it by means of a solution of green-vitriol, or by phosphorus either in solution or vapour ; or by coating the surface with bronze powder or metallic leaf.

The next event of importance in the history of electroplating consisted in the application of magneto-electricity to that object. On August 1, 1842, J. S. Woolrich took out a patent for the use of a magneto-electric machine in electroplating, and this machine was used during many years in several electro-plating establishments, but has since been quite superseded by other magneto-machines of an improved kind. During the same year Mr. Palmer patented his ' improvements in producing printing and embossing surfaces,' and employed electro-deposition for producing the copper plates ; the process was termed ' Glyphography.' Dr. H. R. Leeson also took out a patent on June 1, in the same year, for improvements in electro-metallurgy, and claimed no less

than about 430 (!) different salts for the purposes of deposition.   He suggested ' elastic moulds,' formed of glue alone
or glue mixed with gum, for receiving deposits ; also 'positive wires' led into cavities or under cut parts of the moulds,
and inserted 'conducting wires in wax moulds to facilitate
deposition ; ' and proposed to keep the articles to be coated
constantly in motion 'by means of a roasting-jack,' or to
' agitate the solution.'

The first use of thermo-electricity appears to be that
made by Moses Poole, who took out a patent, dated May
25, 1843, for the use of a thermo-electric pile instead of a
voltaic-battery for depositing purposes.   The method, however, did not come into general use.   During the following
year (viz. 1844) a patent was taken out by Napier, dated
October 22, and one by Parkes, dated October 29, for electro-depositing metals from mineral ores and salts whilst in a
state of fusion.

The next improvement arose in a singular way.   The
surface of silver deposited from the ordinary cyanide of silver
and potassium plating solution has a frosted or snow-white
appearance, and is more or less ' chalky ' and dull, and requires to be burnished or made bright by mechanical means.
This with articles of highly figured design, or which are
hollow and require the interior to be bright, is a great disadvantage, because the process of burnishing is tedious, and
when it has to be applied to the interior of vessels it is also
very awkward to perform.   As with the difficulty of obtaining thick deposits of firm silver during the early period of
the electro-process a little circumstance led to that obstacle
being overcome, so was it with this difficulty, and it happened as follows :—In the process of copying figures for
electro-typing, by moulds composed of a mixture of wax and
resin, the surface of the mould was covered with a film of
phosphorus by means of a solution of phosphorus in bisulphide of carbon.   It was observed by Mr. W. Milward, at
Messrs. Elkington's establishment, that when these moulds

were put into the cyanide of silver-plating solution, for the purpose of receiving a coating of silver, the silver coating upon other articles, such as spoons, forks, &c., which were being plated in the same vats, and especially those nearest to the moulds, acquired a brightness more or less perfect, which occurred sometimes in patches or streaks, and sometimes extended all over the deposited surface, instead of the ordinary snow-white appearance. This circumstance of course attracted attention, and induced Mr. Milward to try the effect of adding bisulphide of carbon alone to the liquid. Considerable success soon resulted; but at this juncture the secret escaped, and in consequence a patent was taken out March 23, 1847, by Mr. Milward and a person of the name of Lyons, who had acquired a krowledge of the secret, for producing bright deposited silver by ' adding compounds of sulphur or carbon,' bisulphide of carbon being preferred, to the cyanide of silver solution. This process has been constantly employed ever since, and is now in extensive use. Bright copper had also been observed about the year 1845, and occurred whenever a large number of phosphorised wax moulds were put into a solution of sulphate of copper to receive an electro-deposit of copper. In the year 1847 also, Professor Silliman copied the iridescent colours of mother-of-pearl, by taking a mould of the shell in fusible alloy, and then an electro-cast from the mould ('Timbs' Year-Book of Facts,' 1847).

The chief improvements which have been made in electro-metallurgy since the year 1847 have been the gradual extension of the process for multiplying printing surfaces, in stereotyping, &c., also the production of works of art, &c., of increased size in copper, until deposits several tons in weight have been attained; the extensive use of nickel as a coating upon harness furniture, &c., the protection of articles of cast-iron, ornamental lamp-posts, &c., from rusting, by a coating of copper; the substitution of magneto-electric machines, and thermo-electric piles, for voltaic batteries; the purification of crude copper in the process of copper smelting; and, quite recently, the economical production of coppered iron rollers for calico printing by means of magneto-electric deposition.

THEORETICAL DIVISION.

PRINCIPLES AND LAWS UPON WHICH THE ART OF ELECTRO-METALLURGY IS BASED.

THE phenomena which occur in electro-metallurgical processes largely involve the principles of dynamic electricity, chemico-electric action, electro-chemical action, and also ordinary chemical affinity ; and, if magneto-electric and thermo-electric apparatus are also employed, the phenomena then include also magneto-electric and thermo-electric action.

Electric conduction and insulation are involved in electro-metallurgical processes, because through all the conductors, electric generators, and depositing liquids, electricity is continually circulating, and requires to be guided in its course, and protected from leakage.

Chemico-electric action, or electric change produced by chemical power, is included, because that action is always occurring in the voltaic batteries employed. Electro-chemical action is also involved because that change is constantly going on in the depositing solutions. In the magneto-machine, magneto-electric induction occurs all the time the machine is at work ; and in the thermo-electric pile, thermo-electric action, or heat producing electric force, operates without cessation. It is evident, then, that a clear perception of the principles of these actions is indispensably necessary to a proper understanding of the subject.

ELECTRICAL PRINCIPLES OF ELECTRO-METALLURGY.

Conduction and insulation.—Conducting and insulating power for electricity, or the capacities of facilitating and hindering the passage of electricity, differ in degree in every different substance. The best conductor is the worst insulator, and *vice-versa*, the degree of the one property being inverse to that of the other in every substance. All bodies conduct electricity, but in different degrees, and all insulate also in different degrees, and the difference of conducting power of the best conductor in relation to that of the worst is enormous ; and similarly with regard to insulating power.

The following is a table in which various common substances are ranged approximately in the order of their relative power of insulating, or hindering the passage of electricity ; it commences with the most perfect insulators, and ends with the best conductors :—

1. Ebonite.	20. Chalk.
2. Shellac.	21. Lime.
3. Caoutchouc.	22. Dry gases.
4. Gutta-percha.	23. Dry steam.
5. Amber.	24. Phosphorus.
6. Resin.	25. Fatty oils.
7. Sulphur.	26. Dry metallic oxides.
8. Wax.	27. Ice at 0° C.
9. Agate.	28. Straw.
10. Glass.	29. Paper.
11. Gems.	30. Marble.
12. Silk.	31. Dry wood.
13. Wool.	32. Alcohol. Ether.
14. Hair.	33. Rain-water.
15. Feathers.	34. Spring-water.
16. Dry paper.	35. Sea-water.
17. Leather.	36. Graphite.
18. Porcelain.	37. Brittle metals.
19. Camphor.	38. Ductile metals.

The difference of insulating power or conduction-resistance between the substances at the two extreme ends of the table is extremely great. If the resistance of silver to the passage of the current be represented by 1, that of a rod of gutta-percha, of equal length and diameter, has been estimated to equal about 850,000,000,000,000,000,000. The insulating substances usually employed in electro-metallurgy are either gutta-percha, glass, or india-rubber, and the wire employed in magneto-electric machines is insulated by being covered with cotton or silk. The large conductors, to supply the main currents to the vats, are covered with gutta-percha or tarred twine. The wires for suspending spoons, forks, &c., in the solutions are enclosed in tubes of glass.

The relative conducting powers of pure metals, &c., according to Dr. Matthiessen, are as follows : —

Silver .	.	.	. 100·0	Tin .	.	.	. 12·4
Copper	.	.	. 99·9	Thallium .	.	.	9·2
Gold .	.	.	. 77·9	Lead .	.	.	8·3
Zinc .	.	.	. 29·0	Arsenic	.	.	4·8
Cadmium	.	.	. 23·7	Antimony .	.	.	4·6
Palladium	.	.	. 18·4	Mercury	.	.	1·6
Platinum	.	.	. 18·0	Bismuth	.	.	1·2
Cobalt	.	.	. 17·2	Graphite .	.	.	·069
Iron .	.	.	. 16·8	Gas coke .	.	.	·038
Nickel	.	.	. 13·1	Bunsen's coke .	.		·025

Copper being the cheapest good conductor, very flexible and ductile, easily obtainable, and not readily oxidised, is nearly always employed for transmitting the current in electro-metallurgical operations. Wire composed of iron or of brass is rarely used for such a purpose.

The conductivity of fused salts varies greatly ; if that of mercury equals 100,000,000, that of chloride of lead is 25,300 ; of iodide of potassium 11,500 ; of nitrate of silver 8,688 ; of common salt 8,660 ; of potassic carbonate 2,150 ; and of chloride of zinc 86 (F. Braun, 'Chemical News,' vol. xxx. p. 207).

The conduction-resistances of liquids are enormous in comparison with those of metals. For instance, if that of copper at 32° Fahr. is equal to 1, those of various liquids are as follows :—

Nitric acid at 55° Fahr.		976,000
Sulphuric acid diluted to $\frac{1}{17}$ at 68° Fahr. . . .		1,032,020
Saturated solution of chloride of sodium at 56° Fahr. .		2,903,538
,, ,, sulphate of zinc . . .		15,861,267
,, ,, ,, copper at 48° Fahr. .		16,885,520
Distilled water at 59° Fahr.		6,754,208,000

According to Dr. Overbeck, if the resistance of water equals 390, that of alcohol is 13,000, of ether 40,000, and of bisulphide of carbon still greater. Small additions of saline matters greatly diminish the resistance of water ('Electrical News,' vol. i., p. 170).

If the conduction-resistance of distilled water is so great in relation to that of copper, we can easily understand, by referring to the previous table, that the resistance of gases must be enormous. The electric conduction-resistance of air heated to redness is 30,000 greater than that of water, containing a 20,000th part of its weight of sulphate of copper in solution. The ordinary unit of electrical resistance employed in this country is termed an 'Ohm' (see Jenkin's 'Treatise on Electricity and Magnetism,' p. 158; also Appendix, p. 387 of this book).

Effects of temperature, purity, &c., upon the electric conductivity of substances. —The conducting power of substances is largely affected by their degree of purity. The least admixture of a foreign substance in a metal has a great effect ; arsenic in copper is very injurious to electric conduction, and may even increase the resistance as much as 66 per cent. ; while one half per cent. of iron increases the resistance of copper as much as 25 per cent.

The conducting powers of bodies are also greatly influenced by temperature. Warming a metal increases its conduction-resistance. Metals usually lose about 30 per cent. of conducting-power by rise of temperature from 0° C. to 100° C. ; german-silver, however, loses only about 4 per cent. by that change. Solid non-metallic substances usually increase in conducting power with rise of temperature ; so also do liquids almost uniformly, and gases very considerably.

Measurements of conduction-resistance are at present rarely made in electro-metallurgical operations, but as science extends they will probably require to be. The degree of conduction-resistance in a wire varies directly as its length, and inversely as the area of its cross-section. Acids and saline liquids obey the same laws of conduction in this respect as metals. The methods of measurement of electric conduction-resistance are fully described in the treatise on 'Electricity and Magnetism,' p. 229, by Fleeming Jenkin.

Heat generated in metals and liquids by conduction-resistance.—As heat can produce electricity, so electricity can produce heat. It is a general truth of electrical science, that wherever an electric current overcomes resistance it produces heat. And it therefore not unfrequently happens that, by the passage of long-continued and energetic currents, the battery, liquids, conducting wires, depositing solutions, the coils of wire in magneto-electric machines, the cooler junctions of thermo-electric piles, &c., become more or less heated ; and these circumstances affect the working.

Electro-chemical action.—The fundamental basis of electro-metallurgy is the fact that, *when a current of electricity passes through a suitable liquid, it produces a chemical change.* This phenomenon is called electro-chemical action, because electricity is the cause, and chemical action the effect. The chief conditions of electro-chemical action are that the substance must be a liquid, a compound body, a conductor of electricity, and traversed by the current. An electrolyte (see p. 33) is usually composed of two elementary substances —the one a conductor, and the other a non-conductor of electricity ; a liquid which is composed either of two conductors or two non-conductors does not usually suffer electrolysis.

The chemical change which takes place consists of a chemical decomposition of the liquid ; and the constituents of the liquid—either primary or secondary—are liberated in a free state. The substances are liberated not in the mass of the liquid, but at the *immediate surfaces of contact* of the liquid with the conductors, by which the current enters and leaves the solutions ; and the distance to which the action extends from the surface of the conductors into the mass of the liquid, is so extremely small, as not to be definitely known.

In all cases of electrolysis the electro-negative elements of the liquid, such as metalloids and acids, either combine with, or are set free, at the surface of the dissolving or positive metal, and the electro-positive elements, such as metals and alkalies, either combine with, or are set free or deposited, at the surface of the receiving or negative metal. For instance, if a piece of silver and a piece of copper are immersed in a solution of sulphate of copper, and the current from a battery be passed into the solution by means of the piece of copper, and out of it by means of the silver, the negative elements of the liquid, viz. the sulphuric acid of the sulphate of copper, will be separated from its associated copper, and will combine with the positive copper terminal, causing it to dissolve in the liquid ; while the positive element of the liquid, viz. the copper of the salt, will be deposited at the surface of the negative or receiving metal, the silver, but will not combine with it. But if we substitute a piece of platinum for the piece of copper, and mercury for the silver, the effects will be reversed ; the acid or negative element will collect around the positive platinum, but will not combine with it, whilst the positive element of the liquid, the copper, will be deposited and combine with the negative mercury.

Nomenclature of electro-chemical action. — The phenomenon of chemical decomposition of a liquid by means of an electric current is, in accordance with the nomenclature adopted by Faraday, who specially investigated it, called *electrolysis.* The liquid in which it occurs is called an *electrolyte.* The conductors immersed in the liquids are termed *electrodes* ; the one by which the current (i.e. the positive electricity) *enters* the liquid is called the *anode*, or positive pole, and that by which it *leaves* the solution is termed the *cathode*, or negative pole. The elements of the liquid which are set free by the action. are called *ions*; those which appear at the anode, or positive pole, are called *anions*, and those at the cathode, or negative pole, *cations*.

Usual phenomena of electrolysis.—The effects which take place at the two electrodes appear in various forms ; in some cases a gas is set free and ascends, or it adheres in bubbles all over the front surfaces, and round the edg: s of the electrode, or is absorbed by the electrode ; in other instances a solid substance is formed, and either adheres to the electrode, falls to the bottom of the liquid, or enters into solution. Usually the anode dissolves and gradually becomes thinner ; sometimes streams of liquid are seen slowly falling from the anode to the bottom, or rising from the cathode to the surface of the solution, caused by the alteration of specific gravity of the layers of liquid in contact with the electrodes, by means of electrolysis. Various special phenomena also occur in particular cases, which will be mentioned under the heads of the deposition of individual substances.

Decomposability of different electrolytes.—Different liquids require very different degrees of electric power to decompose them. Faraday has given the following order with the substances here mentioned, the first named being the most easily decomposed :—Solution of iodide of potassium ; melted chloride of silver ; melted chloride of zinc ; melted chloride of lead ; melted iodide of lead ; hydrochloric acid ; dilute sulphuric acid. Smee, in his ‘Elements of Electro-Metallurgy,’ 2nd edition, p. 185, gives nitric acid as the most easily decomposable liquid, and other substances in the following order :—

Nitric acid.	Dilute sulphuric acid.
Chloride of gold.	Sulphate of cadmium.
Nitrate of palladium.	Sulphate of zinc.
Chloride of platinum.	Sulphate of nickel.
Nitrate of silver.	Sulphate of iron.
Sulphate of copper.	Sulphate of manganese.
Sulphate of tin.	Salts of the alkalies, generally.

I have many times observed that water is decomposed before hydrofluoric acid, and hydrochloric acid in preference to water ; also selenic acid before selenate of nickel ; but the order of decomposability would probably be affected, both by pressure, temperature, and degree of dilution. Much investigation is required in this branch of the subject.

Direction of electrolysis.—In the electrolysis of any given liquid, the substances set free at the anode are always relatively electro-negative to those set free at the cathode, and the latter of course are relatively positive. Anions, therefore, are electro-negative and cations are electro-positive substances. The negative substances appear at the positive pole, or the electrode attached to the copper plate of the battery ; and the positive substances at the negative pole, or the electrode attached to the zinc of the battery. Metals, alkalies, and bases, being electro-positive, are cations, and appear at the negative plate ; and metalloids, peroxides, and acids, being electro-negative, are anions, and appear at the positive one. As the conditions of positive and negative are relative, and no substance is absolutely either, the same body may sometimes appear at the anode, and on other occasions at the cathode ; for instance, sulphur when separated from a positive body, such as a metal, appears at the anode, but when liberated from oxygen, a more negative substance than itself, it appears at the cathode. '

Circumstances which affect the quality of electro-deposited metals —The physical qualities of the deposited substances are largely affected by a number of circumstances ; such as the composition of the liquid, its degree of fluidity, temperature, &c., also the 'density' of the current, or the actual quantity of electricity entering a given magnitude of surface in a given 'time ; the rate of deposit ; and, further, the condition of the receiving surface.

The terms employed to designate the quality of the deposited metal refer usually to its mechanical state and colour.

By 'reguline'[1] metal is meant that which possesses the ordinary or more perfect *metallic* qualities of the metal : thus reguline copper is red, bright, and tough ; reguline antimony is grey, hard, and crystalline ; amorphous antimony is without crystalline structure. The terms, crystalline, sandy metal, black powder, &c., are also frequently used.

Every different solution, and at every different temperature, requires to be electrolysed at a different rate, in order to deposit its metal from it in any particular desired state of aggregation, as crystals, reguline metal, or black powder ; and this is further modified by the form and mechanical state of the receiving surface, a rough surface, or one covered with points, or having ragged edges, requires a slower rate of deposition than a smooth one, in order to obtain reguline metal. A large amount of experimental investigation remains to be made also in this part of the subject.

The circumstance which most affects the quality (and purity) of the deposited metal is the composition of the electrolyte. By far the greater number of solutions yield up their metal only in the form of a black powder, however carefully the process be conducted. From only a very few liquids are metals deposited in their ordinary or reguline state, and only a few of the metals have yet been obtained in that condition in thick masses. Copper, antimony, and silver are the easiest to obtain in thick layers ; I have obtained reguline deposits of antimony $1\frac{1}{2}$ inch in thickness. Zinc, gold, platinum, tin, nickel, cadmium, and lead are usually only obtained in thin reguline films, or in roughly aggregated nodules or crystalline forms ; and most of the other metals have been only obtained in the state of a dark powder. The degree of fluidity and diffusive power of the electrolyte also affects the deposit ; a thick, immobile, non-diffusive liquid will adhere to the cathode after the layer next that electrode has been deprived of its metal, and thus alter its

---

[1] The term ' reguline ' differs from ' regulus ;' the latter means metal which has been reduced from its ores by means of fusion and a flux.

quality. Some solutions also will only yield reguline metal whilst they are hot, or whilst being stirred.

A paper was read at the meeting of the British Association, at Bradford, by Dr. Gladstone, on 'Black Deposits of Metals,' in which he remarks that the allied metals, platinum, palladium, and iridium are generally, if not always, black when precipitated by substitution (i.e. by single immersion process), and bismuth and antimony form black fringes and little else ; similar fringes are also formed by gold ; but it also yields green, yellow, or lilac-coloured metal according to circumstances. Copper, when first precipitated on zinc, whether from a weak or strong solution, is black ; but in the latter case it becomes chocolate-coloured as it advances, or red if the action be more rapid. Lead, in like manner, is always deposited black in the first instance, though the growing crystals become of the well-known dull grey. Silver and thallium appear as little bushes of black metal on the decomposing plate if the solution be very weak, otherwise they grow of their proper colour. Zinc and cadmium give a black coating, quickly passing into grey when their weak solutions are decomposed by magnesium. The general result may be stated thus : if a piece of metal be immersed in the solution of another metal which it can displace, the latter metal immediately makes its appearance at myriads of points in a condition that does not reflect light ; but as the most favourably circumstanced crystals grow, they acquire the optical properties of the massive metal, the period at which the change takes place depending partly on the nature of the metal, and partly on the rapidity of its growth ('Telegraphic Journal,' vol. i., p. 302).

Some metals are especially liable to crack and curl up whilst being deposited ; this is the case in a strong degree with antimony, and less so, but conspicuously, with iron and nickel. Others form very hard and tenacious deposits ; that of iron is extremely hard, and of nickel so much so that it cannot be burnished, and if scratch-brushed it becomes quite yellow by tearing off particles of the brass scratch-brushes. Electro-deposited copper, and even gold, is also sometimes remarkably hard. All electro-deposited metals are more or less crystalline or porous, and can rarely be made to form air-tight, or even water-tight, vessels without annealing, especially if the layer of metal first deposited is scraped or filed away ; electro-deposits therefore do not protect metallic surfaces from oxidation, like a coating of the same metal put upon them by a process of fusion. They are also more porous on their outer surfaces, i.e. the side last deposited, than upon their inner, and if heated in a fire will often become concave on the side last deposited. Seams also cannot be closed (or edges joined together) by electro-deposition ; they always became larger ; even a scratch will gradually become a deep groove.

Influence of density of the current.—Next to the composition of the liquid, the circumstance which most affects the quality of the deposit is what has been termed the ' density of the current,' or the actual quantity of electricity entering a given area of receiving surface in a given time. With a good depositing solution, a non-oxidisable metal, if the density of the current is great, is deposited as a black powder ; if it is small the metal assumes the form of crystals, or is precipitated in the state of a subsalt; and between these two extremes, the soft, reguline, hard, and sandy deposits occur. A saturated solution of sulphate of copper, acidulated with dilute sulphuric acid, yields good reguline metal when the metal is deposited from it at the rate of about half an ounce per square foot of surface per hour. The density of the current affects not only the quality of the liberated cations, but also that of the anions, as has been already observed (see ' Telegraphic Journal,' vol. ii., p. 178) in the instance of oxygen, also with chlorine.

The effect of small density of the current is different with solutions of the noble metals to what it is with those of the base ones, because the latter metals are easily oxidised and the former are not. With solutions of noble metals a current of small density deposits the metal in the form of crystals, but with those of the base ones it deposits an oxide, or even a subsalt, instead of the metal. And as the current is strongest at its first commencement and is then weakened, first by polarisation and subsequently by the deposition of the less-conducting layer, the deposit rapidly changes in many cases from good reguline metal to a non-adherent coating of oxide or salt. Solutions of nickel are very apt to show this change (p. 232).

A perfectly smooth, bright surface is the best to receive a deposit; if the surface is rough the deposit will be rough also. As a deposit increases in thickness it becomes more and more uneven, and therefore, in nearly all cases, the deposit sooner or later, and in most cases very quickly, loses its reguline character and brightness, and becomes a layer of either black or crystalline powder; the edges of the cathode which receive the thickest deposit are the soonest affected.

Formation of metallic crystals.—Every different metal, and even the same metal when deposited from different solutions, gives a different form of crystal. In some solutions— chloride of tin for example—crystals form with great rapidity. The annexed figures represent the forms of crystals of silver, tin, and gold; they are copied from Dr. Gladstone's Lectures, 'Telegraphic Journal,' vol. iii., pp. 29 and 38. Fig. 4 represents silver deposited from a weak solution of nitrate of silver; Fig. 5, ditto, from a strong solution; Fig. 6, from a very strong solution; Figs. 7 and 8, from a nearly exhausted solution; and Fig. 9 from a $2\frac{1}{2}$ per cent. solution. Fig. 10 shows tin, deposited from a solution of chloride of tin; and Fig. 11, gold, from a solution of chloride of gold.

Circumstances which affect the quantity of electro-deposited metal.—In a liquid which yields at the cathode a single metal only, the quantity of the particular metal set free depends entirely upon the amount of electricity which has passed; but with different metals it depends also upon what

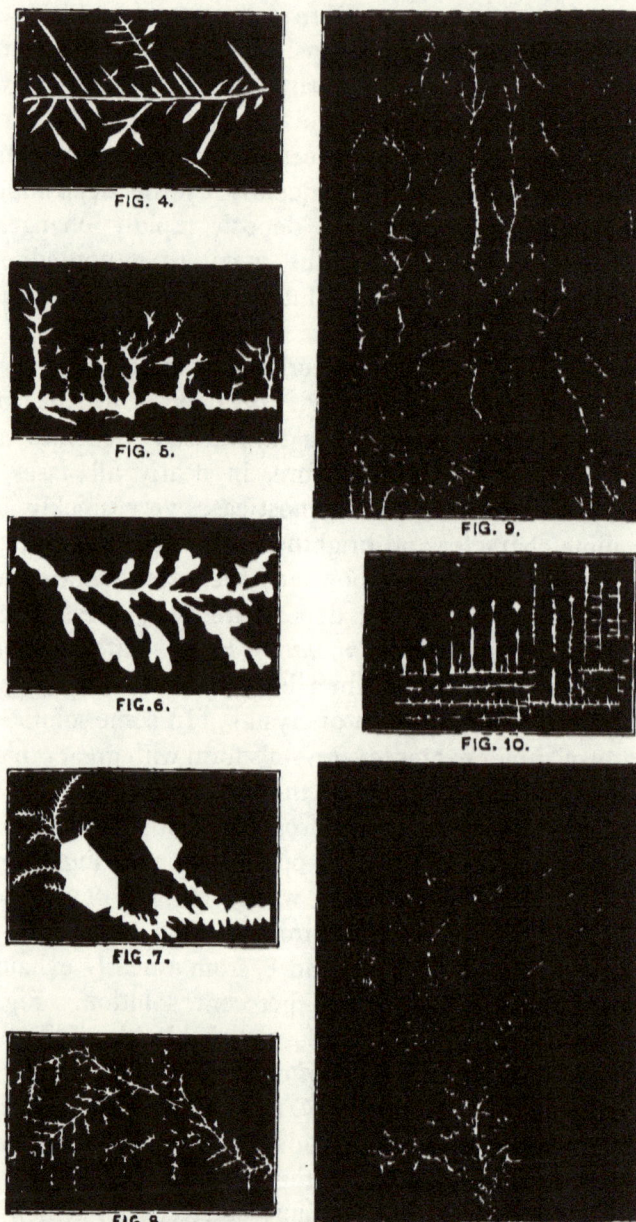

FIG. 4.

FIG. 5.

FIG. 6.

FIG. 7.

FIG. 8.

FIG. 9.

FIG. 10.

FIG. 11.

is termed the valency of the metal and upon its atomic weight; by valency is meant the amount of chemical force associated with an atom.

This part of the subject requires on the part of the student a certain amount of knowledge of chemistry; and as this book is not one on that subject, I assume that the student already possesses a sufficient preparatory acquaintance with chemistry to enable him to enter on the subject of electro-metallurgy; that he is, for instance, acquainted with the chief properties of the commoner elementary substances, and of their chief compounds, of acids. bases and alkalies, of the meaning of the terms specific gravity, molecular weight, combining and equivalent proportions, molecular gaseous volumes, &c.; but, if he is not, I must refer him to Miller's ' Inorganic Chemistry,' published in this series of text-books. I therefore only insert in this treatise such chemical information as I consider must not be omitted.

Relation of chemical value, or valency of atoms, to electrolysis.—In accordance with the doctrines of chemistry, the elementary substances are classed into Monads, Dyads, Triads, Tetrads, &c., as in the following table :—

MONADS	DYADS	TRIADS	TETRADS	HEXADS
Hydrogen	Oxygen	Nitrogen	Carbon	Molybdenum
Lithium	Sulphur	Boron	Silicon	Vanadium
Fluorine	Selenium	Phosphorus	Titanium	Tungsten
Chlorine	Tellurium	Arsenic	Tin	Osmium
Bromine	Barium	Antimony	Zirconium	Chromium
Iodine	Strontium	Rhodium	Thorinum	Manganese
Cæsium	Calcium	Gold	Cerium	
Rubidium	Lanthanum	Bismuth	Iron	
Potassium	Didymium	Aluminium	Nickel	
Sodium	Glucinum	Indium	Cobalt	
Thallium	Magnesium		Uranium	
Silver	Zinc		Lead	
	Cadmium		Ruthenium	
	Copper		Iridium	
	Mercury		Palladium	
			Platinum	

A monad is an elementary substance, one atom of which possesses one equivalent of chemical power; a *dyad* is one, an atom of which possesses two such equivalents; a *triad* three ; a *tetrad* four, &c. One atomic weight therefore of a dyad element is chemically equivalent to two of a monad element ; one such weight of a triad element is chemically equivalent to three of a monad, or one and a half of a dyad ; one of a tetrad is equivalent to four of a monad, two of a dyad, or one and a third of a triad ; and so on.

Equivalency of electro-chemical action.—As substances by chemical union unite in certain definite proportions by weight, it follows as a matter of course that when a chemical compound is decomposed, the separated ingredients must also possess definite proportions by weight. When, therefore, by the passing of a current of electricity through a compound conducting liquid, elementary substances are liberated at the electrodes, they are set free in their chemically *equivalent* proportions by weight. And as the chemically equivalent proportions are either the same as their atomic weights, or are some simple submultiple of them, the following table of atomic weights is inserted for the purpose of reference :—

Symbols and Atomic Weights of Elementary Substances.

Name	Symbol	Atm. Wt.	Name	Symbol	Atm. Wt.
Aluminium	Al	27·5	Copper	Cu	63·5
Antimony	Sb	122·	Didymium	D	96·
Arsenic	As	75·	Erbium	E	(?)
Barium	Ba	137·	Fluorine	F	19·
Bismuth	Bi	210·	Glucinum	G	9·3
Boron	B	10·9	Gold	Au	196·6
Bromine	Br	80·	Hydrogen	H	1·
Cadmium	Cd	112·	Indium	In	113·4
Cæsium	Cs	133·	Iodine	I	127·
Calcium	Ca	40·	Iridium	Ir	197·
Carbon	C	12·	Iron	Fe	56·
Cerium	Ce	92·	Lanthanum	La	92·
Chlorine	Cl	35·5	Lead	Pb	207·
Chromium	Cr	52·5	Lithium	L.	7·
Cobalt	Co	59·	Magnesium	Mg	24·3

Symbols and Atomic Weights of Elementary Substances—continued.

Name	Symbol	Atm. Wt.	Name	Symbol	Atm. Wt.
Manganese .	Mn	55˙	Silver . .	Ag	108˙
Mercury . .	Hg	200˙	Sodium . .	Na	23˙
Molybdenum .	Mo	96˙	Strontium .	Sr	87˙5
Nickel . .	Ni	59˙	Sulphur . .	S	32˙
Niobium . .	Nb	97˙5	Tantalum .	Ta	138˙
Nitrogen .	N	14˙	Tellurium .	Te	129˙
Osmium . .	Os	199˙	Thallium .	Tl	204˙
Oxygen . .	O	16˙	Thorinum .	Th	119˙
Palladium .	Pd	106˙5	Tin . .	Sn	118˙
Phosphorus .	P	31˙	Titanium .	Ti	50˙
Platinum .	Pt	197˙	Tungsten .	W	184˙
Potassium .	K	39˙1	Uranium .	U	120˙
Rhodium .	Ro	104˙3	Vanadium .	V	137˙
Rubidium .	Rb	85˙	Yttrium . .	Y	(?)
Ruthenium .	Ru	104˙2	Zinc . .	Zn	65˙
Selenium .	Se	79˙5	Zirconium .	Zr	89 5
Silicon . .	Si	28˙			

As a monad exerts one equivalent of chemical force, a dyad two, a triad three, &c., and the elementary substances set free at the two electrodes are liberated in their chemically equivalent proportions, it follows that when compounds are decomposed by an electric current, for each atomic weight of a dyad set free at one electrode, two of a monad, or one of a dyad, are liberated at the other ; similarly when an atomic weight of a triad is separated at one pole, three of a monad, or one and a half of a dyad, are separated at the other, and so on ; for instance, in the electrolysis of strong hydrochloric acid, one part by weight of hydrogen is set free at the cathode, and thirty-five and a half parts of chlorine at the anode. In the electrolysis of water two parts by weight of hydrogen appear at the cathode and sixteen parts of oxygen at the anode, and so on.

Frequently, however, the substances which are set free, or appear at the electrodes, are not simple elements, but compound bodies, and even mixtures of compounds ; but in these cases also a similar principle of chemical equivalency operates, and we may say generally, *whatever substance, or mixture of substances, are set free at one electrode, a chemically equivalent weight of substances of opposite electrical nature is set free at the other.*

Definite electro-chemical action.—Not only are the elementary and compound substances which are set free at the two electrodes, by the electrolysis of a given liquid, liberated in definite proportions by weight, which are identical with their chemically combining or equivalent proportions, but also, if the same current is made to traverse a series of different liquids, the chemical actions in all of them are also in chemically equivalent quantities; and therefore produced by the same amount of electricity. This great truth was discovered by Faraday.

The same amount of electricity which will decompose one molecule or eighteen parts of water, setting free two parts by weight of hydrogen and sixteen parts of oxygen in one vessel, will decompose two molecules or seventy-three parts of hydrochloric acid, setting free two parts of hydrogen and seventy-one parts of chlorine in another. If we cause the same current to pass through a solution of cyanide of silver and potassium, then through one of sulphate of copper, and finally through one of antimony, each solution being prepared and acted upon so as to yield only pure metal, we find that for every 108 parts of silver deposited in the first vessel, 31·75 parts $\left(\text{or } \dfrac{63\cdot5}{2}\right)$ of copper are set free in the second one; and 40·66 parts $\left(\text{or } \dfrac{122}{3}\right)$ of antimony in the third one.

By employing this method of passing the same current or amount of electricity through successive solutions, and weighing the products set free, their relative chemical equivalents may be found : for instance, I passed a voltaic current through a solution which deposited pure copper only, and through another which deposited pure antimony alone, and weighed the amounts of deposit. The weight of the copper obtained was 31·7 grains, and of the antimony 40·6 grains, which is equal to one atomic weight or 63·5 parts of copper, as the equivalent of 81·32 parts of antimony, or two-thirds an atomic weight of that metal=121·98 as the full atomic weight of antimony.

Theory of electrolysis.—Faraday considered that electrolytic decomposition is the result of a peculiar corpuscular action developed in the direction of the current ; it proceeds from a force which is either added to the affinity of the bodies present or determines the direction of that force. The electrolyte is a mass of acting particles, of which all that are in the course of the current contribute to the terminal action ; and in consequence of the affinity between the elements being weakened or partially neutralised by the current parallel to its own course in one direction, and strengthened and assisted in the other, the combined particles acquire a tendency to move in different directions. The particles of one element, *a*, cannot travel from one pole to the other unless they meet with particles of an opposed substance, *b*, ready to move in the opposite direction. For in consequence of their increased affinity for these particles, and the diminution of their affinity for those which they have left behind them in their way, they are continually driven forward.

In addition to the law of definite electro-chemical action Faraday has advanced what is termed the binary theory of electrolysis—that ‘ only those compounds of the first order are *directly* decomposable by the electric current, which contain one atom of one of their elements for each atom of the other ; for instance, compounds containing one atom of hydrogen or metal with one atom of oxygen, iodine, bromine, chlorine, fluorine, or cyanogen,’ whilst ‘ boracic anhydride (BO_3), sulphurous anhydride (SO_2), sulphuric anhydride (SO_3), iodide of sulphur, chloride of phosphorus (PCl_3) and (PCl_5), chloride of sulphur (S_2Cl), chloride of carbon (C_4Cl_6), tetrachloride of tin ($SnCl_4$), terchloride of arsenic

(AsCl$_3$), pentachloride of antimony (SbCl$_5$),' are non-conductors of electricity, and incapable of electrolysis.

Secondary effects of electrolysis.—Very frequently the substances which appear at the electrodes are not those originally set free, but are liberated, or set free, by the action of the originally liberated bodies upon the liquid or upon the electrode. Some substances which are not of the simple binary character mentioned are decomposed by current electricity, and yield their positive and negative elements in equivalent proportions at their respective electrodes ; but according to this theory they are *indirectly* decomposed, i.e. they are decomposed by the chemical action of some of the elements set free by the direct action of the current upon other substances present. For instance, 'fused borax (biborate of soda Na$_2$O, 2BO$_3$) yields oxygen gas at the anode and boron at the cathode ; now, since fused boracic acid is not decomposed by the electric current, the separation of the boron must be attributed to indirect action ; the current resolves the soda (Na$_2$O) into oxygen and sodium, and the latter separates boron from the boracic acid ' (Faraday). In a similar way a solution of common salt yields chlorine at the anode, and hydrogen and soda at the cathode. The salt is probably first decomposed into chlorine and sodium, and the liberated sodium decomposes the water, forming soda and setting free hydrogen, for when the cathode consists of mercury, an amalgam of sodium is obtained. In other cases, as that of a solution of sulphate of copper, nitrate of silver, or chloride of gold, the hydrogen produced by the decomposition of the water deoxidises the metallic salts in solution, setting free their metals and reconstituting water. Similar secondary actions take place at the anode ; if that electrode is formed of an easily corrodible metal the elements set free at its surface combine with it, and form a new compound. In other cases where the electrode is not corroded, as an anode of platinum in a solution of nitrate of silver or acetate of lead, the liberated oxygen sometimes converts the salt into

an insoluble peroxide, which adheres to the anode. Secondary effects at the anode may sometimes be avoided by the
employment of an anode of platinum or gas-carbon.

Alloy of deposits with the cathode and with each other.—
When a perfectly clean surface of a metal receives a deposit
by electrolysis, in the great majority of cases, the first portions
of the metal deposited penetrate into the receiving surface,
and form either a mixture or alloy. I have met with some
new instances of this kind. I deposited a thick coating of
copper on the outside of a platinum cup. After heating this
cup to low redness a few times, the coating became loose,
and I tore it all off and digested the platinum surface in
nitric acid, and washed it ; it then looked perfectly free from
copper. On heating it again to low redness its surface became
black with oxide of copper. I cleaned it again in a similar
manner and heated it once more ; the surface again became
black. In this way, even after cleaning and heating six or eight
times, copper appeared ; the copper, therefore, must have
passed deeply into the substance of the platinum, and diffused
again outwards in the process of heating. In another instance a piece of thick platinum foil, upon which I had deposited a film of tellurium in dilute chloride of tellarium, with
a tellurium anode and a current from five Smee's cells, was
scraped very clean from the deposit and heated to low redness. An easily-fusible alloy was formed upon the surface,
and was oxidised and reduced repeatedly in the flame of a
Bunsen's burner, as if the tellurium had soaked into the
platinum.

Sonstadt also observed that a platinum crucible which had
been thinly electro-gilded lost its golden appearance by
heating to a moderate redness, by the gold soaking as it
were into the platinum (Weldon's ' Register,' No. 36, July
1863, p. 498). In the process termed 'pyroplating' also, an
electro-deposited layer of gold upon steel nearly disappears on
heating the steel article ; a second layer behaves similarly but
to a less extent ; a third does not disappear at all (' Chemical

News,' vol. xxvi., p. 137). The first films of one metal electro-deposited upon another frequently form an alloy, even with-out the aid of heat ; for instance, films of zinc or cadmium deposited upon copper impart to its surface a yellow colour, and in other cases similar effects probably occur, but do not happen to be observable. If one of the metals happens to be a liquid or a gas the effect is often more perfect ; thus most metals, even those of the earths and alkalies, when deposited upon mercury are absorbed by it and form amalgams ; hydrogen also, when deposited upon palladium, iron, and the surfaces of various metals, penetrates deeply into them, and alters their properties. This diffusive action of one metal within another operates not only during the process of deposition, but continues afterwards ; thus a yellow film of alloy, obtained by depositing copper upon zinc or cadmium, disappears in a few weeks, as if absorbed by the metallic substratum. This formation of metallic compounds takes place also, and even more completely, when two metals are simultaneously electro-deposited. In this way the less easily reducible metals, such as nickel, and iron, in the electro-deposition of which hydrogen is simultaneously de-posited, are very liable to contain hydrogen ; and in the case of nickel this enclosed gas is said to sometimes cause the deposit to split and curl up, and separate in brilliant films (Sprague's ' Electro-Metallurgy,' p. 305). In the case of antimony, especially that deposited from the bromide and iodide, large bubbles of gas gradually accumulate upon the surface of the deposit, and the deposited metal is full of them. The explosive character of some electro-deposits is also probably due to absorbed hydrogen. I have been in-formed that zinc which had been electro-deposited in a grey-black, spongy mass upon the iron plates of an exhausted battery, consisting of ten pairs of zinc and iron plates, ex-ploded when struck ; but I several times attempted with-out success to obtain such a deposit. Napier, in his ' Electro-Metallurgy,' 5th edition, p. 182, speaks of explosive electro-deposited bismuth. A writer in Dingler's 'Polytechnic Journal'also speaks of electro-deposits of rhodium and iridium exploding when struck ('Journal of the Chemical Society,' vol. ii., p. 1007).

The absorption of electro-deposited hydrogen has a great effect upon the properties of iron and steel. ' If after immersion, say, ten minutes in either sulphuric or hydrochloric acid, a piece of iron or steel be tested, its tensile strength and resistance to torsion will be found to be diminished. Exposure to the air for several days, or gentle heat, will however restore its original strength' (W. H. Johnson, 'Chemical News,' vol. xxvii., pp. 82 and 176; also vol. xxix., pp. 89 and 213 ; also Professor O. Reynolds, p. 118 of the same volume. Compare also the paragraphs on electro-deposition of hydrogen and its absorption by deposited metals, pp. 96, 247).

As steam boilers are occasionally supplied with water containing traces of acids, and the degree of acidity of the water becomes stronger by the evolution of steam, it is reasonable to suppose that the deposition of hydrogen by the simple immersion process, and its absorption by the iron, may in some cases contribute to the bursting of those vessels ; and in such cases the electro-deposition of hydrogen is a circumstance not to be neglected.

Purity of electro-deposited metals.—Electro-deposited metals are by no means necessarily pure ; they rarely are so, and the reason, probably, why the popular notion has arisen that they are very pure is because copper is the metal most frequently deposited, and such copper happens to be an exceptional instance of purity. The degree of purity of deposited metals depends chiefly upon the degree of purity of the solution ; if that is pure the deposit is likely to be so, and will be so unless it unites with the hydrogen liberated simultaneously with it, or with any of the constituents of the liquid, as in the instance of amorphous or 'explosive antimony.' The purity of the solution largely de-

E

pends upon the circumstance whether the anode is pure, and whether its impurities are soluble in the liquid ; if they are not, they cannot be deposited ; if they are soluble, then their deposition or not will largely depend upon the circumstances mentioned in the immediately following and preceding paragraphs. The great purity of electro-deposited copper is largely dependent upon the fact that any lead contained in the anode is insoluble in a sulphate solution, and any zinc contained in it is too electro-positive in an acid solution to be thrown down with the copper.

Electrolysis of mixed liquids.—Of the electrolysis of mixed metallic solutions comparatively little is known ; if, however, a solution contains several dissolved metals of very different degrees of positive electric capacity, the least positive metal is usually deposited first, unless there is an insufficient amount of that metal present to convey the whole of the current ; and if the current is continued till the whole of the metals are deposited, the most positive one is the last to be liberated. It is probable, also, that if the metals are about equally positive in the particular liquid, and their salts possess an approximately equal degree of conducting power, and are not widely different in quantity, the current divides itself between them, and is conducted by each ; but much investigation needs to be made in this part of the subject.

Electro-deposition of alloys.—It is much more difficult to deposit zinc than copper, because zinc is more electro-positive ; it is still more difficult to deposit two metals than one, especially if one of the metals is highly electro-positive to the other, as zinc usually is to copper, and it can only be effected by selecting a liquid in which the one metal is but feebly electro-positive to the other. According to the late eminent investigator, Professor Magnus, the separate and simultaneous deposition of substances from a mixed solution depends : 1st, on the density of the current ; 2nd, on the proportions in which the different substances exist in the fluid ; 3rd, on the nature of the electrodes ; and 4th, on the greater or less facility with which one or the other substance can be carried from stratum to stratum within the fluid ; as well as on the obstacles which stand in the way of this transmission, either in the shape of porous walls or in any other form ('Philosophical Magazine,' 4th series, vol. xii., p. 159).

With solutions in which alloys are to be deposited the most important condition is, that neither of the metals to be deposited shall be electro-positive to the other in that liquid. This is best tested by taking a wire of each metal, connecting them with a galvanometer, and simultaneously immersing their free ends in the liquid ; if either is electro-positive the needle of the instrument will be deflected, while the amount of deflection will indicate the amount of their electric difference in that liquid. It may also be tested by immersing a wire of each metal (not in mutual contact) in the liquid; if either become coated with the other metal in one hour, that one is positive; but if neither becomes coated in six hours, there is probably no considerable electric difference between them.

The following experiments of mine show that if a liquid contains two metals in solution, and a wire or other piece of each of those metals is immersed in the liquid, and one becomes coated with a deposit of metal, while the other does not, the coated one is electro-positive to the other in that liquid, and the solution will only yield by means of a feeble separate current the same metal which is deposited by simple immersion.

First experiment.—With an alloy solution, consisting of equal measures of a strong solution of protochloride of tin, and terchloride of antimony, with an anode either of tin or of antimony (the latter is the more suitable because it does not become coated by simple immersion in the liquid), a copper cathode, and a single cell of small Smee's battery, only antimony was deposited ; the tin became coated with anti-

mony by simple immersion, and was found by the galvanometer to be strongly positive to that metal.

Second experiment.—With a liquid composed of equal
measures of a solution of protochloride of tin and chloride
of bismuth, and either a bismuth or tin anode (the former
is the best), a brass cathode, and a small single cell of
Smee's battery, only bismuth was found to be deposited ; the
tin was positive to the bismuth by the galvanometer, and
became coated quickly with that metal by simple immersion.

Third experiment.—With a mixture of equal measures
of terchloride of antimony and chloride of bismuth, antimony
anode, copper cathode, and a feeble Smee's battery, only
antimony was deposited ; bismuth became slowly coated
with antimony in the solution by simple immersion, and was
found by the galvanometer to be moderately positive to the
latter metal in it.

Fourth experiment.—With 100 grains each of protochloride of tin and chloride of zinc dissolved together in an
ounce of distilled water, tin anode, copper cathode, and one
small Smee's cell, only tin was deposited ; zinc was positive
to tin in this liquid by the galvanometer, and deposited tin
upon itself by simple immersion.

Fifth experiment.—With equal measures of strong solutions of nitrate of zinc, and ternitrate of bismuth, and a
little nitric acid, bismuth anode, copper cathode, and a
feeble one-pair battery, only bismuth was deposited ; zinc
was strongly positive to bismuth in this liquid by the galvanometer, and became quickly coated with that metal by
simple immersion.

Sixth experiment.—With a solution of mixed sulphates
of zinc and copper, copper anode and cathode, and a single
small battery, copper alone was deposited ; zinc was strongly
positive to copper in this liquid by the galvanometer, and
coated itself immediately with copper in it by simple
immersion.

Further, if we take some distilled water, and dissolve

some caustic potash in it, and pass a moderately-strong current through it by platinum electrodes, hydrogen gas will alone be set free at the cathode ; in this case also the least positive of the two elements of the liquid—potassium and hydrogen—is set free or deposited. If we now add a little sulphuric acid to the solution to neutralise and convert it into a solution of sulphate of potash, add some sulphate of zinc besides, and pass a weak current through the liquid, we shall obtain a deposit of zinc on the cathode, but no hydrogen or potassium. In this case we cannot determine by the galvanometer which is the most positive in this liquid, hydrogen or zinc, because the former is a gas ; but it is probable that hydrogen is most positive, because the zinc does not evolve it by simple immersion in this liquid.

If we further add to the liquid a small quantity of sulphate of copper, and treat it as before, neither potassium, hydrogen, nor zinc will be deposited, but only copper ; we also find by the galvanometer that copper is less positive than zinc in such a liquid, and that zinc coats itself with copper in it by simple immersion ; in this case also the least positive of the positive elements is alone deposited. From these and many other experiments which I have made with similar results we deduce the following rule :—If a liquid contains several metals or electro-positive substances, and a *weak* electric current is passed through it, only that substance which is the least electro-positive will be deposited.

With regard to the influence exercised by the proportions of the ingredients of the liquid, and the strength of the current, I may observe that if the liquid contains several metals dissolved in equal quantities, and only one is being deposited by the passage of a weak current, a considerable increase in the strength of current will cause a portion of the next more positive metal to be deposited along with the less positive one ; but this alloy deposit will not be very coherent, because the power required to deposit the second metal in the reguline state will be so great as to deposit the first as a soft powder ; and this holds most true when the difference of electric power required is the greatest. Thus, 1st, if small quantities of sulphate of zinc and sulphate of copper are dissolved together in a large quantity of water, and a feeble current passed through the solution, only reguline copper will be deposited; but if the current passing be considerably increased, the deposit of copper will cease to be reguline, and zinc will be deposited with it. If the power be still further increased, hydrogen gas will also be evolved at the surface of the deposited metals. 2nd, if we dissolve a small quantity of sulphate of copper and a large quantity of sulphate of zinc in a large quantity of water, and pass a strong current through the solution, copper, zinc, and hydrogen will be set free at the cathode. 3rd, if we slightly moisten a lump of caustic potash with pure water, and pass a weak current through it by platinum electrodes, hydrogen alone will be set free at the cathode ; but if a very powerful current is employed, potassium also will be deposited. In each of these cases we find that when the current is least dense the least positive of the positive substances is alone deposited ; but if the power is sufficiently increased, and there is only a small portion of the less positive substance present, the more positive substances, even though they are much more positive, will also be deposited. The weaker affinities are overcome first and to the greatest extent ; the current of electricity exercising its influence first, and in the greatest proportions, upon the salt of the least positive metal.

Polarisation of electrodes.—After an electric current has been passing for some time between two electrodes in a liquid, if the electrodes be separated from the source of the current, and whilst they remain undisturbed in the liquid be connected with a galvanometer, a current occurs, and in a reverse direction to that of the original one. It often also occurs, but to a less extent, if the two electrodes are transferred to another liquid, or if two fresh electrodes are carefully immersed in the corresponding parts of the same liquid.

This phenomenon, known as a 'polarisation,' continually occurs in electro-metallurgical processes, and may be explained as follows :—the various substances, either primarily liberated, or secondarily formed, at the two electrodes, either adhere in a solid state to the electrodes, or fall to the bottom, or dissolve in the liquid, or escape as gas ; but in either case they more or less accumulate about the electrodes, adhere to, or are absorbed by them, and alter their electrical relations. The direction of the current produced by polarisation is opposite to that of the original one, because the latter has liberated electro-negative substances at the positive pole, and positive substances at the negative pole.

Formation of peroxides upon the anode.—Solutions of some metals, when electrolysed with platinum electrodes, are specially liable to form layers of insoluble peroxide upon the anodes by the action of the free oxygen of the water, liberated there by electrolysis. This is the case with those of the nitrates of lead, silver, and bismuth, the nitrate and acetate of manganese, and alkaline solutions of lead, cobalt and nickel (See ' Journal of the Chemical Society,' vol. ix., p. 307). In preparing the peroxides of bismuth, lead, nickel, cobalt, and manganese, by this method, very weak electric currents must be used. That of cobalt is readily prepared, and is permanent; its colours are magnificent, and may find an industrial use in the art of metallo-chromy (See pp. 242, 260 ; Wernicke, ' Chemical News,' vol. xxii., p. 240).

Movements in electrolytes.—As incidental effects of electrolysis I may mention the movements which take place in depositing liquids during the passage of the current. In nearly all cases the layer of liquid in contact with the anode, by dissolving a portion of that body, becomes specifically heavier than the remainder of the solution, and gradually sinks to the bottom ; whilst that in contact with the cathode, by abstraction of its metal, suffers a reverse change ; i.e. it becomes specifically lighter and rises to the top. In this way a layer of heavier liquid accumulates at the bottom of the vessel, and one of less specific gravity collects at the surface. At the same time these two layers slowly diffuse into the intervening strata of the liquid, and thus the whole solution tends to become uniform ; but if this process of diffusion is less rapid than that of separation, there is a constant state of difference of chemical composition and of specific gravity maintained between the upper and lower parts of the liquid. Certain effects result from this, viz., the anode corrodes away very freely at its upper part, the cathode receives a rapid and thick deposit at its lower part ; the current traverses the liquid in an oblique direction downwards; the anode does not dissolve at its lower portion, and the upper end of the cathode receives no deposit. And if the liquid is very dense, and contains much free acid, each electrode behaves like a single metal immersed vertically in two liquids (see p. 84), and generates a current between its upper and lower parts independently of the one which comes from the battery ; this independent current at the anode dissolves the upper portion of that body, and produces a metallic deposit upon its lower end, and the one at the cathode produces similar effects upon that metal ; and thus a deposit upon the upper end of a cathode may actually redissolve and disappear even whilst the battery current is passing. The most effectual way of counteracting these effects is to have the solution sufficiently dilute, without an excess of free acid or other solvent, to electrolyse it with moderate speed, and to stir it continually, or keep the electrodes in motion.

Magneto-electric action.—Electro-chemical and chemico-electric actions, together with the principles and laws which regulate them, constitute the essential part of the basis of electro-metallurgy ; magneto-electric action is only a subsidiary subject, because magneto-electric machines are merely one of the sources which may be employed of the electric current used in the art, and forms no part of the process itself. As also the principles of magneto-electric induction, and the construction and action of magneto-electric machines, are already described in the treatise on ' Electricity and Magnetism ' (pp. 70 and 280) in this Series, it is unnecessary for me to say more than a few words on this part of the subject.

Strictly speaking, magneto-electric action should be termed mechanico-electric action, because mechanical power is the cause and electricity the effect, and the magnetism acts only as an intermediate agent, by means of which the mechanical energy is enabled to produce or be transformed into electricity.   The fundamental fact or principle of magneto-electric action is, *wherever there is varying magnetism, there is an electric current induced in an adjacent closed conducting circuit at right-angles to it.*

Magnetism in the vicinity of a conductor of electricity may be caused to vary by several means, viz. by heating or cooling the magnetised bcdy (I have employed this method, ' Proceedings of the Royal Society,' 1869, No. 108), by otherwise changing the strength of the magnet, by altering the distance of the magnet from the conductor, or by varying their relative positions to each other.   The first of these methods is rarely used, because it is not convenient; but the others are commonly employed, and are very suitable and effective.   The current lasts only during the increase or decrease of the magnetism, and is reverse in direction in the two cases, as indicated in the annexed figures 12 and 13.

The magnetism acting upon the conductor may be increased or decreased, by alternately approaching the magnet towards and withdrawing it from the conductor, or by alternately magnetising and demagnetising a bar of soft iron, upon which the conductor is wound; the latter is the method generally employed, and is usually effected by rotating the iron bar and its surrounding conductor between the poles of a magnet.

The alternate currents produced in the surrounding conductor by the magnetisation and demagnetisation of the iron are opposite in direction, and are caught up and thrown into one uniform course by means of a mechanical arrangement termed a commutator, which is fixed to the iron arma-

FIG. 12.

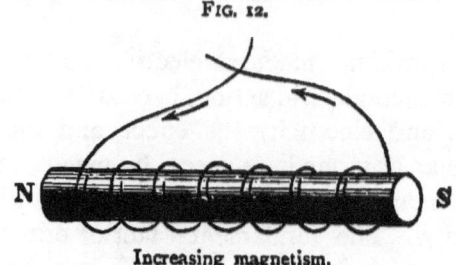

Increasing magnetism.

FIG. 13.

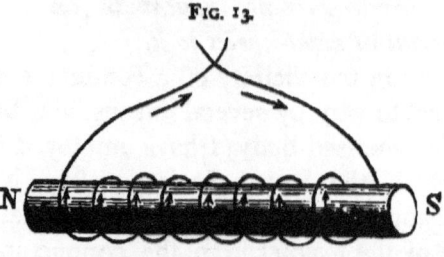

Decreasing magnetism.

ture and revolves with it. The current, therefore, usually obtained by magneto-electric action differs from that resulting from a chemico-electric source, in being a succession of momentary streams of electricity, all flowing in the same direction.

Mr. Henry Wilde, of Manchester, discovered that if the current from the wire of the revolving armature was made to flow through a coil of insulated wire surrounding a large bar of soft iron, a degree of magnetism many times stronger than that of the original magnet might be produced by revolving the armature sufficiently fast ; and that by an extension of this principle of accumulation, magnets of any degree of power might be obtained, limited only by the capacity of the iron to receive magnetism, and the amount of mechanical power expended in rotating the armatures.

In all magneto-electric machines there exists, during their

continuance of action, incessant molecular vibrations in the magnet and armature, caused by the changes of magnetism, and in the conducting-wires, caused by the electrical waves ; and in consequence of these vibrations considerable heat is produced, which differs in amount in different machines. In this way a portion of the mechanical power expended is converted into heat instead of into electricity ; and, in addition to this, the heat is liable to injure the insulation of the wires upon the armature, and damage the bearings of the revolving parts of the machine.

The kinds of magneto-electric machines employed for electro-deposition, together with additional information of a more technical kind, will be illustrated and given in the practical division (p. 345) of this book.

Thermo-electric action.—Thermo-electric action is a much more secondary matter at present than magneto-electric induction in relation to electro-metallurgy, because it is as yet but little used in the art ; but its use will probably be largely extended, and may even supersede that of magneto-electric induction, because it is a much more direct conversion of heat into electric force.

The chief fact of thermo-electric action was discovered by Seebeck in 1823, and is as follows: If we take two bars, A and B (see Fig. 14), of any two metals, especially bis-. muth and antimony, solder their junctions, connect their free ends by wires, and apply heat to the soldered junction,

FIG. 14.

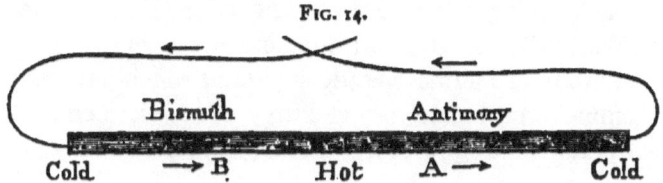

a portion of the applied heat is absorbed and disappears, and an electric current is produced in its stead.

All metals and other conductors of electricity are capable

of producing the current, and they are usually classed into thermo-electro-positive, or those in which the current proceeds from the colder to the warmer portion, as with bismuth; and thermo-electro-negative, or those in which it proceeds in the reverse direction, as with antimony.

In the following table by Dr. Mattheissen, the various metals, &c., are arranged in a series according to their relative degrees of thermo-electric tension :—

Thermo-electro-positive.

Bismuth, commercial +	.	97·	Cobalt	. .	. 22·
,, pure	.	89·	German-silver	.	. 11·75
,, crystal axial	.	65·	Mercury	. .	. ·418
,, ,, equatorial		45·	Lead .	. .	Neutral

Thermo-electro-negative.

Copper, commercial	.	·1	Arsenic	. .	. 13·56
Platinum	. .	·9	Iron, pianoforte wire	.	17·5
Gold .	. .	1·2	Antimony, crystal axial		22·6
Antimony, pressed wire	.	2·8	,, ,, equatorial	.	. 26·4
Silver, pure and hard	.	3·	torial	.	. 26·4
Zinc, pure, pressed wire		3·7	Red phosphorus .	.	29·7
Copper, electro-deposited		3·8	Tellurium .	. .	. 502·
Antimony, commercial, pressed wire	.	6·	Selenium .	. .	. 807·

Not only solid conductors of electricity, but also liquid ones, are capable of yielding thermo-electric currents, and I have devised and employed, in several researches on the subject, various apparatus for the purpose ('Philosophical Magazine,' January 1857; 'Proceedings of the Royal Society,' 1871 ; Watt's Dictionary, 2nd Supplement, p. 457).

The kinds of thermo-electric apparatus which have as yet been employed in electro-metallurgy are described in the special practical section of this book (p. 349).

CHEMICAL PRINCIPLES OF ELECTRO-METALLURGY.

Chemico-electric relations of substances.—Chemico-electric action often takes place in electro-metallurgical processes; it continually occurs in the voltaic-batteries whilst they are in use ; it also takes place under several circumstances when metals are immersed in conducting liquids ; for instance, it occurs when any metal is immersed in an acid solution of a less positive metal than itself, as when steel or iron is dipped into a solution of sulphate of copper, a portion of the steel or iron dissolves, and produces an electric current. The importance of a knowledge of the relation of chemico-electric action to electro-metallurgy may be shewn by the fact, that any liquid, say the one just mentioned, in which the metal to be coated, say iron, is strongly electro-positive to the metal in solution, and with which it is intended to be coated, is usually unfit to be used for coating that particular metal, because if a thick coating is formed upon it the deposit will not adhere firmly.

Chemico-electric series.—The following table exhibits the *usual* relative electrical positions to each other in most liquids of a number of the elementary substances ; the first substance named being the most electro-positive, and the last one the most electro-negative :—

Potassium +	Copper	Phosphorus
Sodium	Silver	Selenium
Magnesium	Mercury	Iodine
Zinc	Platinum	Bromine
Iron	Gold	Chlorine
Aluminium	Hydrogen	Nitrogen
Lead	Antimony	Sulphur
Tin	Carbon	Fluorine
Bismuth	Tellurium	Oxygen —

In this series every substance is usually positive to all those below it, and negative to those above ; consequently none are absolutely positive or negative, and therefore the series cannot strictly be divided into two classes, one consisting wholly of positive and the other of negative bodies; but it is usual, nevertheless, to speak of the metals, especially the alkali and base metals, as positive, and the metalloids as negative substances. Many exceptions might be shewn with regard to the positions of the substances in the above series, because the order varies with every different liquid in which they may be immersed, and therefore the table is only of value for usual guidance.

Electrical relations of metals in liquids.—There are several arrangements of immersing metals in liquids, by means of which electric currents are produced, and they may be classed as follows : 1. By the immersion of two metals in one liquid. 2. Of one metal in two liquids. 3. Of two metals in two liquids, &c.

1. *By the immersion of two metals in one liquid.*—If we connect two pieces, A and B, of different kinds of metal with the two ends of the coil of a galvanometer C (see Fig. 15), and

Fig. 15.

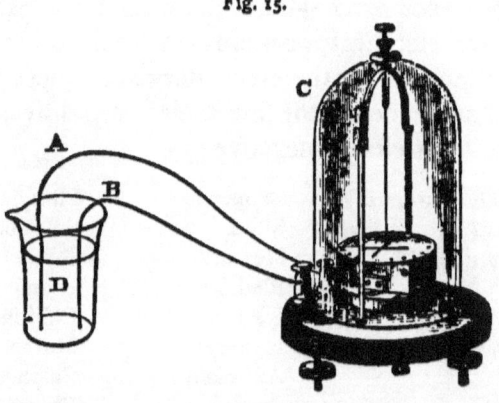

immerse them in a conducting liquid, D, an electric current is generated by chemical action, and the needle is deflected ; and if we examine the behaviour of a number of metals thus in a variety of different liquids, and arrange them in series according to their degrees of electrical power, we find that the electrical positions of the metals usually agree with the order given in the foregoing table ; and we also find that their order is somewhat different in every different liquid.

The electrical orders of metals, &c., in a number of

continually occurs in the voltaic-batteries whilst they are in use ; it also takes place under several circumstances when metals are immersed in conducting liquids ; for instance, it occurs when any metal is immersed in an acid solution of a less positive metal than itself, as when steel or iron is dipped into a solution of sulphate of copper, a portion of the steel or iron dissolves, and produces an electric current. The importance of a knowledge of the relation of chemico-electric action to electro-metallurgy may be shewn by the fact, that any liquid, say the one just mentioned, in which the metal to be coated, say iron, is strongly electro-positive to the metal in solution, and with which it is in-tended to be coated, is usually unfit to be used for coating that particular metal, because if a thick coating is formed upon it the deposit will not adhere firmly.

Chemico-electric series.—The following table exhibits the *usual* relative electrical positions to each other in most liquids of a number of the elementary substances ; the first substance named being the most electro-positive, and the last one the most electro-negative :—

Potassium +	Copper	Phosphorus
Sodium	Silver	Selenium
Magnesium	Mercury	Iodine
Zinc	Platinum	Bromine
Iron	Gold	Chlorine
Aluminium	Hydrogen	Nitrogen
Lead	Antimony	Sulphur
Tin	Carbon	Fluorine
Bismuth	Tellurium	Oxygen —

In this series every substance is usually positive to all those below it, and negative to those above ; consequently none are absolutely positive or negative, and therefore the series cannot strictly be divided into two classes, one con-sisting wholly of positive and the other of negative bodies; but it is usual, nevertheless, to speak of the metals, especially the alkali and base metals, as positive, and the metalloids as

5. *Hydrochloric Acid* (Faraday).

Zinc +	Iron	Silver
Cadmium	Copper	Antimony —
Tin	Bismuth	
Lead	Nickel	

6. 1 *volume of Nitric Acid and* 7 *volumes of Water* (Faraday).

Zinc +	Iron	Copper
Cadmium	Nickel	Silver —
Lead	Bismuth	
Tin	Antimony	

7. *Nitric Acid, sp. gr.* 1·48 (Faraday).

Cadmium +	Iron	Silver
Zinc	Bismuth	Nickel —
Lead	Copper	
Tin	Antimony	

8. *Concentrated Nitric Acid* (De la Rive).

Tin +	Copper	Silver
Zinc	Lead	Peroxide of Iron —
Iron	Mercury	

9. *Pure Dilute Hydrofluoric Acid of* 10 *per cent.* (Gore).

Aluminium +	Lead	Bismuth
Zinc	Silicon	Copper
Magnesium	Iron	Silver
Thallium	Nickel	Gold
Cadmium	Cobalt	Gas-carbon
Tin	Antimony	Platinum —

10. *Pure Dilute Hydrofluoric Acid of* 28 *per cent.* (Gore).

Zinc +	Nickel	Gas-carbon
Magnesium	Cobalt	Platinum
Aluminium	Antimony	Rhodium
Thallium	Bismuth	Palladium
Indium	Mercury	Tellurium
Cadmium	Silver	Osmi-iridium
Tin	Copper	Gold
Lead	Arsenic	Iridium —
Silicon	Osmium	
Iron	Ruthenium	

11. *Aqueous Potash or Soda* (H. Davy).

Alkali-metals +	Copper	Gold
Zinc	Iron	Platinum −
Tin	Silver	
Lead	Palladium	

12. *Solution of Potash or Soda, strong or weak* (Faraday).

Zinc +	Lead	Nickel
Tin	Bismuth	Silver−
Cadmium	Iron	
Antimony	Copper	

13. *Solution of Potash or Soda, sp. gr.* 1·33 (Pfaff).

Tin +	Copper	Steel
Zinc	Gold	Silver−
Antimony	Platinum	
Lead	Bismuth	

14. *Aqueous Ammonia, sp. gr.* ·95 (Pfaff).

| Zinc + | Lead | Copper− |
| Tin | Silver | |

15. 1 *part of Cyanide of Potassium in* 8 *parts of Water* (Poggendorff).

Amalgamated Zinc +	Nickel	Iron
Zinc	Antimony	Platinum
Copper	Lead	Cast-iron
Cadmium	Mercury	Coke −
Tin	Palladium	
Silver	Bismuth	

16. *Dilute Yellow Sulphide of Potassium* (Faraday).

Zinc +	Silver	Bismuth
Copper	Lead	Iron−
Cadmium	Antimony	
Tin	Nickel	

17. *Dilute Hydrosulphate of Potassium* (H. Davy).

Zinc +	Bismuth	Gold
Tin	Silver	Charcoal −
Copper	Platinum	
Iron	Palladium	

F

18. *Colourless Solution of Sulphide of Potassium* (Faraday).

Cadmium +	Antimony	Nickel
Zinc	Silver	Iron —
Copper	Lead	
Tin	Bismuth	

19. *Solution of Sal-ammoniac* (Poggendorff).

Zinc +	Magnetic Iron	Copper pyrites
Cadmium	German-silver	Tellurium
Manganese	Cobalt	Gold
Lead	Bismuth	Galena
Tin	Antimony	Coke
Iron	Arsenic	Platinum
Steel	Chromium	Plumbago
Uranium	Silver	Peroxide of Man-
Brass	Mercury	ganese —

20. *Solution of Common Salt* (Fechner).

Zinc +	Antimony	Gold
Lead	Bismuth	Platinum —
Tin	Copper	
Iron	Silver	

21. *Fused Boracic Acid* (Gore).

Iron +	Platinum	Silver —
Silicon	Gold	
Carbon	Copper	

22. *Fused Phosphoric Acid* (Gore).

Zinc +	Copper	Platinum —
Iron	Silver	

23. *Fused Potassic Hydrate* (Gore).

Silicon +	Iron	Platinum
Aluminium	Lead	Silver —
Zinc	Carbon	

24. *Fused Potassic Carbonate* (Gore).

Silicon +	Carbon	Platinum —
Iron	Copper	
Zinc	Silver	

25. *Fused Potassic Chloride* (Gore).

Aluminium +	Iron	Silver
Zinc	Copper	Platinum —

26. *Fused Potassic Fluoride* (Gore).

Palladium +	Platinum	Iridium —
Gold		

27. *Fused Ammonic Nitrate* (Gore).

Magnesium +	Silver	Silicon
Zinc	Tin	Carbon
Lead	Aluminium	Platinum —
Copper	Iron	

Additional information respecting the chemico-electric relations of metals in aqueous solutions will be found scattered throughout the book under the headings of the respective metals, and for additional tables of the electrical relations of metals in fused substances see 'Philosophical Magazine,' June 1864; 'Chemical News,' vol. ix., p. 266.

The earliest kinds of voltaic batteries were composed of two metals immersed in one liquid, and from the outset zinc appears to have been almost the only metal employed as the positive element. The currents obtained from them were stronger in proportion as the two metals were further asunder in the general series (p. 61); thus they were stronger when silver was substituted for copper, and platinum or carbon for silver, as the negative element; and they were obtained still stronger by selecting from the special series the most suitable liquid in which to immerse them. It was practically by selecting metals and liquids in accordance with these series, and especially those which were the most durable and least expensive, that the earlier kinds of batteries were invented. No powerful battery is composed of two metals which lie close together in the series, such for instance as zinc with cadmium, tin, lead, or iron in dilute sulphuric acid. The chief batteries of this class, i.e. of two metals immersed in one liquid, are the old zinc and copper one, Cruickshank's, Wollaston's, and Smee's.

The value of cyanide solutions, for the purposes of electro-plating, is also largely dependent upon the electrical relations of the baser metals to gold and silver in such liquids. If we refer to Tables 1 to 12 (pp. 63, 64) we may perceive that in aqueous acids the baser metals are high up in the lists, and more electro-positive, and gold, silver, and copper either lower or a long way down ; but in aqueous cyanide of potassium (Table 15), or sulphide of potassium (Tables 16, 17, 18), each of them strongly alkaline liquids, copper and silver are much higher up, and iron much lower down ; and in consequence of this the anode of nobler metal is more easily dissolved, and a receiving surface of iron is less corroded.

The electrical relations of metals and carbon, &c., in fused substances have not yet found to any notable extent similar practical applications, but it is not improbable that some will be found, because they hold out a prospect of obtaining electric currents by means of the combustion of coke. 'The discovery of some suitable fused salt or mixture, in which carbon is highly electro-positive at a high temperature to iron, nickel, or other infusible and also in other respects suitable conductor, would probably prove a cheap and powerful source of electricity ; cheap, because of the low chemico-electric equivalent of carbon in relation to that of zinc, and the low price of coke and gas-graphite ; and powerful, because of the intense affinity of carbon for oxygen at high temperatures, sufficient indeed to set the alkali metals free from their oxides. The nearest approach to this object in these experiments' (in experiments with metals in fused substances) 'was obtained with carbon and nickel, immersed in a fused mixture of soda, lime, and silica, i.e. in a species of glass' ('Philosophical Magazine,' June 1864 ; 'Chemical News,' vol. ix., 1864, p. 266).

2. *By immersion of one metal in two liquids.*—If two pieces of the same metal are immersed in two different conducting liquids, the liquids being in contact with each other through the medium of a porous partition or other suitable means, such as placing the two liquids in the two legs of a bent glass tube, &c., an electric current is produced. A large number of instances of currents generated by means of this arrangement are described in Gmelin's 'Handbook of Chemistry,' vol. i., p. 397, and one or two voltaic batteries have been constructed according to it, but have not come into extensive use.

3. *By immersion of two metals in two liquids.*—Any two metals immersed in any two conducting liquids which are in mutual contact will also produce an electric current, and the strongest voltaic currents are obtained by means of this arrangement, because there is not only an electrical and chemical difference of metal, but also of liquid. Daniell's, Bunsen's, Grove's, and many other batteries, are of this kind.

The particular liquid in which the most positive metal is the most positive is not necessarily the same liquid as that in which the most negative metal is the most negative, and therefore a one-liquid battery does not give the strongest current ; but this arrangement of two metals and two liquids enables us to combine the advantages of both liquids, and thus obtain a more powerful current.

A very large number of batteries have been devised and constructed in accordance with one or other of these arrangements, and many are in practical use. The particular kinds of them which have been employed for the purposes of electro-metallurgy will be described in the practical division of this book.

Voltaic currents.—As the origin of the currents in voltaic batteries, the terms *tension, potential, intensity of current, quantity of current, electro-motive force,* &c., are fully explained in another book of this series, viz. the 'Treatise on Electricity and Magnetism,' by Fleeming Jenkin, F.R.S., I shall say no more than is requisite on those points, but refer the reader to that treatise for fuller information.

The electrical relations of metals in liquids are the chief source of all voltaic currents, and the main fact upon which the action of all voltaic batteries depends. If two different metals are immersed in a conducting liquid, or in two such liquids, the liquids being in mutual contact, and the metals united by a wire, a constant state of electric difference is produced in them by their mutual contact, and by a difference of chemical action of the liquid or liquids upon them; and this is the commencement of all voltaic action, and the origin of *electro-motive force.* The particular kind of chemical action which is the main source of the current in nearly all such batteries is the union of the zinc with the oxygen of the liquid in contact with it.

Electro-motive force.—The electro-motive force, or strength of the current to overcome resistance, depends upon the degree of difference of strength of chemical affinity of the two metals for the electro-negative constituents of the liquid. The farther asunder, therefore, the two metals are in the chemico-electric series (p. 61), the greater usually is the difference of intensity of chemical action of ordinary acid liquids upon them, and the stronger also is the electro-motive force. This general truth may be easily verified by connecting pieces of platinum and copper with a galvanometer, and immersing their ends in dilute nitric acid whilst watching the needle; then make a similar experiment with copper and zinc in dilute sulphuric acid; also with zinc and magnesium in extremely dilute sulphuric acid; and it will be found that in each case the current proceeds from the metal which is most acted upon, through the liquid to the other metal. The electro-motive force evidently depends also upon a more fundamental point, which as yet has been but comparatively little studied, viz. upon the *kind* of chemical affinity exercised between the positive metal and the negative constituents of the liquid ; for instance, copper is powerfully corroded by nitric acid, but although the intensity of chemical attraction is great in this case, the electro-motive force generated is comparatively feeble. I have made some experiments upon this point. The ordinary unit of electromotive force employed in this country is termed a ' volt ' (see Jenkin's 'Treatise on Electricity and Magnetism,' p. 159 ; also Appendix, p. 387 of this work).

Potential and tension.—Previous to the completion of the circuit and formation of an unimpeded current, the free ends of the polar wires attached to the two metals are charged with the two kinds of electricity in an accumulated or free static condition, and are in a state of *electric potential*, i.e. possessing a capability of doing electric work. These accumulated electricities in the wires may be detected by means of a very delicate electroscope. The free electricities are also in a state of *tension*, constantly tending to escape and unite ; and their degrees of tension may be measured by means of an electrometer ; the degree of tension, however, in a single voltaic cell is extremely small, and has been estimated to be about ten million times less than that of an ordinary frictional electric machine.

Current; strength of current.—The continual union of the two electricities through the connecting wire, or other conductor, constitutes an electric *current*. Any given voltaic battery can only yield a given maximum strength of current. The strength is the amount or quantity of electric force which flows through any given section of the circuit in a given period of time. It depends upon two conditions, viz. the electromotive force of the battery, and the total amount of resistance in the circuit. The strength of the current is equal to the electro-motive force divided by the resistance ; this

is known as Ohm's law; it is directly proportional to the
electro-motive force, and inversely proportional to the resistance ; if the resistance remains the same, and the electro-
motive force varies, the strength is directly proportional
to the electro-motive force ; and if the electro-motive
force remains the same, and the resistance varies, it is inversely proportional to the whole of the resistance in the
circuit (see Appendix, p. 387).

Resistance ; 'intensity' of current.—The total resistance
in the circuit is usually divided into *internal,* or that in the
battery itself; and *external,* or that in the connecting wires
and other portions of the circuit outside the battery.   If the
external resistance is much less than that of one cell, as
when the poles of a single voltaic cell are connected by a
short and thick copper wire, any addition to the number of
cells in the battery will produce no perceptible increase of
current, because by that addition we augment the internal
resistance as fast as we increase the electro-motive power.
But if the external resistance is much greater than the resistance in the battery, any addition to the number of cells
will produce a nearly proportionate increase in the quantity
or strength of the current, because we then increase the
electro-motive force much faster than we augment the total
amount of resistance.

A current which is but little diminished in amount by the
introduction of a given external resistance is, in common
language, said to possess great 'intensity'; but the difference of effect produced by means of a current from one cell
and that from many, does not arise from any real difference
in the nature of the currents in the two cases, but from the
difference of proportion of external to internal resistance.
No difference has hitherto been recognised in any two currents of equal quantity per minute, obtained from different
voltaic sources.

Measurement of current.—The quantity of electricity
circulating may be measured by the amount of electric work

performed, and the strength of the current by the amount of such work done in a given period of time.

The instruments usually employed for this purpose are either a galvanometer or a voltameter. The 'construction, action, and mode of using a galvanometer are already fully described in the 'Treatise on Electricity and Magnetism' of this series, p. 187. One suitable for use in electrolytic experiments should offer but little resistance to the passage of the current, and what is termed 'a tangent' one may be conveniently employed, because the magnetic action of the current upon the needle is then much less powerful.

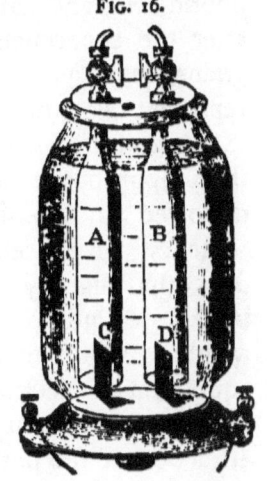

FIG. 16.

Voltameters are of different kinds; that originally employed by Faraday contained a mixture of sulphuric acid and water as the electrolyte. A water voltameter is shown in the annexed sketch, Fig. 16. Two graduated glass tubes, A and B, open at the bottom, and provided with stop-cocks at their upper ends, are inverted over two large plates of platinum, C and D, which are connected to the binding screws, by means of wires beneath. The outer glass jar is nearly filled with a previously-cooled mixture of about $3\frac{1}{2}$ or 4 measures of distilled water and 1 measure of pure sulphuric acid, the acid being added to the water, *not the reverse.* On passing the current to be measured from C to D, oxygen collects in C and hydrogen in D, and the quantity of electricity passed varies directly as the amount of gas evolved.

In this form of voltameter a little error arises in consequence of a small portion of the gas being dissolved in the water, and also from unequal pressure of the liquid upon different quantities of gas ; if also the current to be measured is one of low electro-motive force, its amount is largely diminished by the resistance in the voltameter itself.

An arrangement which offers less conduction resistance, and which is very convenient, consists in passing the current through two large electrodes (the cathode being a thin one) of pure and clean sheet copper, immersed in a nearly saturated solution of sulphate of copper, to which has been subsequently added about one-sixth its bulk of pure dilute sulphuric acid, and carefully weighing the cathode before and after the experiment. Each atomic weight, 63·5 parts in grains, may, as copper is a dyad (see p. 41), be said to represent two equivalents of electricity.

Amounts of electricity produced by different metals.—As the proportionate number of atomic weights of a substance dissolved or deposited by electrolysis depends upon the 'valency' of the elements (see p. 41), so in like manner does the quantity of the current generated in a voltaic battery depend upon the same condition. Thus one atomic weight (say in grains) of a monad element will produce one equivalent of electricity, a dyad two, a triad three, a tetrad four, &c. Whilst it is the degree of *intensity* of chemical attraction of the positive metal by the negative element of the liquid in the battery which determines the *electro-motive force* of the current, it is the *quantity* of the substances attracted which determines its *amount*.

Relation of the quantity of chemical action in the battery, to that in the depositing vessel.—The relations of the electric current, both to the quality and quantity of electrolytic effect, have already been described (p. 35 to 45); and those relations are the same whether the current producing the electrolysis is generated in the electrolytic vessels themselves, or in separate batteries, or other electro-motors. And, as a chemico-electric equivalent of metal dissolved in the battery generates an equivalent of electricity, and an equivalent of electricity deposits an equivalent of metal in the deposit-cell (pp. 42, 44), we may now say that, usually, *for each chemical equivalent of substances dissolved, set free, or formed, in each cell of the battery, a chemical equivalent of other substance is dissolved, set free or formed, in each depositing vessel, in the same circuit*; and that this equivalency of action throughout the circuit is due to the fact that each chemical equivalent of any substance has associated with it an equal amount of electricity. This is the law of definite electro-chemical action.

Faraday established that law. He found that 'if the current of a battery be passed through a voltameter containing dilute sulphuric acid and platinum electrodes, and thence by means of a platinum wire entering the upper end, and conveying positive electricity into a glass tube containing fused protochloride of tin, and having inserted into its lower end a platinum wire, which serves as the negative electrode, then for every 9 parts of water decomposed in the voltameter 58·53 parts of tin are deposited on the last-mentioned wire' (the atomic weight of tin is 118). 'When fused chloride, iodide, oxide, and borate of lead were treated in a similar manner, the quantity of lead obtained was too small in proportion to the water decomposed, viz. to 9 parts of water, 100·8, 89·, 93·2, and 101·3 lead, whereas the atomic weight of lead is 207' (or 103·5 × 2, lead being a dyad). 'The cause of the deficiency is, probably, that a portion of the precipitated lead was redissolved by the anion; a kind of circumstance which occasionally happens. When two silver wires are introduced as electrodes into fused chloride of silver, the weight of the positive electrode diminishes almost exactly 108·1 parts for every 9 parts of water decomposed in the voltameter, whilst that of the negative electrode increases by the same quantity. Chloride and iodide of lead treated in the same manner, lead being used as the positive electrode, gave 101·5, and 103·5 lead for every 9 parts of water.'

According to Quincke, the force tending to separate the constituents of an electrolyte is proportional to the density of the current, i.e. to the strength of the current per unit of sectional area of the liquid; it also increases with the electromotive force of the current employed, and is inversely pro-

portional to the length, but independent of the cross-section
and of the conductivity of the liquid, if the resistance of the
remainder of the circuit is small in comparison with that
of the electrolyte ('Journal of the Chemical Society,' vol. x.,
p. 208).

PRACTICAL DIVISION.

SECTION A.

General methods of depositing metals.—There are various
methods which either have been or are still employed in
depositing metals from their solutions for practical purposes,
and they may be classed as follows—1st. *By immersing one
metal in one liquid,* as, for instance, by immersing steel or iron
in a slightly-acidulated solution of sulphate of copper. 2nd.
By immersing two metals in one liquid, as by immersing the
article to be coated in contact with zinc or other sufficiently
positive metal in the particular metallic solution. 3rd. *By
immersing one metal in two liquids,* as, for instance, if a deep
glass vessel be half filled with a saturated aqueous solution
of cupric sulphate, the vessel be then nearly filled with water
containing a small quantity of sulphuric acid, poured in
quietly so as not to mix with the copper solution, and a
bright rod of copper, as deep as the vessel, be allowed to
remain in a vertical position in the liquid during twenty-four
hours without disturbance; the upper half of the rod will
slowly corrode and dissolve, whilst the lower half will receive
a deposit of copper.   4th. *By immersing two metals in two
liquids,* as in the ordinary 'single cell' electrotype apparatus.   5th. *By the separate current plan,* as when a voltaic
battery, magneto-machine, or thermo-electric pile is employed with a separate depositing vessel.

Under each of these classes will be mentioned a number
of experiments made by the author, and it is desirable that

the student should repeat a few of them, in order to fix the general principles more firmly in his memory.

Method No. 1.—*Deposition by immersing òne metal in one liquid* (see Fig. 17). With aqueous solutions of the following salts I obtained the effects mentioned. In *hydrochlorate of terchloride of antimony* (a solution of terchloride of antimony in hydrochloric acid), as prepared for pharmaceutical purposes, zinc, bismuth, tin, lead, brass, and german-silver became coated with antimony; whilst antimony, nickel, silver, gold, and platinum did not.

FIG. 17.

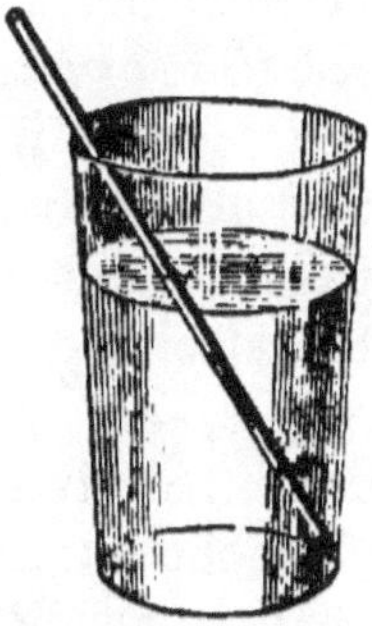

Chloride of bismuth.—Zinc, tin, lead and iron deposited the bismuth upon themselves, whilst antimony, bismuth, copper, brass, german-silver, gold, and platinum did not.

Tetrachloride of platinum.—Platinum was deposited from a solution of its chloride by arsenic, antimony, tellurium, bismuth, zinc, cadmium, tin, lead, iron, cobalt, nickel, copper, brass, german-silver, mercury, and silver; but not by gold or platinum.

Gold solutions.—From an acid solution of terchloride of gold, the base metals, likewise mercury, silver, platinum, and palladium, deposited gold in the metallic state; arsenic rapidly deposited gold in this solution; antimony, tellurium, and bismuth became gilded; zinc, cadmium, lead, iron, cobalt, mercury, silver, platinum, and palladium deposited the gold. In a solution of the *double cyanide of gold and potassium*, zinc quickly became gilded, and copper, brass, and german-silver slowly, whilst antimony, bismuth, tin, lead, iron, nickel, silver, gold, and platinum did not.

Silver solutions.—The following metals, viz. manganese, arsenic, antimony, bismuth, zinc, cadmium, tin, lead, iron, copper, and mercury deposited silver from its solutions in the metallic state. An *aqueous solution of nitrate of silver* yielded its metal to manganese, arsenic, antimony, bismuth, zinc, tin, lead, iron, nickel, copper, brass, and german-silver ; but not to silver, gold, or platinum ; lead and tin deposited the silver most quickly ; then followed the other metals in this order : cadmium, zinc, copper, bismuth, antimony, arsenic, mercury. Arsenic deposited silver from the *alcoholic solution of nitrate of silver* ; antimony received a coating of silver either in the aqueous sulphate or alcoholic nitrate ; bismuth deposed silver from the alcoholic nitrate, but not from the aqueous sulphate ; zinc received a silver deposit in the alcoholic nitrate ; tin became silvered in the alcoholic nitrate, but more quickly in the aqueous sulphate ; iron deposited silver from the sulphate of silver, but not from the alcoholic nitrate ; copper deposited it from the aqueous sulphate or alcoholic nitrate ; brass and the alloys of silver, with zinc, tin, or lead, deposited silver from silver solutions completely. In a solution of the *double cyanide of silver and potassium* (the ordinary silver-plating liquid), zinc, lead, and copper became silvered ; also brass and german-silver, but more slowly ; whilst antimony, bismuth, tin, iron, nickel, silver, gold, and platinum did not.

Mercurous salts.—Solutions of mercurous salts have their metal deposited by arsenic, antimony, bismuth, zinc, cadmium, tin, lead, iron, copper, and brass, also by the alloys of silver with zinc, tin, lead, or copper.

Nitrate of mercury.—A solution of nitrate of mercury yielded its metal to bismuth, zinc, cadmium, lead, iron, or copper, and if acidulated with nitric acid, to antimony also, but not to silver, gold, or platinum.

Acetate of mercury.—Iron deposited mercury from a solution of this salt.

Sulphate of copper.—In a solution of sulphate of copper, zinc, tin, lead, and iron became coated with copper, whilst antimony, bismuth, nickel, copper, silver, gold, and platinum did not.

Cupric chloride.—In a solution of chloride of copper,

bismuth, zinc, tin, lead, and iron received a copper deposit ; whilst antimony, nickel, copper, silver, gold, and platinum did not.

Nitrate of copper.—In a solution of nitrate of copper, zinc, tin, lead, and iron became coated ; whilst antimony, bismuth, nickel, copper, silver, gold, and platinum did not.

Ammonio-chlorides of copper.—With a solution of sub-chloride of copper in liquid ammonia, or of black oxide of copper in a solution of sal-ammoniac, zinc received a deposit ; whilst antimony, bismuth, tin, lead, iron, nickel, copper, silver, gold, or platinum did not.

Ferrous sulphate.—'Zinc,' as Fischer says, 'immersed in a perfectly neutral solution of ferrous sulphate (protosulphate of iron), contained in a stoppered bottle, throws down metallic iron, which is deposited partly on the zinc ;' but, in my experience, with this solution neither antimony, bismuth, tin, lead, iron, nickel, copper, brass, german-silver, silver, gold, or platinum received any metallic deposit.

Hyponitrite, nitrate, or acetate of lead.—In a solution of hyponitrite, nitrate, or acetate of lead, zinc received a coating of·lead ; whilst antimony, bismuth, tin, lead, iron, nickel, copper, brass, german-silver, silver, gold, and platinum received no deposit.

Stannous chloride.—In a solution of stannous chloride, zinc and lead become tinned ; whilst antimony, bismuth, tin, iron, nickel, copper, brass, german-silver, silver, gold, and platinum receive no deposit.

Sulphate, chloride, nitrate, or acetate of zinc.—In a solution of either sulphate, chloride, nitrate, or acetate of zinc, neither antimony, bismuth, zinc, tin, lead, iron, nickel, copper, brass, german-silver, silver, gold, or platinum became coated with zinc.

Observations upon class of instances No. 1.—In reviewing all these instances, we may make the following observations: 1st, that various metals by mere immersion in solutions of other metals, at the ordinary temperature of the atmosphere,

sometimes become coated with a deposit of metal, and sometimes not; 2nd, that a metal rarely becomes coated by mere
immersion in a solution of the same metal; for instance,
zinc does not become coated with zinc in a solution of sulphate of zinc (see p. 79), copper with copper in a solution of
its sulphate, gold with gold in its chloride, &c.; 3rd, that the
baser metals, especially zinc, cadmium, tin, lead, and iron,
become coated more frequently than the noble metals,
especially gold and platinum; 4th, that solutions of base
metals, especially of zinc and iron, yield their metal less
frequently than those of the noble metals, especially those of
gold and platinum; 5th, that of all the ordinary metals mentioned in the foregoing instances, zinc deposits metal from the
greatest number of solutions, and appears to have the strongest depositing power; 6th, that the coherent and adhesive
deposits obtained are in all cases exceedingly thin; and 7th,
that oftentimes the deposited metal, whatever its kind may
be, has the appearance of a black or dark-coloured powder on
its surface, especially when it has been deposited very rapidly;
and that sometimes it exhibits its ordinary colour and
appearance, especially if its outer portion is rubbed off.

By the simple immersion process a thin coating only of
metal is usually obtained, and even that is imperfect, because
the surface to be coated and the coating of metal act elec-
trically as two different substances, the former being electro-
positive and the latter electro-negative. In consequence of
this electrical difference there is set up a voltaic action at
minute points all over the surface; this action is not perceptible at first because it is of microscopic minuteness, but
it gradually spreads from those points all over the surface,
and causes the metal beneath the coating to dissolve, and
the deposit to become loose and full of spots.

It is, however, possible to coat a metal perfectly with
another metal by means of simple immersion by adopting
the following rule, now I think for the first time published.
Take an electro-positive metal, A (say copper), dip it into a

solution of a less positive metal, B (say mercury), its surface then dissolves and a film of B is deposited upon it. Now dip it into a solution of a third and still less positive metal, C (say gold) ; the film of B and also any non-coated particles of A then dissolve, and a film of C is deposited in their stead. Now re-dip the metal A into the solution of B, and any still non-coated particles of it are dissolved. and deposit B in their place ; then dip again into the solution of C, and a similar effect takes place as before. By thus alternately dipping the metal A, into the solutions of B and C, the number of its non-coated particles becomes less and less, until every one is coated with the metal C (see p. 128). The rule is applicable to various metals to which it has not yet been applied ; and its application offers an opening for new inventions.

To the simple immersion mode of depositing, belongs the process of tinning brass articles (wash tinning), by boiling them in water containing a salt of tin

FIG. 18.

and bitartrate of potash; the process of silvering brass nails, buttons, hooks and eyes, buckles, &c., by rubbing them with any of the well-known silvering compositions moistened with water; also the water-gilding process (see pp. 265, 152, 127).

Method No. 2. *Deposition by two metals and one liquid* or 'by simple contact process' (see Fig. 18). The following instances belong to the class of deposition by two metals and one liquid, the two metals being either in mutual contact (touching each other either above or beneath the surface of the liquid), or connected together by a wire.

In chloride of antimony.—On immersing a piece of antimony in contact with a piece of zinc in a solution of the ordinary chloride of antimony, it received a coating of antimony, and on immersing a piece of platinum in contact with a piece of tin in this liquid, it received a deposit of antimony; but on immersing a piece of antimony in contact with a piece of platinum, or a piece of platinum in contact with a piece of silver in this liquid, it received no metallic deposit.

Chloride of bismuth.—In a solution of this salt, brass in contact with a piece of zinc, copper in contact with tin, or german-silver with iron, received a deposit of bismuth; but brass in contact with a piece of gold, gold in contact with silver, or german-silver with platinum, received no deposit.

Tetrachloride of platinum.—In a solution of this salt, platinum in contact with zinc became coated with platinum, but in contact with gold it received no such coating.

Nitrate of silver.—In a solution of argentic nitrate, gold in contact with zinc received a deposit of silver, but in contact with platinum it did not.

Nitrate of mercury.—In a solution of this salt, silver in contact with either zinc or iron, or platinum in contact with copper, received a metallic deposit; but platinum in contact with silver did not.

Sulphate of copper.—In a solution of cupric sulphate, brass in contact with zinc, or tin, german-silver, silver, or platinum, in contact with iron, received a deposit of copper; whilst silver in contact with antimony, or platinum in contact with brass, received no deposit.

Oxide of copper in ammonia.—In a solution of oxide of copper in ammonia, platinum in contact with zinc received a deposit; but silver in contact with iron did not.

Chloride of nickel and ammonium.—In a solution of the double chloride of nickel and ammonium, copper in contact with zinc received a deposit of nickel, but in contact with silver it did not receive such a deposit.

Protosulphate of iron.—With a saturated solution of this salt, platinum in contact with zinc received a deposit of iron; but in contact with copper it received no metallic deposit.

Hyponitrite of lead.—With a solution of hyponitrite of

lead, either tin, copper, or brass, in contact with a piece of zinc, received a deposit of lead ; but copper in contact with tin or lead, or brass with platinum, received no deposit.

Nitrate of lead.—With a solution of plumbic nitrate, either copper, brass, or silver, in contact with zinc, received a coating of lead ; but copper in contact with iron, brass with tin, or silver with copper, received no such coating.

Stannous chloride.—With a solution of this salt, either antimony, tin, or copper, immersed in contact with zinc or lead, received a coating of tin ; but antimony in contact with tin, tin with silver, copper with iron, or either gold or platinum with copper, did not receive a deposit.

Sulphate, chloride, or nitrate of zinc.—With a solution of either sulphate, chloride, or nitrate of zinc, no metal of any pair of metals, selected from amongst the following, received a deposit of zinc : antimony, bismuth, zinc, tin, lead, iron, nickel, copper, mercury, silver, gold, platinum, or palladium.

Observations upon class of instances No. 2.—The following general observations may be made upon the foregoing facts : firstly, that in some instances deposition does, and in others it does not, occur ; secondly, that a metal will not usually cause another metal to be coated by this method, unless it can coat itself in the same liquid by simple immersion—for instance, zinc cannot coat itself with zinc in solutions of that metal, neither can it usually cause other metals to become coated with that metal in those solutions (see above) ; copper cannot usually coat itself with zinc in a solution of sulphate of zinc, or with tin in a solution of chloride of tin, neither does it usually cause silver, gold, or other metal, to become coated with zinc or tin in those liquids (see Raoult's experiments, pp. 268, 273, 276); thirdly, that one of the two metals which receives a deposit by this method, derives its power of receiving it by virtue of its contact with the other metal ; fourthly, that any metal which has the power of coating itself by simple immersion in a given liquid can, by this method, cause other metals which do not coat themselves by simple immersion in that liquid, to become coated; for instance, zinc, tin, and iron, coat themselves with copper, by simple immersion in a solution of sulphate of copper, but silver, gold, and platinum do not ; but if either of the former metals be connected with either of the latter, the two being immersed together in that liquid, the latter metals, as well as the former, will become coated with copper ; fifthly, that base metals, and especially zinc, have generally the power of causing other metals to become coated by this method, whilst the noble metals, and especially gold and platinum, rarely possess this power; sixthly, that by this method, metal is deposited much more frequently from solutions of the noble metals, than from those of the base ones ; and, finally, that thick deposits of metal may be obtained by this method, provided the action is continued sufficiently long, and the liquid properly renewed.

Method No. 3.—Deposition by one metal and two liquids (see Figs. 19, 20).—The following instances belong to deposition by the immersion of one metal in two liquids, separated by a porous diaphragm, the metal being either in two pieces, connected together by a wire or wires, or in one piece, and bent so as to dip into both liquids, or the diaphragm may be dispensed with, as already explained, by pouring the lighter liquid carefully above the other, and placing the piece of metal vertically in the two liquids (see Fig. 19).

Antimony in chloride of antimony.—Two pieces of antimony, connected together by a wire or wires, were immersed, one in dilute nitric acid, and the other in a solution of chloride of antimony; the piece in the dilute acid dissolved, whilst that in the chloride solution received a metallic deposit.

Iron in chloride of antimony.—With iron in dilute sulphuric acid on one side, and in a solution of chloride of antimony on the other, the end in the metallic solution received a deposit of antimony, whilst that in the dilute acid dissolved.

Antimony in chloride of bismuth.—Two pieces of antimony were immersed in the previous manner, one in hydro-

FIG. 19.   FIG. 20.

chloric acid, and the other in a solution of chloride of bismuth ; that in the acid dissolved, and the other received a coating of bismuth.

Bismuth in chloride of bismuth.—With bismuth in hydrochloric acid on one side, and in a solution of chloride of bismuth on the other, a free deposit of bismuth was soon obtained.

Bismuth in nitrate of bismuth.—With bismuth in dilute nitric acid, and in a solution of acid nitrate of bismuth, a thin deposit of that metal was found in twelve hours.

Silver in sulphate of copper, and in cyanide of silver plating liquid.—With silver in either dilute sulphuric or dilute nitric acid on one side, and in a solution of sulphate of copper on the other, no deposit of copper took place in twelve hours · but with silver in a solution of cyanide of potassium on one side, and in the double cyanide of potassium and silver on the other, a free deposit of silver occurred upon the end or piece in the latter solution.

Antimony in sulphate of copper.—With antimony in dilute hydrochloric acid on one side, and in a solution of sulphate of copper on the other, a deposit of copper was obtained.

Platinum in nitrate of copper.—With platinum in aqua regia on one side, and in either a solution of nitrate of copper, the ordinary cyanide gilding solution, or a solution of tetrachloride of platinum on the other, no deposit of copper, gold, or platinum occurred.

Brass or copper in sulphate of copper.—With brass or copper in dilute sulphuric acid on one side, and in a solution of sulphate of copper on the other, a deposit of copper was obtained in twelve hours; similarly with copper in dilute hydrochloric acid, and in a solution of chloride of copper, a metallic deposit occurred.

Tin in chloride of tin.—With tin in dilute hydrochloric acid on one side, and in a solution of stannous chloride on the other, a deposit of tin was obtained.

Copper in sulphate of zinc.—With copper in dilute sulphuric, or dilute nitric acid, on one side, and in a solution of sulphate of zinc on the other, no deposit of zinc occurred in twelve hours.

Iron in sulphate of zinc, and in sulphate of iron.—With iron in dilute sulphuric acid on one side, and in a solution of sulphate of zinc on the other, no deposit of zinc was obtained in twelve hours; similarly with iron, dilute sulphuric acid, and a solution of protosulphate of iron, no deposit occurred in twelve hours.

Zinc in chloride of zinc.—A piece of zinc was bent so as to dip into dilute hydrochloric acid on one side, and into a neutral solution of chloride of zinc on the other ; a free deposit of zinc was found upon the end in the metallic solution, after a period of twelve hours.

Zinc in sulphate of zinc.—With zinc in dilute sulphuric acid, and in a solution of sulphate of zinc, a free deposit of the metal occurred in twelve hours.

Zinc in acetate of zinc.—With zinc in a solution of acetate of zinc on one side, and in dilute sulphuric acid on the other, that in the dilute acid dissolved, whilst the other end received a metallic deposit.

Observations on class of instances No. 3.—1st, it appears that in this class also, we obtain negative as well as positive instances ; 2nd, that by this arrangement, unlike the previous classes, almost any metal may cause the same metal to be deposited—for instance, zinc may deposit zinc, copper deposit copper, and silver deposit silver ; 3rd, that by it even a noble metal may cause the deposition of a base metal, provided we have a suitable combination of liquids—for instance, if a piece of gold or silver be immersed in a strong solution of cyanide of potassium on one side, and in a solution of sulphate of copper or chloride of antimony on the other, the end in the free cyanide solution dissolves, whilst that in the copper or antimony one receives a deposit ; 4th, that the metal or end which receives a deposit, derives that power from its contact with the metal in the other liquid ; 5th, that, as a general rule, base metals have a greater power of causing deposition by this method than the noble ones ; 6th, that the noble metals are more readily and more often deposited than the base ones ; and, 7th, that we may produce thick and coherent deposits.

Method No. 4.—*Deposition by two metals and two liquids,* or 'single-cell process' (see Fig. 20, also Fig. 1, p. 18). The following instances belong to the class of deposition produced by the immersion of two metals in two liquids, the metals being in mutual contact, or connected together by a wire, and the liquids separated by a porous partition.

Zinc depositing antimony.—A piece of antimony was immersed in a solution of its chloride, and a piece of zinc in dilute sulphuric acid, and the two metals being connected together by a wire, a free deposit of antimony took place in twelve hours.

Iron depositing antimony.—With iron in dilute hydrochloric acid, and antimony in a solution of its chloride, a copious deposit of antimony was formed in twelve hours.

Copper and chloride of antimony or chloride of tin.— With copper in dilute hydrochloric acid, and antimony in its chloride, or tin in chloride of tin, no deposit of antimony or tin occurred in twenty hours.

Bismuth and chloride of antimony.—With bismuth in dilute hydrochloric acid, and antimony in chloride of antimony, no deposit of the latter occurred in twenty-four hours.

Zinc depositing copper.—With zinc in dilute sulphuric acid, and brass in a solution of cupric sulphate, copper was deposited.

Iron and chloride of tin.—With iron in dilute hydrochloric acid, and tin in a solution of its chloride, no deposit of tin took place in eighteen hours.

Tin depositing zinc.—With tin in hydrochloric acid, and zinc in a neutral solution of its sulphate, a deposit of zinc was obtained in the metallic solution.

Observations on class of instances No. 4.—1st, it appears that negative as well as positive instances occur in this arrangement in common with the others ; 2nd, that by using suitable metals and liquids, deposition may be effected more rapidly by this method than by the preceding ones ; 3rd, that the metal which receives the deposit derives its power from its contact with the other metal ; 4th, that base metals in strong acids have the greatest power of causing a deposit upon the other metals, and noble metals the least ; 5th, that the noble metals are more readily deposited than the base ones ; and, 6th, that thick and coherent deposits may be obtained.

In all the above instances, instead of using one vessel divided into two parts by a porous diaphragm, it will be found convenient to put one of the liquids in an unglazed earthenware porous cell, and immerse the cell in the other liquid (see Fig. 1, p. 18). In this case either liquid may be in the outer vessel.

FIG. 21.

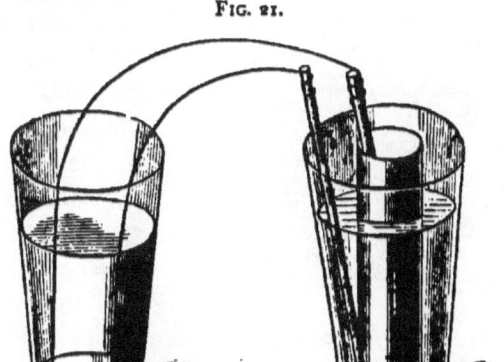

Method No. 5.—Deposition by a separate current, or 'battery process' (see Fig. 21; also Fig. 2, p. 24). The next class of instances are those in which any of the foregoing arrangements, except the first, may be connected by wires with two pieces of similar metal immersed in a separate liquid. In this class of instances, the method or arrangement differs from the three preceding ones, simply by the wire which connects the two pieces of metal being cut in two, and its free ends either immersed in a separate liquid, or connected with two pieces of metal dipping into that liquid. It is not necessary to have the depositing vessel separate ; it may even be attached to the same piece of apparatus, provided the liquid in it is perfectly separated from the other liquids and metals. The pieces of metal in the separate liquid, possess no power of deposition of themselves in that liquid (unless they coat themselves by simple immersion), even if they are connected together, but wholly derive their power of dissolving and receiving a deposit, from the other metals and liquids by means of the current passing through the wires.

Compound depositing vessels.—In each of the foregoing arrangements the deposition is limited to a single vessel, but any number of depositing vessels may be connected together in a series (see Fig. 22, also Fig. 3, p. 24) so that solution of

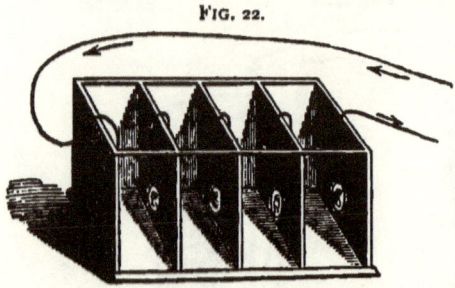

FIG. 22.

the anode, and deposition upon the cathode, may be simultaneously obtained in every one of them by means of the same current. It was at one time imagined that this was a very economical process, because by this means, with the aid of one equivalent of electricity, and at the expense of but one equivalent, each of acid and zinc, several equivalents of metal were dissolved and deposited; but it was soon found that the process was rendered so slow, as to neutralise the other advantages, and this arrangement therefore is but rarely employed. This practical result would have been anticipated by anyone who could have interpreted the chemical equivalent of electrical energy. It has been recently patented by E. Casselburg (see p. 381).

General remarks.—In each of the foregoing arrangements, the size or shape of the containing vessels, the volume or depth of the liquids, or the size or positions of the metals, have no material impression upon the production or non-production of a deposit ; the temperature, however, is an important condition, and in all the experiments I have described, this was about 60° Fahr.

Practical points to be observed.—In practical working, according to any one of these methods, it is necessary to attend to a number of points based upon the principles already given in the theoretical division. We must see that the depositing liquid is really an electrolyte, that it has a proper chemical composition, and contains the requisite amount of water, free acid, free alkali, or is neutral, as the case may be; that it has also the proper degree of fluidity, and is at the right temperature; that the strength of the electro-motor is suitable; that the electrodes are of the proper sizes and forms; that all the substances through which the current has to pass will conduct electricity; that all the points of contact of the wires and screws, and the surfaces of the immersed metals, are well cleaned; that the circuit is really complete; that the electrodes are in proper positions; that the current is passing in the right direction; that the different parts of the liquid are maintained of uniform composition by stirring, &c.

There are many metallic solutions, such, for instance, as the anhydrous terchloride of arsenic, pentachloride of antimony, the tetrachloride of tin, &c., which do not conduct electricity, and cannot therefore be used as electrolytes. Some others are unsuitable from other causes; for instance, those containing nitric acid or a nitrate, are not usually good, and sometimes will not yield a metallic deposit at all, because of the highly-oxidising character of the liberated acid; chlorates, bromates, and iodates, are also rarely used. Iodides are liable to liberate free iodine. Selenates and phosphates, also, do not usually form good solutions; the former are apt to have their selenium set free by the deposited hydrogen. Sulphides cannot often be employed, because most of them are insoluble; and those which are soluble have a very offensive odour, and are decomposed by the atmosphere. Aqueous solutions of chlorides, sulphates, and cyanides, are the most usually suitable liquids; fluorides and bromides, are also often available.

Methods of forming a depositing solution.—There are two ways of forming a depositing solution, one termed the battery process, and the other the chemical one. In making a solution by the former method, the particular metal which the liquid is to contain and deposit, is employed as an anode, and a current from any suitable source passed through the liquid by means of it, until a smooth and clean cathode of suitable metal, receives a sufficient and proper metallic deposit. Some operators, in making a solution by the battery process, fill a porous cell with the liquid, put the cathode in it, and place the porous cell in the larger bulk of liquid ; but this is an unnecessary precaution. The chemical method consists in preparing the ingredients by chemical means, and uniting them in the proper proportions, to form the desired solution ; in the use of this method it should always be remembered that freshly-prepared precipitates in a wet state, usually dissolve much more quickly, th.:n those which have been long prepared, or have become dry.

Method of using a depositing solution (see also p. 341).—If a solution contains a very large excess of uncombined acid, or other solvent, metallic deposition will sometimes not occur, especially if the metal to be deposited is a highly positive one ; for instance, in a solution of sulphate of zinc, the presence of a large amount of free sulphuric acid will prevent the deposition of zinc. If on the other hand, a depositing solution contains no free combining substance, deposition will either proceed very slowly, or be entirely stopped, in consequence of an insoluble salt (often a non-conducting one) being formed upon the anode ; for instance, when an electric current is sent through two silver plates immersed in a solution of the pure double cyanide of silver and potassium, the dissolving plate becomes covered with a white insoluble layer of cyanide of silver, which first impedes, and then stops the current.

If a depositing solution is diluted with water to a very large extent, deposition will be greatly retarded, but if, on the other hand, it contains a great insufficiency of water, crystals of metallic salt will collect upon the anode, especially at its lower part, and gradually stop the current. This happens in a saturated aqueous solution of cupric sulphate, containing copper electrodes and plenty of free acid.

It is often a great advantage to raise the temperature of a depositing solution, because the strength of affinity of the liberated negative elements for the anode, increases by elevation of temperature, whilst that of the different elements of the electrolyte for each other diminishes. Rise of temperature also increases the electric conductivity of an electrolyte, and decreases that of the metal plates immersed in it, but, as the latter are by far the best conductors, the final effect of heating a depositing liquid, is a considerable increase of the amount of current passing.

I have repeatedly observed, that with some solutions used hot for depositing, if the cathode was immersed in the liquid at 60° Fahr, and the liquid then heated, no conduction or deposition occurred ; nor did it take place if the cathode was removed, washed in cold water, and re-immersed. But if the liquid was first heated, and then the cathode immersed, deposition occurred freely, and the liquid might be cooled down considerably without stopping the action. In coating iron with tin in some liquids, if the iron was immersed before heating the solution, no deposition occurred even at 150° Fahr, but if the liquid was first heated, even only to 90° or 100° Fahr., deposition took place. I have not examined whether this was due to what has been termed 'the passive state.'

Usually it is not necessary to screen electro-depositing solutions from the direct action of light. In some cases, however, light decomposes a liquid, and renders it unfit for deposition ; this is the case with a liquid formed by dissolving hyposulphite of silver in a solution of hyposulphite of sodium ; cyanide of silver-plating liquids are also affected, and turn brown, by the influence of light, but not in such a way as to render them unfit for depositing ; the light only affects the 'free cyanide' in them.

The rapidity of deposition is affected by the superficial area of the electrodes, the length and transverse sectional area of the intervening solution, and of the connecting wire. The larger the immersed surface of the electrodes, the shorter the length, and the greater the transverse section of the solution, and of the connecting wire, the more rapid is the process.

The various other conditions, especially that of proper density of current, &c., necessary to be secured in order to obtain a metallic deposit, and the desired quality of metal, have been already described in the theoretical section (see also p. 341).

DEPOSITION OF INDIVIDUAL METALS.

CLASS I. GASEOUS METALS.

HYDROGEN.

As THERE are many persons, students, electro-platers, inventors and others, who wish to make experiments for themselves, and require to know what has already been done in the separation of particular metals by means of electrolysis, the following abstract is given of that portion of the subject.

1. **Hydrogen.**—Electro-chemical equivalent weight = 1. As the separation of hydrogen is a very common result in electro-metallurgical operations, and this element is considered by most chemists to be a metal, although a gaseous one, and has been called 'hydrogenium' in order to indicate its metallic nature, I include it amongst the metals. Water, and all acids may be regarded as salts of hydrogen, and this element is set free by electrolysis in many solutions which contain water or an acid, in some cases by direct action, and in others as a secondary product, being in the latter case produced by the action of more electro-positive substances, such as the alkali metals, liberated at the cathode, decomposing the water or acid, taking the oxygen, etc. to themselves, and setting the hydrogen free.

Deposition of hydrogen by simple immersion process.— The liberation of hydrogen by contact of the alkali-metals with water, is one of the most familiar and striking phenomena of modern chemistry. The metals of the alkaline earths also, usually evolve hydrogen slowly from water, and nearly all the base metals also behave similarly, if the water is acidulated. Even finely-divided silver, gold, and platinum, set it free from a hot concentrated solution of potassic cyanide (H. St. C. Deville).

Magnesium liberates hydrogen from water; and its amalgam with mercury does so with violence (C. N. Hartley, 'Chemical News,' vol. xiv., p. 73). Magnesium sets free hydrogen from water, especially if the water contains common salt, salammoniac, or some acid (Roussin, 'Chemical News,' vol. xiv., p. 27). Hydrogen is always evolved, when a metal is precipitated from an aqueous liquid by means of magnesium (Commaille, 'Chemical News,' vol. xiv., p. 196). Finely divided iron (but not either cobalt or nickel), decomposes water, slowly at 16° C. but rapidly at 100° C. and liberates hydrogen (Troost and Hautefeuille, 'Chemical News,' vol. xxxi., p. 196).

I have observed, that magnesium does not evolve hydrogen in dilute hydrofluoric acid, and but little in an aqueous solution of chloride of potassium; but that it evolves it freely, in a mixture of the two liquids. Similarly with the same acid, and a solution of potassic chlorate. It did not evolve the gas, in a mixture of the same acid and perchlorate of potassium. It liberated hydrogen from a mixture of the acid and a solution of bromide of potassium, but not from either alone. Similarly with the same acid and iodide of potassium; but not in a mixture of the acid and a solution of potassic iodate. In a mixture of hydrofluoric acid and a solution of potassic sulphate, magnesium set free hydrogen, but not in either liquid singly. According to Deville, even silver deposits hydrogen violently, and forms argentic iodide in liquid hydriodic acid ('The Chemist,' New Series, vol. iv., p. 329).

Deposition of hydrogen by separate current process.—
Concentrated hydrochloric acid yields chlorine at the anode, and hydrogen at the cathode, as direct results of the action of the current ; but, according to Bourgoin, the oxygen and hydrogen obtained, on passing an electric current by means of platinum plates, through distilled water acidulated with pure sulphuric acid, are probably not results of an action of the current upon the water, nor even results of the action

of liberated electrolytic products upon the water, but of direct decomposition of a hydrate of sulphuric acid (see 'Telegraphic Journal,' vol. i., p. 91; March 15, 1875).

I have electrolysed on many occasions, pure dilute hydrofluoric acid with electrodes of numerous metals ; and also the extremely dangerous liquid, anhydrous hydrofluoric acid, with electrodes of palladium, platinum, gold, gas-carbon, &c. ; hydrogen was always deposited at the cathode ; the numerous other effects obtained with that acid, will be found described under the heads of the respective metals.

According to Brester, when nitric acid does not liberate any hydrogen gas at the surface of a cathode of platinum or charcoal, by the passage of an electric current, the acid is reduced to the state of ammonia ('Chemical News,' vol. xviii., p. 144). Bloxam also, has shewn, that the hydrogen evolved from a platinum cathode immersed in dilute nitric acid, or in a solution of nitrate of potassium, contained in a porous cell, immersed in dilute sulphuric acid containing the anode, converts a portion only (not more than one-half), of the nitric acid of either of those liquids, into ammonia ('Chemical News,' vol. xix., p. 289).

The electrolysis of concentrated formic acid by platinum electrodes, yields carbonic acid and oxygen at the anode, but that of dilute acetic acid gives pure oxygen ; aqueous benzoic acid yields oxygen at the anode and hydrogen at the cathode, and the latter sometimes acquires a black deposit, which disappears on exposure to light; it is probably a hydride of platinum. A saturated aqueous solution of oxalic acid, yields twice the bulk of gas at the cathode, to that at the anode ; the latter is a mixture of two volumes of carbonic anhydride, and one of oxygen. A saturated solution of tartaric acid, gives oxygen at the anode, and hydrogen gas with hydride of platinum at the cathode (Brester, 'Chemical News,' vol. xviii., p. 145).

Absorption of hydrogen by electro-deposited metals.—As hydrogen is often deposited from a solution by electrolysis, simultaneously with other metals, electro-deposits frequently contain it. Various experimentalists have observed that deposits of palladium and nickel absorb it ; pure tin also in a less degree has the same property ; but cadmium, zinc, aluminium, copper, lead, silver, mercury, bismuth, gold, and platinum (?) do not. The correctness of these statements, however, depends largely upon the kind of liquid electrolysed. If a piece of palladium, nickel, cobalt, or tin, has a wire of aluminium twisted round it, and is then immersed for a few minutes in dilute acid, it absorbs sufficient hydrogen to exert a slightly reducing action upon a solution of ferri-cyanide of potassium. According to Böttger, a palladium plate coated with palladium black, absorbs the hydrogen more quickly, and when taken from the electrolyte, and dried quickly by blotting-paper, becomes red hot in the air in a few seconds. I have repeatedly observed that the steel blade of a knife, or a steel wire, becomes much more brittle after having been made the cathode and evolved hydrogen in an electrolyte, and that this occurs not only with a dilute acid but also with an alkaline liquid. It is not improbable that steam boilers are sometimes weakened by a similar absorption of hydrogen, when the water employed in them is decomposed by the iron. The explosive variety of antimony formed by electrolysis is also said to contain hydrogen.

CLASS II. BRITTLE NEGATIVE METALS.

ARSENIC, TELLURIUM, ANTIMONY, BISMUTH.

2. **Arsenic.**—Elec.-chem. eq. $= \dfrac{75}{3} = 25$. The commonest salts of arsenic are arsenious acid, i.e. the common white oxide known as 'arsenic ;' arsenic acid ; and the compounds of those two acids with potash and soda. Metallic arsenic itself is a brittle substance, and an inferior conductor of electricity. Arsenious acid is soluble in warm hydrochloric acid ; also by heating it to dryness with strong

nitric acid it is converted into arsenic acid, which is a deliquescent substance, readily soluble in water.

Deposition of arsenic by simple immersion process.—
This element is easily deposited by the simple immersion
process, by dissolving arsenious acid in warm and somewhat
dilute hydrochloric acid, and stirring the solution with a strip
of bright copper. This experiment is well known in toxico-
logical chemistry as being an extremely delicate test for
arsenic, devised by Reinsch. According to Roussin, from
solutions of arsenic, magnesium deposits arseniuretted hydrogen, but no arsenic in the metallic state ('Chemical
News,' vol. xiv., p. 27).

*Deposition of arsenic by the simple contact of another
metal.*—All the arsenic may very easily be extracted from
arseniferous substances by placing a solution of them in a
platinum vessel, and immersing in it a piece of zinc in contact with the vessel; the arsenic appears on the platinum.
By prolonging the action the whole of the arsenic is extracted
('Cosmos,' Second Series, vol. i., p. 595; and 'Chemical
News,' vol. xii., p. 3).

Deposition of arsenic by separate current process.—I have
made many experiments of electrolysis of solutions of arsenic,
and have obtained from the aqueous fluoride small portions
of a scaly deposit, which appeared to exhibit in a feeble
degree the peculiar explosive property of the amorphous
variety of electro-deposited antimony.

3. **Tellurium.**—Elec.-chem. eq. $= \dfrac{129}{3} = 43 \cdot 0$. Very little
has been done in the electro-deposition of this metal, probably in consequence of the great cost of the substance.
Ritter could only obtain a pulverulent deposit of it from its
solutions; and both he and Sir H. Davy found that, in electrolysing water by means of an anode of this metal, the water
surrounding the anode acquired a purple colour, and precipitated a brown powder; Magnus shewed that this powder
was metallic tellurium. I have electrolysed a pure solution

of its chloride by means of large and smooth platinum electrodes and a very feeble current, but obtained only a jet-black deposit, the inner portion only of which was adherent ; the anode was not corroded. I have also electrolysed pure dilute hydrofluoric acid with an anode of pure tellurium, by a current from a single Smee's cell, and have obtained by very slow action most excellent deposits of bright reguline metal, of grey colour, and brittle crystalline structure.

4. **Antimony.**—Elec-chem. eq. $= \dfrac{122}{3} = 40 \cdot 66$. Its most common salts are the oxide, sulphide, terchloride, and potassic-tartrate (tartar-emetic). The acid terchloride is the ordinary chloride of antimony as prepared for pharmaceutical purposes; it is formed thus.—Take one pound of black sulphide of antimony, add to it four pints of hydrochloric acid, gently heat the mixture, until the gas decreases, then boil it slowly down to two pints, keeping it partly covered all the time; cool it, filter it through calico, and keep it in a stoppered bottle. It is now a yellowish red liquid (the colour being due to iron in the sulphide), of specific gravity about 1·35 to 1·50.

A similar solution may be made by the battery method ; this consists in passing an electric current from several cells through pure and strong hydrochloric acid, by means of a large anode of antimony, until a good deposit is obtained upon a cathode of platinum of equal surface ; this solution is nearly colourless, nearly free from iron, and much more pure than the other. A very good solution may also be easily made by saturating ten ounces, by measure, of strong hydrochloric acid with freshly-precipitated teroxide of antimony. (N.B. Not that made by oxidising antimony by nitric acid, nor that which has been long exposed to the air.) Then add about five ounces more of the acid to the clear portion and stir the mixture. About three ounces of the oxide will be required. An excellent solution may also be made by dissolving an avoirdupois ounce of oxychloride of anti-

H 2

mony in five ounces of pure hydrochloric acid of specif
gravity 1·12.

The acid chloride of antimony is an excellent conduct(
of electricity ; it dissolves the anode freely, yields plenty (
bright reguline metal if the battery power is not too stron;
and its depositing power does not deteriorate by exposure t
light or air.   It is decomposed more or less readily by zin(
tin, lead, iron, brass, copper, and german-silver, each of whic
coat themselves with antimony in it by simple immersio1
Articles immersed in it require to be washed with hydr(
chloric acid before washing them with water, otherwise th
latter decomposes the adhering film of liquid, and cove1
the articles with a white insoluble powder.

The mixed chlorides of antimony and ammonia form
very good depositing liquid.   It may be made either b
the battery process, or by mixing two measures of a saturate
solution of sal-ammoniac with two measures of hydrochlori
acid and one measure of water, and dissolving antimony i
it by means of a large anode of that metal and a stron
battery current, or by simply mixing together equal measure
of a saturated solution of sal-ammoniac and commerci;
chloride of antimony.   This solution conducts well, yield
plenty of metal of good quality, and does not act so strongl
upon base metals as chloride of antimony alone, but in oth(
respects it is like the chloride.

The potassic tartrate of antimony (tartar-emetic) is mo;
conveniently obtained by purchase.   It is a salt not ver
soluble, it requires about fifteen times its weight of water t
dissolve it ; its aqueous solution is a very bad conductor (
electricity, and is not to be compared with the chloride f(
depositing purposes ; even with a very feeble electric curre1
the deposited antimony consists only of a small quantity (
a perfectly black powder.   On the other hand, however, thi
salt is very freely soluble in a mixture of two volumes of h5
drochloric acid and one of water.   The solution may b
made by mixing together about two pounds of water, fou

hloric acid, and eight pounds of the potassic
ter proportion of water may be added if
xture forms an excellent one for depositing
good conductor of electricity, it is not in-
tinued working, or exposure to light or the
re deposited antimony from it constantly
nths) ; it will bear a very strong current
it being caused to pass into the state of a
yields reguline metal rapidly, and in coat-
d thickness (I have obtained deposits from
nch thick). Deposits of about one-twelfth
tness may be obtained in about three days
articles which are wet with this solution
ean in water alone, without requiring to be
with hydrochloric acid.
timony by simple immersion process (see also
y may easily be deposited from an acid solu-
ide by the simple immersion process. In
t of chloride of antimony is used for impart-
ɔ articles of brass. A large quantity of water
ll quantity of chloride of antimony, which
ite precipitate of oxychloride of antimony;
led until the whole is nearly redissolved,
ed to the solution, and again boiled in like
ing filtered, this clear liquid is raised to the
the articles of brass, previously cleaned, are
ley immediately deposit a film of antimony
upon themselves by the simple immersion
illowed to remain a greater or less length
to the tint required. They are then well
vater, dried in hot sawdust in the usual
cted from alteration of colour by lacquering.
imony is also used for bronzing gun-barrels.
der, deposited by simple immersion process
by means of zinc, is used for imparting an
y cast-iron to figures of plaster-of-Paris.

According to Roussin, magnesium deposits antimoniurette
hydrogen, but no metallic antimony, from solutions of ant
mony ('Chemical News,' vol. xiv., p. 27).   I have observe
that crystals of silicon did not become coated with antimor
in a solution of fluoride of antimony containing free hydr
fluoric acid ; that zinc deposited antimony as a black powd
by simple immersion in a solution of the mixed fluorides
antimony and potassium ; and that the oxide of iron w:
rapidly dissolved from a rusty iron wire in a mixture of equ
measures of solution of terchloride of antimony and a sat
rated solution of sal-ammoniac.

Watt coats copper with antimony by simple immersic
thus :   Dissolve one ounce of chloride of antimony in or
pint of spirit of wine, and add hydrochloric acid until tl
liquid is clear.   Immerse the clean article in it during abo
half an hour ; it receives a bright coating.

Gold in contact with antimony, in a solution of cold (
boiling salt of antimony, does not acquire a coating of met
(Raoult, ' Chemical Society's Journal,' vol. xi., p. 465).

Deposition of antimony by separate current process (see al:
pp. 81, 87, 89).—In depositing antimony by the battery pr
cess, the metal may be obtained not only in a state of loo
black powder, but also in two distinctly different, coherer
reguline conditions, viz., as a very brittle metal of
grey-slate colour, and hard crystalline structure ; and al:
as a highly lustrous steel-black deposit, of amorphous stru
ture, and somewhat less hard than the pure variety ; whi(
retains its colour and brightness without oxidizing for a loi
time.

A satisfactory solution for obtaining the pure grey met
is composed of :—

Distilled water	12 ounces.
Tartar-emetic	1 ounce.
Tartaric acid	1 ,,
Pure hydrochloric acid	$1\frac{1}{2}$,,

It is not a good conductor, and should be worked slow

:h two Smee's elements, at such a rate as to deposit about
e-eighth of an inch thick of the metal in four weeks.

Whilst engaged in depositing antimony from an acid solu-
n of the terchloride by the separate current process, in
:tober 1854, I observed a singular development of heat
the deposited metal when scratched or rubbed, and pub-
hed a brief account of it in the 'Philosophical Magazine' for
nuary 1855. I afterwards investigated the phenomenon
)re fully; and the following account, condensed from the
'ransactions of the Royal Society' contains all the leading
:ts relating to it.

The best solution for forming the strongly-thermic
:iety of deposit is composed of one avoirdupois ounce of
oxide of antimony or oxychloride of antimony, dissolved
five or six ounces of hydrochloric acid of sp. gr. 1·12 ; or
nay be made by saturating two measures of hydrochloric
.d with oxide or oxychloride of antimony, and then add-
; one measure more of the acid.

If, instead of the terchloride solution, a solution of either
'bromide or teriodide of antimony is employed, the de-
sited coating possesses a similar property of evolving heat,
t in a much less conspicuous degree, especially the deposit
)m the teriodide; and if a solution of fluoride of antimony
employed, the deposited metal is of a grey colour, perfectly
/stalline, and entirely destitute of the peculiar heating
)perty. Under some circumstances this crystalline variety
deposit may also be obtained by electrolysis from a weak
lution of terchloride of antimony, especially if the battery
wer is very feeble, or the liquid is employed in a dilute or
ated state.

In common with electro-deposits generally, the inner
d outer surfaces of both the black and grey deposits are in
equal states of cohesive tension, frequently in so great a
gree as to rend the deposit extensively, and raise it from
e cathode in the form of a curved sheet, with its concave
le towards the anode. This state of tension is most mani-

fest with rapidly-formed, thin deposits, especially upon extended flat surfaces.

It is worthy of notice that if the speed of deposition was gradually diminished to about 0·5 grain per square inch of cathode per hour, when it attained a certain degree of slowness (about 0·7 grain per square inch per hour), the character of the metal depositing suddenly changed from the amorphous black to the crystalline grey variety without exhibiting the slightest gradation between, and the two layers of active and inactive metal might be readily separated by means of a knife. With deposits very rapidly formed, the fractured surface was coarse and less black, and the thermic change was found to be very strong, shattering the metal with almost explosive violence.

Faint crackling sounds frequently issued from the depositing metal, evidently caused in most instances by the cohesive action just mentioned, and in other cases they were due to the sudden expulsion of bubbles of gas from holes in the deposited metal, especially with the bromide variety, or by depositing upon an iron cathode in the terchloride solution. The deposit obtained in the bromide solution was frequently perforated with holes all over its surface, and had the appearance of a metallic sponge caused by the numerous bubbles of gas. The thermic property of the deposit from the terchloride gradually disappears, the substance in a state of powder loses its power in six months ; fragments one-sixteenth of an inch thick lose their power in the course of twelve months, whilst others a quarter of an inch thick still possess a portion of their heating power at the end of three or four years.

Each of the varieties of active antimony is fragile and easily broken; that from the iodide solution is extremely so. Thin pieces, one-sixteenth of an inch, of the chloride variety may be broken in the air at 60° Fahr. without discharging their heat, if broken with care ; thicker pieces should be broken under the surface of cold water, by gentle blows with wood, or other substance not very hard. Very thin pieces may with care be reduced to fine powder in a mortar under a mixture of ice and water, and the powder so produced, after drying in a thin layer in a slightly warm place, possesses all the heating properties of the original solid mass.

Heating the chloride variety to 212° Fahr. for one hour in boiling water, or keeping it at a somewhat lower temperature (185° or 190° Fahr.) for a longer period in an air-bath, causes it gradually to evolve its heat, and lose its peculiar heating power.

A cylindrical bar of the chloride variety, about half an inch in diameter, formed upon a rod of grain tin one-eighth of an inch thick, when changed by the momentary contact of a heated wire, evolved sufficient heat to melt the tin completely, and the tin ran out through a crack in the antimony, and remained liquid a short time.

By applying momentary heat to the ends of deposits formed upon heliacal copper wires, the action was gradually transmitted to the opposite ends at a speed varying from twelve to thirty feet per minute, the rapidity of progress depending chiefly on the absence of cooling influences, cracks in the metal, and portions of grey crystalline deposit.

The specific gravity of the active chloride variety varied from 5.739 to 5.944; but after having been discharged suddenly of its heat, its specific gravity varied from 5.748 to 6.029. The specific gravity of the inactive or pure crystalline variety varied from 6.369 to 6.673.

Their electro-chemical equivalents, determined by electrolysing their solutions simultaneously in the same circuit with a solution of sulphate of copper, and weighing the deposits, were 42.30 to 43.81 parts of the active variety, and 40.41 to 40.79 parts of the crystalline kind for every 31.7 parts of copper deposited in the copper solution.

The peculiar change in the active chloride variety is attended by alterations in the colour, cohesion, and fractured surfaces in the substance; from a bright steel colour and glassy fracture, it passes to a dull grey colour and granular fracture, and its cohesive power greatly increases. These changes occur whether the heat has been evolved suddenly or very gradually by long lapse of time. Similar effects are observed, but in different degrees, with the bromide and iodide deposits ; straight bars of the active chloride variety suddenly discharged become curved by the heat, the outer side, or that last deposited, invariably becoming concave; this is similar to the effect of annealing upon electro-deposited metals generally.

The heat evolved by the peculiar change in the chloride variety is not due to cohesive action ; for it has been found that the amounts of heat evolved by similar weights of the substance in a single solid mass, in small pieces and in fine powder, in a calorimeter, were not sensibly different. Nor is the heat due to alteration of the specific heat of the substance during the change.

The temperature to which the active chloride variety must be raised, either locally or throughout its mass, to produce the sudden discharge, varies according to several circumstances, but is generally about 200° or 210° Fahr. in an air-bath. The discharge is not limited to one particular temperature, but commences between 170° and 190°, and gradually increases in rapidity by rise of temperature to some point about 200° or 210° Fahr., when it attains its maximum, and discharges all its remaining heat suddenly. A rod of the substance may be gradually discharged of its heat at one end, without discharging the opposite end, by immersing that end for one hour in nearly boiling water.

The total amount of heat evolved by the sudden discharge of the chloride variety was considerable, and was sufficient in most instances to raise the temperature of an equal weight of ordinary antimony of specific heat 0·0508, about 650 or 700 Fahr. degrees, and in one instance 705·89 Fahr. degrees, above the atmospheric temperature (60° Fahr.) at which it was discharged.

When the active chloride variety is suddenly discharged of its heat, there is invariably evolved from it a small quantity (generally about 3·5 per cent.) of vapour, consisting almost entirely of terchloride of antimony. This evolution of vapour is not a cause but an effect of the heat.

The following are the results of two analyses of specimens of amorphous antimony, obtained from an acid solution of the pure chloride :—

No. 1.			No. 2.		
Sb	. . .	93·36	Sb	. . .	93·51
SbCl₃	. .	5·98⎫	SbCl₃	. .	6·03⎫
HCl	. .	0·46⎭ = 6·44	HCl	. .	0·21⎭ = 6·24
		99·80			99·75

The second variety of active antimony may be obtained as follows : dissolve one part of teroxide of antimony in ten parts of hydrobromic acid, of specific gravity about 1·3 ; filter the solution, and electrolyse it by means of three Smee's elements, and an anode of antimony, at a speed of deposition of about 3 to 5 grains per square inch of receiving surface per hour.

This variety is of a lighter colour than that from the chloride, and is generally quite dull in aspect. It exhibits less of the cohesive cracking action than the first kind, and is less hard. Its specific gravity at 60° Fahr. varies from 5·415 to 5·472.

By momentary contact of a red-hot wire, it exhibited a similar molecular and thermic change, but the action did not spread throughout the mass unless it was previously raised to a temperature of about 250° Fahr. ; if then touched with the wire it evolved all the heat instantly, with explosive violence and projection of pieces of the substance. Scratching the heated substance by a steel pointer did not cause it to discharge its heat. Pieces heated upon mercury, or melted fusible alloy, discharged themselves suddenly and powerfully when the bath attained a temperature of about 320° Fahr.

By fusion in a bent tube of refractory glass, it was found to consist of 79·52 per cent of metal and 20·48 per cent. of volatile matter—a colourless, buttery substance, slightly semi-fluid at 60° Fahr., which doubtless consisted of terbromide of antimony and a little aqueous hydrobromic acid. Other specimens treated thus gave respectively 18·42 and 20·40 per cent. of volatile matter; the two specimens being part of a single deposit, the first being from the upper and the second from the lower part of the deposit, as it was suspended in the electrolyte.

The electro-chemical equivalent of this variety was determined by depositing it simultaneously by the same current with the chloride variety, and ascertaining the relative weights of the two deposits. In two experiments of this kind there were obtained respectively 50·09 and 50·11 parts of this variety for every 42·5 parts of the active chloride variety, or 32·2 parts of zinc consumed. And in two other determinations 51·2 and 51·4 parts of bromide deposit were obtained. Each of these quantities of deposit contained the same amount of metallic antimony, viz. 40 parts or one-third of an atomic weight; the remainder being the associated salt of antimony. These results indicate that the pure metal alone is deposited by the current.

The third variety of heat giving electro-deposited antimony was obtained as follows : Dissolve one part by weight of teroxide of antimony in fifteen parts of hydriodic acid, of specific gravity 1·25, and electrolyse it at a speed not exceeding one grain per square inch of cathode per hour.

The deposit is scaly-grey, dull in appearance, very friable, and much less metallic in character than either of the other kinds, unless it has been deposited with extreme slowness. The specific gravity of a slowly-formed specimen was 5·27. On immersing dry pieces in water a hissing sound, as of strong absorption, occurred, and numerous bubbles of gas issued from all parts of its surface during a few seconds. The tendency to evolution of hydrogen gas in the solution by this variety is so great as frequently to disintegrate the deposit completely.

Pieces one-ninth of an inch thick required to be heated upon mercury to 338° Fahr. before the contact of a red-hot wire would cause a discharge of heat ; it then discharged but feebly, with evolution of red vapours of iodide of antimony.

By fusing the unchanged substance in a glass tube, it yielded 77·76 per cent. of metal, and a solid, red, easily fusible sublimate, together with a little moisture, evidently teriodide of antimony, and a little aqueous hydriodic acid.

Its electro-chemical equivalent was determined in the same way as the previous kind. With slow action (0·5 grain per square inch of cathode per hour) 50·39 parts of deposit were obtained, and with very slow action (0·2 grain per square inch per hour) 48·07 parts were obtained for every 42·5 parts of deposit in the chloride solution.

The explosive kind of antimony electro-deposited from the chloride solution has been several times rediscovered in America and other places by different persons.

Amorphous antimony is one of the easiest of metals to deposit in a firm, coherent state. Its appearance when deposited from the chloride, or from the solution of the potassic-tartrate in hydrochloric acid, by means of the current from two or three of Smee's elements and an anode of antimony, is very beautiful, and when deposited at a proper speed it has much the appearance of highly-polished steel. The process should be continued until the coating is about one-twelfth of an inch in thickness on each side of a thin sheet of bright copper, employed as the cathode : this will occupy about three days and nights if the current is suitably strong. The solution should be stirred with a rod of gutta-percha each morning and evening during the action. Sometimes a deposit explodes in the liquid during its formation.

When the deposit is sufficiently thick, transfer the coated

sheet to a wooden or gutta-percha bowl into which a stream of
cold water is freely running, and clean the metal by first pouring
dilute hydrochloric acid over it, and washing it in the cold
water with the aid of a soft brush.   By bending the sheet of
metal very slowly in the water, the antimony falls off in large
plates, which may be broken into smaller pieces upon a concave surface of wood under cold water by a gentle blow with
the end of a wooden rod.   Each fragment after washing and
drying (without the aid of heat) should be wrapped in cotton
wool, and kept in a cool place.   It gradually, during many
months, loses, more or less, its heating property and brightness, and acquires an acid reaction.

A singular phenomenon sometimes occurs in depositing
explosive antimony.   As the solution is a very dense one,
if it is rapidly worked, the exhausted liquid rises to the top
and lies in a layer upon the surface, and, if the solution is
not occasionally stirred, a film of deposited metal forms
around the cathode upon the surface of the liquid in the
form of a button one and a half inch in diameter.

I have frequently deposited collections of shining grey
crystals of the pure variety of antimony from a saturated
neutral solution of the fluoride, by means of a current from
six Grove's or ten Smee's elements ; also from a dilute solution containing free hydrofluoric acid, by a current from two
Smee's cells.   As these crystals have a beautiful appearance
and do not oxidize, some practical use might probably be
made of them for the purpose of ornamentation.

For a full account of the properties of electro-deposited
antimony see 'Phil. Trans. Roy. Soc.' 1857, 1858, and 1862 ;
'Chemical News,' vol. viii., pp. 257 and 281 ; also 'Journal
of the Chemical Society.'

Both the black and red sulphides of antimony dissolve
in hydrosulphate of ammonia, and the resulting solutions
conduct very freely with an antimony anode, and one Smee's
element, but yield no metal even with a current from twenty-
five cells in series.   Aqueous solutions, either of caustic

potash, tartrate or oxalate of potassium, scarcely conduct at all with an anode of antimony, and the current from one or two Smee's elements. Cyanide of antimony dissolved in a solution of cyanide of potassium has been proposed as a depositing solution, but I have found a solution of cyanide of potassium to be a very bad conductor with an anode of antimony.

A solution composed of ten litres of water, 500 grammes of finely-powdered sulphide of antimony, and 2000 grammes of carbonate of sodium, dissolved by boiling, filtered whilst hot, and electrolysed at a boiling temperature, has also been recommended (Roseleur's ' Galvanoplastic Manipulation,' p. 282) for depositing antimony.

Antimony used as an anode in water becomes covered with oxide. Fused oxide of antimony yields antimony at the cathode, but antimonic acid is formed at the anode, and stops the current. According to Faraday, fused terchloride of antimony conducts badly and is but little decomposed. I have observed that electro-deposited antimony did not spread over the blackleaded surface of gutta-percha.

5. **Bismuth.**—Elec.-chem. eq. $= \dfrac{210}{3} = 70$. The commonest salts of bismuth are the basic nitrate (pearl white or mineral cosmetic), the acid nitrate, and the chloride. The basic nitrate is formed by treating the acid nitrate with abundance of water; it is a white powder, soluble in nitric or hydrochloric acid. The acid nitrate is made by digesting the metal in warm dilute nitric acid, evaporating and crystallising the solution. The chloride may be made by dissolving bismuth in a mixture of four measures of hydrochloric acid and one of nitric acid, and expelling all excess of acid by evaporation.

Deposition of bismuth by simple immersion (see also p. 77). — Magnesium deposits pure metallic bismuth from solutions of bismuth salts (Commaille, ' Chemical News,' vol. xiv., p. 188). ' To coat articles of tin with bismuth by simple immersion :

dissolve ten grains of nitrate of bismuth in a wine-glass full of distilled water, to which two drops of nitric acid have been added. Immerse the articles ; the bismuth will at once begin to be deposited upon them in very small, shining plates.'

Deposition of bismuth by a separate current (see also pp. 82, 85, 88).—I have deposited this metal by means of an extremely feeble current, from a solution of the nitrate in water, with nearly the minimum amount of free acid. The metal was then reguline, and appeared very beautiful, white, with a faintly pinkish tint, and with a fine silky lustre, but the coating was rather thin ; the deposit would not spread over a blackleaded surface in the liquid. According to some writers, such a deposit is liable to explode when struck. I have also deposited it from a solution of iodide of bismuth and iodide of potassium, and obtained an extremely bulky, jet-black powder, which contained iodine after persistent washing, and slowly oxidized and became greyish white in the air after many months.

Pure dilute hydrofluoric acid, with an anode of bismuth, and a current from a single Smee's element, conducted very badly indeed, and yielded only a black film upon a copper cathode in thirty hours.

A cyanide solution has been recommended for depositing this metal, but a bismuth anode does not dissolve readily in a hot solution of cyanide of potassium.

The current from two Daniell's cells passed through a solution of basic nitrate of bismuth, and tartrate of sodium, gave a deposit of hydrated peroxide of the metal upon an anode of platinum (W. Wernicke, ' Journal of Chemical Society,' vol. ix., p. 307; ' Chemical News,' vol. xxii., p. 240).

Fused oxide of bismuth, electrolysed with copper electrodes, deposits bismuth upon the cathode ; with platinum electrodes the cathode forms a very fusible alloy with the deposited metal (P. Buckhard, ' Chemical News,' vol. xxi., p. 238).

According to M. A. Bertrand, metallic bismuth may be deposited upon copper or brass from a solution composed of thirty grains (grammes?) of the double chloride of bismuth and ammonium, dissolved in a litre of water slightly acidified with hydrochloric acid, by means of a current from a single Bunsen's cell. He states that antimony may be deposited in a similar manner ('Athenæum,' April 22, 1876, p. 570). He recommends its use for artistic decorations, instead of platinum-black.

CLASS III. NOBLE METALS

OSMIUM, RUTHENIUM, RHODIUM, IRIDIUM, PALLADIUM, PLATINUM, GOLD, SILVER, MERCURY.

6. **Osmium.**—Elec.-chem. eq. $= \dfrac{199}{6} = 33 \cdot 16$. Scarcely anything has been done in the electro-deposition of this metal. F. Wöhler employed it as an anode in the electrolysis of dilute sulphuric acid, with a current from two Bunsen's cells, and found it freely converted into osmic acid ($Os\,O_4$), but with a solution of caustic soda as the electrolyte, the liquid became of a deep yellow colour, and metallic osmium was deposited upon the cathode ('Chemical News,' vol. xix., p. 10.) Smee electrolysed a solution of osmic acid, and obtained a black deposit.

7. **Ruthenium.**—Elec.-chem. eq. $= \dfrac{104 \cdot 2}{4} = 26 \cdot 05$. According to the same authority, ruthenium behaves like osmium.

8. **Rhodium.**—Elec.-chem. eq. $= \dfrac{104 \cdot 3}{3} = 26 \cdot 07$. Smee deposited this metal from a solution of its sodio-chloride, and obtained a brittle white deposit by a current from ten cells with platinum electrodes ; with a stronger current the deposit was a black powder. According to a writer in Dingler's 'Polytechnic Journal,' electro-deposited rhodium (and iridium) detonate when heated (see 'Journal of Chemical Society,

I

vol. xi., p. 1007), probably in consequence of their containing hydrogen.

9. **Iridium.**—Elec.-chem. eq. $= \dfrac{197}{4} = 49\cdot25$.   Smee states that he has reduced this metal in a bright reguline state on a small scale.   According to F. Wöhler, osmi-iridium is readily dissolved as an anode in a solution of caustic soda ('Chemical News,' vol. xix., p. 10).

10. **Palladium.**— Elec.-chem. eq. $\dfrac{106\cdot5}{4} = 26\cdot62$.   A suitable depositing solution may be made by precipitating and redissolving chloride of palladium by an excess of a solution of potassic cyanide, or by sufficiently saturating a solution of cyanide of potassium with the metal by the battery process.   The solution dissolves a considerable quantity of the metal, and is said to yield thick metallic deposits in a white reguline state.

The ammonio-chloride is also a good salt for the purpose, and should be worked with a palladium anode, and a current from two or three cells ; the current is a little impeded in this solution by a bright golden yellow powder forming upon the anode. Palladium nitrate is less fitted for electro-metallurgy, because it acts more freely upon base metals, and is apt to yield the metal as a black powder ; it is however a good conductor of the current.   Iodide of palladium dissolved in a solution of iodide of potassium is still less satisfactory.   M. A. Bertrand recommends a perfectly neutral solution of double chloride of palladium and ammonium, for electro deposition of palladium, either with or without the use of a battery ('Chemical News,' vol. xxxiv. p. 227).

Palladium used as an anode in dilute sulphuric acid, with a current from two Bunsen's cells, becomes slowly covered with an almost black film of peroxide of palladium ($Pd O_2$) (F. Wöhler, 'Chemical News,' vol. xix., p. 10).

. I have electrolysed pure dilute aqueous hydrofluoric acid containing about 30 per cent. of the anhydrous acid, by means of a current from six Smee's elements, and a small sheet of palladium as anode, in a large platinum cup as cathode. Conduction was free, much gas was evolved from each electrode, and there was a strong odour of ozone. A dark red-brown film quickly formed upon the anode, but did not dissolve after fifteen hours of action; the liquid was black by being filled with floating particles of metallic palladium. After six days' action the anode was greatly corroded. In the electrolysis of aqueous hydrofluoric acid, by means of a palladium anode and platinum cathode, two effects occur: First, the water is decomposed, oxygen being evolved at the anode and hydrogen at the cathode. Second, hydrofluoric acid is decomposed also, the fluorine uniting with the anode, and the hydrogen escaping at the cathode; and the stronger the acid, the less is the proportion of the water decomposed to that of the acid.

As this process offered a very likely means of obtaining fluorine itself, I electrolysed pure anhydrous hydrofluoric acid on several occasions, with a thick sheet of palladium as the anode, and the platinum-containing vessel as the cathode. This process was difficult and very dangerous; in each case it was conducted in the opening of a chimney, and the platinum vessel containing the acid was immersed in a large bulk of a freezing mixture, composed of ice and chloride of calcium. Notwithstanding the low temperature employed, and the vessel being closely covered by a lid of paraffin, the acid volatilised rapidly, partly in consequence of the escaping hydrogen, and the heat evolved by the passage of the current. The coldness of the vessel and the intense attraction of the acid vapour for moisture caused drops of water to condense upon the lid of the vessel, and made it difficult to preserve the acid in a perfectly anhydrous state; the lid was therefore made so as to overhang the outer edge of the vessel, and had laid upon it a layer of cotton wool to absorb the moisture.

In the three first experiments the platinum cup was usually about three and a quarter inches deep and one and three-quarters inch wide, and the whole of its bottom part inside was occupied by a shallow dish of paraffin. The battery consisted of from six to thirty Smee's elements; with twenty and upwards the conduction was copious. In each experiment, with a strong current, the anode quickly became coated with a dark, red-brown, brittle crust, which was of a redder colour on the side next the anode. As the coating entirely covered the immersed part of the anode, and did not greatly diminish the current, either it or the acid in its pores freely conducted electricity. The crust was scraped off, at intervals of half an hour or an hour, into a platinum dish, standing upon a heated slab of iron, in order to dry it quickly, and at once transferred to a perfectly closed platinum bottle. The heated dish should not be at a temperature much exceeding 300° or 400° Fahr. In each instance, some black powder collected upon the cathode and in the paraffin dish, and was found to be metallic palladium.

From the small amount of deposit upon the cathode, and the absence of colour in the liquid, after eleven hours' action, it was evident the crust was but little if at all soluble in the anhydrous hydrofluoric acid; the crust found in each experiment was nearly black when dry, and shewed signs of metallic particles when rubbed between smooth surfaces of agate; it is probable that the crust was partly decomposed by contact with hydrogen from the cathode.

The most perfect forms of experiment to exclude the reducing effect of the hydrogen were made with a platinum cup two and three-quarters inches wide and three and a quarter inches deep, divided into two equal parts by a well-fitting, thin, vertical plate of paraffin, extending to within half an inch of the bottom, and covered by two half circles of that substance. The palladium anode and platinum cathode were each about four inches long and one inch wide, and firmly fixed in slits in the two halves of the cover. With five and a half ounces of the perfectly anhydrous acid, and a current from twelve one-pint Grove's elements, the conduction was copious, and in five minutes the part of the anode in the acid had acquired a deep-brown colour. The electro-lysis was continued during five hours, the anode being taken out and scraped each half-hour. The crust was hard, and a few sparks were produced on some occasions by particles of the hot crust being decomposed by the heat of friction in removing it. A hissing sound was heard during the whole of the electrolysis, but the fume of the acid prevented any effervescence being seen. 10·46 grains of black powder was found upon the cathode and adjacent parts of the con-taining-vessel and partition, and yielded 10·11 grains of metallic palladium. The anode had lost 37·90 grains in weight, and 54·13 grains of the dry brown crust was obtained. 1·37 grain of the crust, gently heated in a platinum lid under a glass cover, generated a red heat in itself, emitted sparks, also a vapour which attacked glass powerfully, and left a black powder weighing 1·14 grain, which by heating to full redness evolved a pungent acid odour, and left about ·87 grain of pinkish metallic palladium. These numbers are in the proportion of 61·2 parts of expelled substance for each 106·5 parts of residual palladium.

I also found that a palladium anode was very rapidly caused to corrode by the passage of a current from three to six Grove's elements through pure fluoride of potassium in a state of fusion; and finely divided palladium was found in the saline residue.

I electrolysed strong nitric acid by means of a current from fifty Smee's elements, with a palladium anode and a platinum cathode. Copious conduction, and rapid decom-position of the acid, with abundant evolution of red fumes, took place. Much gas was evolved from the anode, but none from the cathode, until after a short time. The anode was not at first visibly corroded, but after half-an-hour's action the palladium slowly dissolved, forming a red liquid. No deposit formed upon the cathode. With either strong or partly diluted hydrochloric acid, instead of nitric, there

was instant and rapid action, copious evolution of hydrogen
at the platinum cathode, and chlorine at the palladium anode,
and the latter dissolved, forming a blood-red liquid, and
there quickly appeared a black deposit of palladium upon the
cathode. With dilute sulphuric acid the conduction was
copious, and a deposit of splendid colour, red, purple, &c.
formed upon the anode, but no odour of ozone was evolved
unless the anode dipped only a very small distance into the
liquid. By making the sheet of palladium the cathode for a
short time, the now well-known phenomenon of bending
by absorption of hydrogen took place, and on taking it out
and bending it by mechanical means it suddenly evolved
much heat.

11. **Platinum.**—Elec.-chem. eq. $= \dfrac{197}{4} = 49\cdot25$. The
only common salt of platinum is the tetrachloride, made by
dissolving scraps of platinum in a hot mixture of one volume
of nitric acid and two and a half volumes of hydrochloric
acid, until the liquid acquires a deep-red colour, and then
evaporating the solution nearly to dryness, and allowing it to
cool and solidify; it is a deep-red salt, very freely soluble
in water. The other salts of platinum are usually made
from it.

There are two names applied to the electro-deposition
of platinum, viz. platinising and platinating; by the former
is usually meant its deposition as a black powder or film,
and by the latter its precipitation as white reguline metal.

For platinising either by simple immersion, or more
rapidly by the aid of a separate current or battery, we may
use the tetrachloride dissolved in water, containing one-
fourth its volume of nitric acid, or dissolved in water alone.
Smee appears to have been the first to platinise sheets of
silver for the negative plates of voltaic batteries.

Deposition of platinum by simple immersion (see p. 77).—
Nearly all the common metals become coated with platinum
by simple immersion in solutions of platinum.

A deposit of platinum in the reguline state is more difficult to obtain than one of either copper, silver, gold or nickel. Bright platinum, but of a dark colour, may be obtained upon clean copper articles by immersing them in a boiling solution, composed of 100 parts by weight of distilled water, twelve of caustic soda, and ten of the tetrachloride of platinum ; forty parts of sodic carbonate may be substituted for the twelve parts of caustic soda.

Or add to a strong solution of the platinum chloride, carbonate of soda in fine powder, until effervescence ceases ; then add some glucose, and afterwards as much chloride of sodium as will produce a whitish precipitate. Place the articles of copper or brass which are to be coated in a zinc colander, and immerse them thus for a few seconds in the mixture, which should be at a temperature of 60° C. ($=140°$ Fahr.) Then wash the articles, and dry them in hot sawdust (Les Mondes, ' The Chemical News,' vol. xix. p. 226). Magnesium deposits pure platinum from a solution of platinic chloride (Commaille, ' Chemical News,' vol. xiv. p. 188). I have observed that crystals of silicon did not deposit platinum from a solution of tetrachloride of platinum.

Deposition of platinum by separate current (see pp. 82, 89). —Probably the best solution for obtaining thick reguline deposits is that employed by Roseleur, and obtained as follows. Convert ten parts by weight of platinum into dry tetrachloride in the manner already described, dissolve it in 500 parts of distilled water, and filter if necessary. (N.B. If there is much sediment, the salt has been dried either too much or carelessly.) Dissolve also 100 parts of crystalline phosphate of ammonia in 500 parts of distilled water, and add it, with stirring, to the platinum solution; this produces a copious precipitate ; then add at once, with stirring, a previously prepared solution of 500 parts of phosphate of soda crystals in 1000 parts of distilled water. Boil the mixture until the odour of ammonia ceases, and the liquid which was previously alkaline begins to redden blue litmus paper; the yellow liquid will then become colourless, and is fit for depositing. This solution is suitable for depositing upon copper or its alloys, but not upon zinc, tin, or lead, because they decompose it or platinise themselves in it by simple immersion. The liquid is used hot, and requires a strong battery current. As an anode of platinum would not be corroded in such a liquid, the solution will of course be gradually deprived of its metal by the process of deposition unless some of the chloride be occasionally added.

A solution made by dissolving the chloride of platinum in one of cyanide of potassium, in the proportion of about twenty pennyweights of the metal to one gallon, has also been employed. It is used warm with a feeble current, but it also has the disadvantage of not dissolving the anode, and therefore requires a stronger electric current, as well as renewal of platinum salt.

Other solutions, composed of the double chloride of platinum and sodium ; or of this salt dissolved in a solution of oxalic acid, and made strongly alkaline by means of caustic soda, have been recommended, but it is difficult to obtain reguline white platinum from them.

Böttger states that for platinising he takes a boiling solution of the ordinary chloride of platinum and chloride of ammonium, to which he adds a few drops of solution of ammonia. The solution contains very little metal, and requires to be occasionally supplied with fresh metallic solution (' Pharmaceutical Journal,' vol. iii. p. 358).

In my experiments, a platinum anode in pure dilute hydrofluoric acid of 10 per cent. was not corroded by the passage of a current from either six Smee's or six Grove's cells during many hours. Very free conduction took place ; a powerful odour of ozone, and a gas which re-ignited a red-hot splint was evolved at the anode, but no deposit of platinum occurred. A solution of pure fluoride of potassium gave precisely similar effects.

Also with a current from twenty-four pairs of magnesium and platinum, excited by a solution of one and a half ounce of common salt dissolved in fifty ounces of water, and the current passed through pure dilute hydrofluoric acid of 40 per cent., gas was evolved from both electrodes of platinum, but no corrosion of the anode took place in eighteen hours. I also electrolysed pure hydrofluoric acid, containing 80 per cent. of the anhydrous acid, with platinum electrodes and a current from ten Smee's cells. Abundant conduction and evolution of hydrogen and ozone occurred; the anode corroded and lost 16·58 grains in weight during thirty-six hours, and became covered with a blackish crust which partly dissolved in the liquid to a brownish solution, but no electro-deposit of platinum occurred. I further electrolysed anhydrous hydrofluoric acid in a similar way with platinum electrodes to that described with palladium (p. 116); with a current from forty Smee's elements the anode corroded rapidly, but did not dissolve in the liquid; it acquired a dark red-brown crust, which rapidly deliquesced in the air, and formed a blood-red liquid, also some basic salt, by solution in water.

I electrolysed during sixteen hours pure dilute hydrofluoric acid of 30 per cent., mixed with an equal volume of nitric acid; gases were freely evolved, but scarcely any platinum dissolved, and none was deposited. Also, when mixed with an equal volume of strong hydrochloric acid, hydrogen and chlorine were evolved, but in four hours the anode was but little corroded. When mixed with an equal volume of sulphuric acid, after many hours' action, the anode corroded very slowly. And with much selenious acid dissolved in it, selenium containing traces of platinum was freely deposited, and gas was evolved as before. With phosphoric anhydride dissolved in it, the anode was slowly corroded, and gas was evolved (See 'Philosophical Transactions of the Royal Society,' 1869, p. 200).

By electrolysing fluoride of potassium or of lithium in a

melted state, a platinum anode was rapidly dissolved, and the resulting salt of platinum simultaneously decomposed, and its metal set free: and by electrolysing pure double fluoride of hydrogen and potassium, in a fused state, the platinum anode was rapidly dissolved, and imparted a colour to the fused salt. The fused fluorides of silver, copper, lead, manganese, uranium, or the fused silico-fluoride of potassium, electrolysed by a current from six Smee's elements, did not corrode the platinum anode.

12. **Gold.**—Elec.-chem. eq. = $\dfrac{196\ 6}{3}$ = 49·15. The commonest salt of gold is the terchloride ; and this is the one from which other salts of the metal are usually prepared. In addition to this one, there have been used for electro-depositing purposes, the oxide, bromide, iodide, sulphite, hyposulphite, cyanide, and double cyanides.

Finely-divided gold is also sometimes employed in electro-metallurgical operations ; it is precipitated by adding a clear solution of protosulphate of iron (green vitriol) to a warm solution of chloride of gold, until it ceases to produce a cloud. It is a brown powder, which assumes the metallic lustre on being burnished. One part of gold requires about five parts of the crystallised sulphate to precipitate it. Oxalic acid produces a similar precipitate of gold. A solution of sulphurous anhydride (sulphurous acid H_2SO_3), or a current of the gas, also precipitates the gold completely as metal. Many organic substances, if wetted with a solution of chloride of gold, reduce it to metal, and hence one's fingers, paper, wood, a feather, calico, linen, &c., become stained of a purple colour by contact with the liquid and subsequent exposure to light.

Preparation of gold salts.—Terchloride of gold is formed by dissolving metallic gold in a warm mixture of one measure of nitric acid, and from two to three measures of hydrochloric acid ; the mixture is called aqua-regia. The gold dissolves slowly with evolution of gas ; when it is all dissolved, evaporate the solution by gentle heat, with stirring, until it is reduced to a small bulk, and solidifies on cooling ; the residue should be entirely soluble in water. If it contains a white substance which will not dissolve, it is chloride of silver, derived from traces of silver in the metal ; if there is a small amount of yellow or brown residue, some of the salt has been overheated ; such residue should be redissolved in a little aqua regia, and evaporated to dryness again. One ounce of gold, if it is in small fragments, or thin sheet, will re-quire about four ounces of aqua-regia to dissolve it. Chloride of gold is a yellow salt and dissolves in one and a half its weight of water. If it is properly made it contains one atomic weight (=196·6 parts) of gold and three atomic weights (= 106·5 parts) of chlorine, and its composition is represented by the formulæ $AuCl_3$. One troy ounce of gold will make 1 oz. 164½ grains of the chloride.

Oxide of gold is obtained by digesting a solution of the chloride with an excess of calcined magnesia, washing the precipitate first with dilute nitric acid, and then with water only ; if caustic potash or soda be used instead of magnesia the oxide is liable to contain some of the alkali.

The terbromide of gold may be formed by digesting oxide of gold in hydrobromic acid, and evaporating the solution by a gentle heat, with stirring, until it solidifies on cooling.

The oxide of gold forms on addition of aqueous ammonia or of solutions of carbonate sulphate, or chloride of am-monia, a dark olive-brown substance, called fulminate of gold, aurate of ammonia, or ammoniuret of gold. The same substance is also formed on adding ammonia, or a solution of a salt of ammonia, to a solution of terchloride of gold. It is *an extremely dangerous substance when dry, and detonates with the least friction or percussion.* To form ammoniuret of gold, which is sometimes used in electro-gilding baths, convert ten parts by weight of gold into the solid chloride. Dissolve that salt in water, and add to the solution fifty parts by weight of the strongest aqueous ammonia, and stir the mixture, an abundant precipitate of the ammoniuret, otherwise called fulminate of gold, is produced in the form of a yellowish brown powder. When it has subsided, pour off the supernatant liquid, and fill up again with water, and repeat this several times, until the precipitate no longer smells of ammonia. The water contains a little gold, and is reserved for recovery of that metal. As the yellow-brown precipitate, *when in a dry state, is highly explosive,* it should never be allowed to get dry, and ought not to be prepared until the time of forming a gilding solution with it. Particles of it also should not be allowed to dry upon the edges of the vessels nor upon filters through which the wash-liquids have been passed. To remove the solid salt from articles, we may dissolve it in a solution of cyanide of potassium. Freshly precipitated wet oxide of gold dissolves in a solution of caustic potash, to form aurate of potassium; the solution is yellow, and may be employed for electro-gilding.

Sulphide of gold is obtained by passing a current of sulphuretted hydrogen gas through a solution of chloride of gold, as long as a precipitate occurs; it is a blackish-brown powder.

'Cyanide of gold is formed by cautiously adding a solution of cyanide of potassium in six parts of water, to a neutral solution (i.e. not containing any free acid) of terchloride of gold, as long as a yellow precipitate settles down; if more cyanide of potassium is added the precipitate becomes dirty yellow, and is more quickly deposited; a still larger quantity renders it orange yellow and redissolves it. It is a crystalline powder, permanent in the air; by ignition it is resolved into gold and cyanogen gas; it is not decomposed by sulphuric, hydrochloric, or nitric acid, or by aquaregia, unless freshly precipitated, and then only slowly. It is not decomposed by sulphuretted hydrogen; hydrosulphate of ammonia dissolves it slowly but completely, forming a colourless solution, from which, by the addition of acid, sulphide of gold is precipitated. It dissolves in aqueous solution of ammonia, hyposulphite of soda, or alkaline cyanides, but not in water, alcohol, or ether.'

'Gold precipitated from a solution of chloride of gold by protosulphate of iron, dissolves in a boiling solution of cyanide of potassium ; a hot solution of cyanide of potassium will also dissolve ordinary metallic gold if air be present. Both oxide of gold and aurate of ammonia dissolve completely in a solution of cyanide of potassium, and form double cyanide of gold and potassium. Cyanide of gold requires twenty-three parts of cyanide of potassium dissolved in water to dissolve it. For every one part of gold to be dissolved by the battery-process, six parts of cyanide of potassium, dissolved in two to four times the quantity of water at 100° Fahr., is required ; two electrodes of gold being connected with a suitable battery, and immersed in it, until the required quantity of gold is dissolved.' 'The crystallised cyanide of gold and potassium dissolves in seven parts of cold and in half a part of hot water' (Himly), 'in four parts of cold and in o·8 part of hot water' (Glassford and Napier). 'It dissolves very sparingly in alcohol. Its aqueous solution gilds copper and silver by simple immersion, especially if hot, and the copper and silver dissolve in it.'

A gold anode was speedily dissolved in dilute hydrochloric acid, or in a saturated solution of chloride of barium, sodium, or ammonium, by a current from twelve Wollaston's cells ; but most rapidly in the sodic chloride. It was also slowly dissolved in a solution of chlorate of potassium, by a current from twenty such cells (H. Bartlett, 'Chemical News,' vol. xvi. p. 257). Runspaden has also shown that a gold anode in dilute sulphuric acid is considerably oxidised, and a definite hydrated oxide of gold formed ('Chemical News,' vol. xx. p. 179).

Electro-deposited gold is not necessarily pure, because other metals are often thrown down with it, in order to obtain the particular shade of colour required ; for instance, white and green gold contain silver, red gold contains copper, and pink gold contains both copper and silver.

Electrolysis of fluorides with a gold anode.—I electrolysed pure, dilute hydrofluoric acid with a gold anode and a platinum crucible cathode, during many hours, with a current from six Smee's elements. Conduction was copious, and much gas was evolved from each electrode, and an odour of ozone came from the anode. In a few hours the anode became covered with a red-brown film, which did not dissolve but fell off. The liquid remained colourless, and was not discoloured by addition of sulphuretted hydrogen water to it. No deposit of gold occurred upon the cathode. I also electrolysed some stronger pure hydrofluoric acid with the same battery current, and also with a current from fifty pairs of magnesium and platinum, excited by a dilute solution of common salt, and obtained similar effects. The deposit upon the anode appeared to be metal, because it was insoluble in nitric acid, and looked like gold on being burnished with agate.

Pure anhydrous hydrofluoric acid at 10° Fahr. would not transmit any current from ten Smee's elements and a large gold anode ; but with forty Smee's cells, as in the experiment with palladium (see p. 116), it conducted very feebly, and, by continuing the action for one and a half hour, the anode acquired a dark reddish-brown film, and a few crystals, at first of a green colour, appeared upon its edges ; the crystals became yellow and then red by exposure to the air.

A solution of pure fluoride of ammonium, containing free ammonia and a gold anode, conducted freely the current from six Smees elements. Much gas was evolved from the anode, and a bright lemon-coloured powder, insoluble in the liquid, formed upon the anode. No deposit of gold occurred.

I also electrolysed pure fluorides of lithium and potassium in a melted state, with a gold anode, and a current from both three and six Grove's elements. The anode was very rapidly corroded, and metallic gold separated.

Solutions for gilding.—There are many solutions for electro-gilding, some being formed by chemical means, and some by the separate current or battery process ; but the best for thick deposits are those formed with pure cyanide of potassium and cyanide of gold, either by the battery process or by chemical means.

Separate gilding solutions are kept for different purposes, some for gilding by simple immersion process, and some by separate current ; others for gilding pale, yellow, pink, &c. Some are employed cold, and others hot ; some for gilding iron, steel, and the baser metals in general.

Deposition of gold by simple immersion (see p. 77).—Acid solutions of gold deposit their metal upon surfaces of phosphorus, silver, mercury, copper, and nearly all the base and brittle metals, by simple contact with those substances. According to Commaille, magnesium deposits pure gold from an aqueous solution of the terchloride. ('Chemical News,' vol. xiv. p. 188). I have observed that crystals of silicon did not deposit gold from a solution of its terchloride, but that by contact of the terchloride in aqueous solution with benzine, 'petroleum ether,' amylene, and a number of other liquid hydrocarbons, films of metallic gold gradually separated.

For gilding articles of copper, bronze, or brass, by simple immersion, the following solution of Roseleur's may be used. Dissolve 800 parts by weight of pyrophosphate of soda in 10,000 parts of distilled water, and add eight parts of strong hydrocyanic acid. Convert ten parts by weight of gold into soluble dry chloride, dissolve it in a reserved portion of the water to which nothing has yet been added, and mix the resulting liquid with the cold solution of pyrophosphate. The mixture is used hot ; it is yellowish, but must become colourless when heated ; if it becomes red, a little prussic acid must be added, with stirring, until the liquid is colourless. If too much of the acid is added it will prevent the articles becoming gilded, and this may be corrected by adding a small quantity of chloride of gold solution. The articles to be gilded must be previously dipped in a very dilute solution of nitrate of mercury; and whilst being gilded they must be kept in continual motion. To gild most successfully by this process, the articles should receive a first coating of gold in a nearly-exhausted solution of the same kind, a second in a less exhausted one, and a third in a more freshly-prepared one, to impart a proper colour. The gilding occupies only a few seconds in each bath. To obtain ' green ' and ' white ' gilding in such a liquid, a solution of nitrate of silver is added, drop by drop, with stirring, until the desired colour is obtained ; before gilding green or white, it is best to gild the articles yellow, then dip them quickly in the nitrate of mercury solution, and then into the bath containing the nitrate of silver.

Gilding by simple immersion is also employed for putting an exceedingly thin deposit of gold upon large articles of bronze previous to proper gilding ; then with a thicker deposit in a cyanide solution by the battery process. The solution employed is composed of 180 parts of caustic potash, twenty parts of carbonate of potash, and nine parts of cyanide of potassium, dissolved in 1000 parts by weight of water, in which has been previously dissolved as much chloride of gold as is formed from one part by weight of the metal. The mixture is used nearly at a boiling temperature. The articles to be gilded do not require to be previously dipped in a mercurial solution. As the solution loses its gold, chloride of gold must be added, but after four or five such additions the other salts must also be added with it, in the above proportions. The solution may thus be kept in order for any length of time.

It is possible to gild copper and brass articles perfectly by simple immersion, by employing the artifice of ' quicking ' the surface before each immersion, by dipping it alternately into a solution of nitrate of mercury and into the gilding liquid ; and this plan is often adopted with large articles. It is said that copper may be gilded so perfectly by this method as to resist for several hours the corrosive action of concentrated acids: The secret of the action is, that each film of mercury, being electro-positive to the gold, dissolves in the auriferous solution, and deposits a film of gold in its place (see pp. 77, 80).

A solution for gilding by simple immersion was at one time extensively used by Messrs. Elkington. It was prepared as follows :—Convert one part of gold into terchloride, and expel all excess of acid; dissolve it in a small amount of water, and add gradually to it thirty-one parts of acid carbonate of potassium ; then mix the liquid with a solution of thirty parts more of the acid carbonate dissolved in 200 parts of water, and boil the mixture for two hours. During the boiling the yellow solution becomes green, and is then ready for use. The previously cleaned trinkets of brass or copper are immersed for about half a minute in the hot liquid. To gild articles of german-silver, silver, or platinum, in this bath, they must be immersed in contact with wires of copper or of zinc. Chlorate of potash is formed in the solution by the gilding process, and a black powder is precipitated, containing carbonate of copper and a little purple of Cassius (see Miller's ' Chemistry,' vol. ii., 3rd ed., p. 814).

The two following liquids have also been used for gilding by the simple immersion or ' water-gilding' process. Convert five troy ounces of gold into chloride ; dissolve it in four gallons of distilled water, add twenty pounds of pure bicarbonate of potassium, and boil it during two hours. The articles to be gilded are immersed in the warm liquid from a few seconds to one minute, according to the degree of quickness of the action. For gilding articles of silver : dissolve equal weights of corrosive sublimate and sal-ammoniac in nitric acid, add some pure grain gold to it, evaporate the liquid to half its bulk ; and apply it whilst hot to the surface of the silver article.

C. D. Braun gilds zinc, by immersing it in a solution of sulphide of gold dissolved in a solution of sulphide of am-

monium, excluded from the atmosphere ('Chemical News,'
vol. xxix. p. 230). W. Kirchmann gilds clean iron by first
applying to it sodium amalgam, which coats it with mercury.
He then applies to the mercurialised surface a strong solution of chloride of gold ; and finally heats the object to redness in a muffle (' Chemical News,' vol. xxvii. p. 268).

Gilding by contact with zinc.—Joseph Steele's patent,
dated Aug. 9, 1855. It consists substantially in immersing the articles to be gilt, in connection with a piece of
zinc, in a hot solution, formed by adding chloride of gold to
a solution of cyanide of potassium. It is not an economical
process, because much of the gold is deposited upon the zinc.

Solutions for gilding by means of a separate current
(see p. 89).—The electric current employed is usually derived either from a Bunsen's battery, or a Clamond's thermo-
electric pile, that from a magneto electric machine being
found to be less suitable. Many solutions have been tried,
but none have succeeded like the double cyanide of gold
and potassium. They may be formed either by chemical
methods, or by means of the battery process.

A solution may be formed by chemical means as follows.
Convert a weighed quantity of gold into solid chloride, dissolve it in water, then add a solution of cyanide of potassium
to it as long as, but no longer than, it produces a precipitate,
stirring the mixture on each addition, and allowing it to subside. As it is difficult and tedious to hit the exact neutral
point, and an excess of either chloride or cyanide causes
some gold to remain in solution, the wash-waters must be
carefully preserved, and the gold in them recovered. When
the neutral point is attained, allow the precipitate to subside,
pour off the clear liquid, and fill up with water again, again
allow to subside, and so on five or six times ; then pour the
sediment into a filter, and complete the washing by addition of water, and allow it to drain thoroughly. The precipitate should not be allowed to dry because it is liable to
contain a little fulminate of gold, derived from ammonia

produced from cyanate of potash present in the cyanide. The wet substance should now be added to a small quantity of a strong solution of cyanide of potassium, and just sufficient additional cyanide of potassium added with stirring, to dissolve the whole, a note being kept of the amount of cyanide consumed. About one-fifth or one-fourth more of cyanide of potassium should now be added, and dissolved to constitute what is termed 'free cyanide ; ' and sufficient water be added, with stirring, to form a solution containing the requisite amount of gold per gallon. The total amount of cyanide required will depend upon the quality of that salt, which is very variable, and with the freedom or otherwise of the gold salt from an excess of acid. Instead of dissolving cyanide of gold in the cyanide of potassium, the oxide, the ammoniuret, or even the chloride of gold may be added, and will be converted into cyanide and dissolve, if sufficient cyanide of potassium is present. But the disadvantage of this method is, that these salts of gold introduce impurities into the liquid; the chloride is the most objectionable, because it leads to the formation of chloride of potassium, which interferes with the perfect working of the solution.

The proportions of cyanide of gold, cyanide of potassium, and water, in an electro-gilding liquid, may vary very greatly without detriment to the process, as will be perceived from the varied proportions used by different persons. A very good proportion is an ounce of gold, sixteen ounces of cyanide of potassium, and one gallon (= 160 ounces) of water, or an ounce of gold converted into cyanide, seven or eight ounces of cyanide of potassium, and 100 ounces of water ; or four ounces of gold, thirty-two ounces of cyanide, and 160 ounces of water. The proportion of gold in solutions used by the separate current process, in large electro-gilding establishments, varies as much as from ten pennyweights to fifty troy ounces per gallon. Moderately dilute gilding solutions yield a better quality of metal, though at a slower rate, than stronger ones.

K 2

Gilding solution of M. De Ruolz—'Dissolve ten parts of cyanide of potassium in 100 parts of distilled water; filter the liquid and add one part of cyanide of gold, prepared with care, well washed, and dried out of the influence of light; keep the mixture in a closed glass vessel at the temperature of 60° to 77° Fahr. for two or three days, out of the presence of light, with frequent stirring.'

Formulæ of M. J. L.—'First, take thirty-one grammes (see p. 381) and twenty-five centigrammes of oxide of gold, five hectogrammes of cyanide of potassium, and four litres of water, and boil them together half an hour. The resulting solution must be worked hot, and may be used to gild copper, brass, and silver.'

'Second, dissolve ten parts of ferro-cyanide of potassium and one part of dry terchloride of gold in 100 parts of water; oxide of iron will be precipitated. Boil the solution two or three hours in a porcelain or glass vessel, until a precipitate collects at the bottom, and the supernatant liquid is transparent, and of a canary-yellow colour; filter the solution and dilute it with three times its volume of water.'

M. De Briant's process.—'Dissolve thirty-four grammes of gold in aqua-regia and evaporate the solution until it becomes neutral chloride of gold; then dissolve the chloride in four kilogrammes of warm water, and add to it 200 grammes of magnesia; the gold is precipitated. Filter and wash with pure water, digest the precipitate in forty parts of water mixed with three parts of nitric acid to remove magnesia, then wash the remaining oxide of gold with water until the wash-water exhibits no acid reaction with test-paper. Next dissolve 400 grammes of ferro-cyanide of potassium and 100 grammes of caustic potash in four litres of water, add the oxide of gold, and boil the solution about twenty minutes. When the gold is dissolved there remains a small amount of iron precipitated, which may be removed by filtration, and the liquid, of a fine gold yellow colour, is ready for use; it may be employed either hot or cold.'

M. Becquerel's gilding liquid.—' Dissolve one part of terchloride of gold and ten parts of ferro-cyanide of potassium in 100 parts of water ; filter the liquid to remove the separated iron ; add 100 parts of a saturated solution of ferru-cyanide of potassium, and dilute the mixture with once or twice its volume of water. In general the tone of the gilding varies according as this solution is more or less diluted ; the colour is most beautiful when the liquid is most dilute, and most free from iron. To make the surface appear bright it is sufficient to wash the article in water, acidulated with sulphuric acid, rubbing it gently with a piece of linen cloth.'

M. Levol's solution for gilding silver.—' Dissolve neutral chloride of gold in water, then add an aqueous solution of sulpho-cyanide of potassium, until the precipitate first formed is re-dissolved. The liquid will retain a slightly acid reaction ; if it has lost it, it must be renewed by adding a few drops of hydrochloric acid.'

Gilding liquids by M. Fizeau.—First, dissolve one part of dry chloride of gold in 160 parts of distilled water ; then add, little by little, a solution of carbonate of potash in distilled water, until the liquid begins to become cloudy. We may use this liquid immediately. And, second (used by M. Lerebour), dissolve one gramme of chloride of gold and four of hyposulphite of soda, in one litre of distilled water.'

Mr. Wood's gilding solution.—' Dissolve four troy ounces of cyanide of potassium and one of cyanide of gold in one gallon of distilled water, and use the solution at about 90° Fahr. with a current from at least two cells.'

Making gilding solutions by the battery process.—Excellent solutions for gilding may be made by this method ; and if the quantity of liquid required is not very large, this plan is by far the most convenient and simple one, and is unattended by the risk of loss of metal, which occurs in the processes of solution and precipitation of gold by chemical means. To prepare a cyanide gilding solution by this plan, simply dissolve some cyanide of potassium in hot distilled water, in an earthenware vessel, in the proportion of from one to two pounds to each gallon. Immerse two large electrodes of pure sheet gold in the liquid, and pass the current from about three Smee's, or two Daniell's cells, stirring the liquid occasionally, until a clean and bright cathode of german-silver (substituted a short time for the gold one) receives a proper coating. The liquid should be kept at a temperature of about 150° Fahr. during the process, by immersing the vessel containing it in an outer vessel of hot water with a lamp beneath. The quantity of gold dissolved from the anode is ascertained by weighing, and is not of material consequence, provided the deposit is good. In this process a portion of the cyanogen from the cyanide unites with the gold, and leaves potash in the solution, and after a time, being exposed to the atmosphere, absorbs carbonic acid, and thus brings carbonate of potassium into the liquid, but the presence of this salt is not objectionable. A very good gilding solution made by this method consisted of one gallon of water, one and a half pound of cyanide of potassium, and fifty pennyweights of gold.

Gold anodes should be suspended in the liquid, either by gold wires protected by tubes of gutta-percha, india-rubber, or glass (see p. 171), or by means of platinum wires, because the gold is liable to be cut through by electro-chemical action at the surface of the liquid.

Cold electro-gilding solutions for the separate current process. —Gilding in cold solutions is usually employed in cases where the objects to be gilt are massive, such as chandeliers, clocks, &c., which would otherwise require large volumes of liquid to be heated. Both bright and dead gilding in cold liquids is practised in large electro-gilding establishments. The articles to be gilded are coated with a film of brass or copper by electro-depositing process, before gilding them in a cold solution. With a good solution the gilding is quickly effected.

The following is the composition of cold gilding solutions

in general use, and recommended by Roseleur as giving satisfactory results.

1st Solution.

Distilled water	1000 parts.
Aqueous ammonia	50 ,,
Cyanide of potassium of 70 per cent. .	30 ,,
Gold	10 ,,

Convert the gold into solid chloride, dissolve the salt in water, add the ammonia with stirring; the precipitate is aurate of ammonia, and *highly explosive* (see p. 123); wash it by decantation and subsequent filtration; preserve the wash-waters, as they contain a little gold. Dissolve the cyanide of potassium in nearly the whole of the water, add the solution to the aurate of ammonia, and stir; the aurate dissolves quickly; wash any residue of aurate into the liquid by means of the remainder of the water, with the aid of a feather. Boil the mixture about one hour to expel excess of ammonia.

As this bath is liable to become weaker in dissolved gold by the process of gilding, it is replenished as follows : Convert some gold into aurate of ammonia, add 100 parts of water for each ten parts of gold, and then gradually dissolve cyanide of potassium in the mixture until the liquid is colourless. A little of this latter mixture is added to the gilding solution as occasion requires.

2nd Solution.

Distilled water	1000 parts.
Ordinary cyanide of potassium . .	30 or 40 ,,
(Or, Pure cyanide of potassium . . .	20 ,,)
Gold	10 ,,

Convert the gold into solid chloride, and dissolve it in 200 parts of the water; dissolve the cyanide in the remaining 800 parts of water. Mix the solutions and filter if necessary. Boil the liquid a short time before first using it. To replenish the solution, add occasionally, as required, solid chloride of gold one part, and pure cyanide of potassium from one to one and a half parts, each dissolved in a little water.

The deposit of gold from these solutions is often yellow; if it is dark red or black, it indicates either an excess of gold in the liquid or too strong a current; and if it takes place very slowly and is of a grey colour, or if one portion of a gilded surface becomes ungilded, whilst that nearer the anode is receiving a deposit, it indicates either that the electric current is too weak, or the presence of too much cyanide of potassium.

' *Solid*' *deposition of gold.*—A process, or branch of trade, termed 'solid depositing' has gradually extended itself. It consists in making solid articles of gold and silver, by electro-deposition, upon gutta-percha or other moulds; such, for instance, as watch and clock faces, ornamental snuff-boxes, and other articles elaborately chased or engraved, or which have very complex or undercut ornaments upon them; the expense of multiplying these by the electro-process being less than by the ordinary means. Mr. Alexander Parkes took out a patent dated March 1841, for a solution for depositing solid articles in gold; it is formed thus :—Dissolve one ounce of pure gold in aqua regia, and evaporate the solution to dryness; then add two gallons of water and sixteen ounces of cyanide of potassium, and work the resulting liquid at a temperature of about 120° or 130° Fahr.

Gilding in hot solutions by separate current process.—This is the best method of gilding in the great majority of cases. For rapid gilding of small articles of silver, copper, bronze, or brass. Roseleur employs a solution composed of—

Distilled water	1000 parts.
Crystallised phosphate of sodium . .	60 ,,
Bisulphite of sodium	10 ,,
Cyanide of potassium (pure) . . .	1 part
Gold	1 ,,

The phosphate of sodium is dissolved in 800 parts of the water made hot. The bisulphite of sodium and cyanide of potassium are dissolved together in 100 parts of the water. The gold is converted into solid chloride, and dissolved in the remaining portion of water, and the solution poured slowly, with constant stirring, into the *cold* one of phosphate of sodium ; and into this mixture, which is greenish yellow, is at once poured the solution of bisulphite and cyanide. The entire liquid soon becomes colourless, and is then ready for use. If the solution of phosphate is not cold, some of the gold is precipitated as a metallic powder.

If the articles to be gilded are composed of iron or steel and require to be gilded directly, i.e. without previously coating them with copper or brass, he employs a liquid composed of—

Distilled water	1000 parts.
Phosphate of sodium	50 ,,
Bisulphite of sodium	12½ ,,
Cyanide of potassium (pure) . . .	½ part.
Gold	1 ,,

and prepares the bath in a similar manner.

These baths are used at a temperature which varies from 50° to 80° C. ; they gild rapidly, and the gilding only occupies a few minutes.

In gilding articles of steel, they are, after previously cleaning, dipped in a very hot bath, with a powerful current at the commencement, and the current then gradually diminished by raising the anode until it is nearly out of the liquid.

Roseleur also employs a solution composed of—

Distilled water	300 parts.
Cyanide of potassium (pure) . . .	5 ,,
Gold	1 part

The gold is converted into solid chloride, and dissolved in one portion of the water, and the cyanide of potassium in the other portion, and the two solutions then mixed.

This liquid may be employed at almost any temperature, but is liable to give a yellow deposit on the upper part of an article, and a red one at the bottom ; and even to un-gild the distant parts, whilst the near parts are receiving a deposit. Both these defects may, however, be diminished by keeping the articles in brisk and continual motion.

Coloured gilding.—To obtain red gold, add to either of the foregoing solutions a sufficient proportion of either of the acetate of copper liquids (see pp. 207, 208), or we may gild red in an old gilding bath in which a great many copper articles have been gilt, taking care to use a strong electric current. To obtain green or white gold, add to one of these baths a sufficiency of either a dilute solution of argentic nitrate, or better, one of double cyanide of silver and potassium. To obtain what is called 'pink' gold, the articles are first gilt yellow, and then red, and afterwards a mere blush of silver is deposited upon the red in a cold silver bath; they are finally burnished. If the proper pink colour is missed, strip off the coating of silver and copper, by immersing the articles during a few seconds in a mixture of five parts sulphuric and one of nitric acid (the yellow gilding then reappears), and try again.

To obtain a rich deep gilding from a cyanide liquid, the solution should be strong, and the articles should either appear of a dark yellow or a rich orange colour, approaching to brown, on coming out of the liquid ; then, on scratch-brushing them, they will acquire the desired appearance. A very rich dead gilding may also be obtained in a cyanide solution, by adding a little wet aurate of ammonia to the liquid just before gilding. And a bright clear yellow gilding may be obtained by adding to an ordinary cyanide gilding solution a small quantity of caustic soda. If a deposit of gold appears of a blackish colour when coming out of the liquid, it will not have a satisfactory appearance imparted to it either by brushing or burnishing.

Sometimes a defective colour in gilding is improved by immersing the article, until its colour is white, in a solution

f nitrate of mercury, then volatilising the mercury by heat,
nd scratch-brushing the article. 2nd. Or dip the article in
trong sulphuric acid, then heat it until white fumes are freely
volved, and immerse it at once in very dilute sulphuric
cid. 3rd. Or make into a wet paste, with a little water, a
nixture in powder, of two parts nitrate of potassium, one of
ulphate of zinc, one of alum, and one of common salt, and
mear the article with a layer of the paste ; then heat it to
lackness upon an iron plate, over a clear fire, and plunge it
nto water. By varying the mixture, different tints of colour
ıre obtained. 4th. Or smear the article with a thick magma
ıf powdered borax and water, heat it until the borax
ındergoes *igneous* fusion, and plunge it into dilute sulphuric
ıcid.

Copper or brass trinkets, which require the gold to have
ı dead appearance upon them, are dipped for a moment in a
nixture of equal parts of sulphuric and nitric acids, to which
ı little common salt is added.

The colour of the deposit may also be largely regulated by
he size of the anode. If the anode dips but slightly into
he solution, and the deposit is then of a pale yellow colour,
t will become full yellow with deeper immersion of the
ınode, and of a red colour if the anode is wholly immersed.
:t is also largely affected by motion of the cathode ; if the
iolution gilds of too dark a colour, keeping the article in
notion will remedy the defect. The colour may also be
ıreatly varied by depositing other metals with the gold. In
ıll electro-gilding establishments alloys of gold, with silver
ınd copper, are deposited, in order to obtain the requisite
:olour, and the general method adopted for regulating the
:olour of electro-gilding is as follows :—After having pre-
ıared the solution, work it with a large copper anode until
he deposited metal begins to deteriorate in colour ; then
eplace the copper by a small gold anode. With the copper
ınode can be obtained a full rich colour, becoming deeper
ıs the temperature of the liquid is higher ; to produce a

paler yellow, use a small gold anode with the liquid at a lower temperature.

With cyanide of potassium solutions, containing various dissolved metals, silver goes down first by the electric-current, then silver plus gold, then gold alone, then gold and copper, then copper alone, then copper plus zinc. If salts of lead are added to the gilding liquid, lead will first be precipitated alone, and then the lead will cause the gold to be deposited in a bright condition.

Necessity of free cyanide of potassium.—In all cyanide gilding liquids it is necessary to add much more cyanide of potassium solution than is sufficient to dissolve all the salt of gold ; this excess is termed 'free cyanide.' The proportion of free cyanide employed by different electro-gilders varies somewhat, but does not exceed certain limits, otherwise the liquid will not work well ; from one-fourth to one-half of the quantity required to dissolve the salt of gold is a usual proportion. If too much free cyanide is employed the gilding is apt to assume what is termed a 'foxy' appearance, the anode also is dissolved whilst the current is not passing, and the solution is more prone to decompose and become of a dark colour. The reason why free cyanide is necessary in gilding operations, is because the cyanide liberated from its gold at the cathode, requires some time to get across the solution to the anode, and meanwhile the anode must be supplied with some, in order to render soluble the cyanide of gold formed upon it. The actual quantity of cyanide of potassium required, both to dissolve the gold salt, and to form free cyanide, differs very greatly, because the quality of commercial cyanide is very variable. The larger the proportion of actual cyanide present in the salt, the smaller the proportion of the salt necessary both to dissolve the gold salt and to form free cyanide. It must not be forgotten that commercial cyanide is liable to contain from about 30 to 98 per cent. of actual cyanide, and that all the other salts in it are almost useless as solvents of gold, and some of them are positively detrimental to electro-gilding operations.

Management of electro-gilding solutions.—Cyanide gilding solution is generally contained in a glazed iron vessel, and heated either by a stove or by gas-jets beneath ; or it is contained in a stoneware or glass pan immersed in boiling water. On account of the usual smallness of the articles to be gilded, the thinness of the deposit, and the rapidity of the action in a hot liquid, the objects only require to be immersed in the solution for a few minutes ; when a thicker deposit is desired, in order to maintain a proper condition of deposit, they should be taken out several times, brushed, and re-immersed. The strength of battery used for gilding is generally about two Bunsen's cells, of different sizes, according to the magnitude of the articles to be gilded. The loss of water by evaporation is generally made good by adding a little distilled water, after having finished gilding.

Gilding done in hot solutions, is superior in several respects to that done in cold ones. The deposits are smoother and cleaner, and, what is more important, the same thickness of deposit is more durable when made in a hot liquid than when effected in a cold one. This may be partly explained by the fact that in the subsequent cooling the pores of the film of metal contract. Gilding done in hot solutions is also generally more uniform, and of a deeper or richer colour than that done in cold ones.

Cyanide gilding solution deteriorates by long-continued working, in consequence of silver getting into it from the anodes, and depositing with the gold, gradually increasing its paleness of colour. Gold may be obtained in a pure state from impure anodes, &c. by dissolving the anode in mercury (see p. 358) ; then immersing the amalgam in warm dilute nitric acid ; this will gradually dissolve all mercury, silver, copper, and lead, and leave the gold in a state of fine metallic powder. Before adding the amalgam to the acid, the excess of mercury may be removed by squeezing it through wash-leather, but a little of the gold passes through also.

In all cyanide solutions, and especially in those containing a large excess of cyanide of potassium, and exposed to heat and sunlight, a small portion of the alkaline cyanide is continually being transformed, by contact with the air, into carbonate of potassium and cyanide of ammonium. To remedy this, most electro-gilders add occasionally a small quantity of cyanide of potassium to the liquid, others add a little hydrocyanic acid; the latter is rather the more suitable, because it re-converts the carbonate into cyanide, but it has the disadvantage of containing only about 5 per cent. of the acid, and of being a deadly poisonous liquid, and requiring to be kept in a dark and cool place.

Cyanide gilding liquids are also liable to vary in composition, in consequence of the quantity of gold deposited being sometimes greater and sometimes less than that dissolved; if it is greater, the cathode liberates an excess of free cyanide, and if it is less, the reverse. The gold anode should always appear clean; if it has a crust upon it there is a deficiency of cyanide, but if it is black and has a slimy appearance, and especially if it evolves gas, the solution is deficient of gold; this may be remedied by employing for a time a very large anode and a small receiving surface. By attending properly to the indications, a cyanide gilding solution may be used for any length of time.

Gilding base metals.—Before gilding articles composed of antimony, lead, tin or zinc, Britannia metal, pewter, type metal, &c. it is better to coat them with a film of copper or brass, in a cyanide solution, or to begin the gilding in a hot and nearly exhausted gold bath, and to carefully scratch-brush them.

A weaker solution is recommended for gilding articles of iron or steel than for gilding copper, viz. one consisting of about one measure of the ordinary cyanide gilding liquid, and four or five measures of water containing about 1 per cent. of cyanide of potassium. The solution should only be moderately warm, and a feeble current employed. In gild-ing german-silver or brass articles, also the solution should be weak, and a feeble current employed, because both those alloys are rather strongly electro-positive to gold in a strong cyanide solution, and will therefore coat themselves by simple immersion in it, and if this action is too strong it will prevent the adhesion of the deposited metal.

Previously prepared and cleaned articles of zinc, are first coated with a thin film of copper or brass in a warm cyanide solution, then 'scratch-brushed,' and a thicker coating of copper put upon them in a cold liquid, composed of a mixture of ten volumes of water and one of sulphuric acid, saturated with sulphate of copper, and then two or three volumes more of water added. If the deposit of copper appears patchy, either the original cleaning process or the first coating of copper was imperfect ; and the articles must be thoroughly scratch-brushed, and both the coatings be repeated. The articles should now be 'quicked' by rapidly dipping them into a solution, composed of distilled water 1000 parts, sulphuric acid two parts, and nitrate of mercury one part (see also p. 323), then rinsed, and slightly silvered by dipping into a mixture of water 1000 parts, cyanide of potassium forty parts, and nitrate of silver ten parts. They are then ready for gilding by means of a current, strong at first so as to cover the articles quickly, and then gradually diminished so as to give a good deposit.

Gilding the insides of vessels.—The insides of vessels are gilded, by filling the vessel with the gilding solution, suspend-ing a gold anode in the liquid, and passing the current. The lips of cream-jugs, and the upper parts of vessels of irregular outline, are gilded by passing the current from a gold anode, through a rag wetted with the gilding solution, and laid upon the part.

Ungilding of articles of silver and iron.—Make them the anode, in a solution composed of one part of cyanide of potassium dissolved in ten parts of water.

Recovery of gold from wash-waters.—Wash-water from gilding operations, or from the making of salts of gold, should never be thrown away without previously testing it in a proper manner. The greater the quantity of free acid, alkali, or alkaline salts, in such a liquid, the more capable is it usually of dissolving some of the gold. All such liquids have their gold thrown down from them completely by immersing in them clean plates of zinc, provided they are made slightly acid, and sufficient time is allowed. Water containing free chloride of gold, may be perfectly precipitated by means of a solution of green vitriol.

Recovery of gold from cyanide solutions.[1]—Add an excess of hydrochloric acid, carefully avoiding the poisonous fumes of prussic acid ; heat to boiling; a yellowish-green precipitate forms, but some gold still remains dissolved ; cool the liquid, this separates more of the gold. Decant or filter the clear portion, heat the liquid, and add some filings of zinc; in an hour or two, all the remainder of the gold will be precipitated. Decant the liquid, boil the residue with dilute hydrochloric acid ; wash it and add it to the other portions. Ignite and fuse the mixture in a platinum crucible, with an equal weight of acid sulphate of potassium. Dissolve the saline residue in boiling sulphuric acid, wash it then with water ; perfectly pure gold will remain (R. Huber, 'Chemical News,' vol. viii. p. 31).

'Cyanide of gold and potassium gilding liquid, when mixed with sulphuric, hydrochloric, or nitric acid, slowly deposits cyanide of gold ; and when boiled with hydrochloric acid, it is completely resolved into cyanide of gold and chloride of potassium. Similar effects are produced by sulphuric or nitric acid, and even by oxalic, tartaric, and acetic acids. When heated with sulphuric acid it gives off hydrocyanic acid gas, and, after ignition, leaves a mixture of gold and sulphate of potassium. Iodine sets free cyanogen gas, forms iodide of potassium, and throws down the cyanide of gold.

[1] See also p. 192.

'If we pour hydrochloric acid into a pure solution of gold in cyanide of potassium, there is slowly formed at ordinary temperatures, and immediately on the application of heat, a yellow precipitate, which is cyanide of gold ; the filtered liquid which has given this precipitate still contains a little gold in solution. On evaporating the liquid to dryness, fusing, dissolving, and filtering afresh, there remains upon the filter the remainder of the gold.

'Crystallised double cyanide of gold and potassium fuses and effervesces by heat, and is resolved into cyanogen gas, ammonia, and cyanide of potassium, if air be present ; its complete decomposition requires a strong red heat. When it is strongly ignited, mixed with an equal weight of carbonate of potash, a button of metallic gold is obtained.

'To obtain the remaining gold from gilding solutions which have become inactive, they should be evaporated to dryness, the residue finely powdered, and intimately mixed with an equal weight of litharge, fused at a strong red heat, and the lead extracted from the alloy button of gold and lead by warm nitric acid ; the gold will then remain as a loose, yellowish-brown, spongy mass' (Böttger, 'J. Pr. Chem.' xxxvi. 169 ; Elsner, Redtel Hessenberg, 'J. Pr. Chem.' xxxvii. 477 ; xxxviii. 169, 256).

Recovering gold or silver, by M. Bolley.—'Cyanide of gold, dissolved in an excess of cyanide of potassium, resists all the means which we have tried to separate them ; and hydrosulphuric acid, for example, does not produce a precipitate. By the wet way we cannot always precipitate the gold completely, and for that reason MM. Böttger, Hessenberg, Elsner, and others, propose to evaporate the liquid to dryness ; mix the residue with its own weight of litharge, fuse the mixture at a strong red heat, then dissolve the lead from the alloy by boiling it a long time with dilute nitric acid, which leaves the gold in the form of a light sponge.

'The following process is applicable on the small scale, with a spirit-lamp and a crucible of platinum. Evaporate the solution to dryness, mix the saline mass with its own weight of sal-ammoniac, and heat it gently; ammoniacal salts decompose, as we have said, the metallic cyanides, and form cyanide of ammonium, which is itself decomposed by the heat and volatilised, whilst the acid of the ammoniacal salt (the body which salifies the ammonia) combines with the metals (passed to the state of oxides), which were previously united to the cyanogen. The sal-ammoniac then in this case forms chloride of potassium and chloride of gold, and, if the salt contains ferro-cyanide of potassium, chloride of iron in addition. The chloride of gold is easily decomposed; the chloride of iron is partly decomposed, and leaves oxide of iron in beautiful crystalline spangles. The undecomposed portion of the chloride of iron, like the chloride of potassium, may, after the decomposition is finished (which only requires a low red heat), be washed away by water, leaving the gold in the form of a light coherent mass, and the iron in small spangles, which may be removed by mechanical means.

'If we fear that a little of the gold remains mixed with the iron in a pulverulent state, we may dissolve it in hot aqua regia, and precipitate the gold from the resulting solution by adding to it a solution of protosulphate of iron; but this appears superfluous; and I am assured, by evaporation of given volumes of the same solution of gold, the evaporation and calcination of the sal-ammoniac, and other operations, that we have collected in a sufficiently exact manner all the gold of these solutions.

'The same process is applicable to the solution of silver, and, independently of the oxide of iron (of the ferro-cyanide of potassium), we obtain chloride of silver, which is soluble in aqueous ammonia.'

13. **Silver.**—Elec. chem. eqt.=108. As silver is the most prominent metal in electro-depositing processes, I shall speak of it the most fully. The commonest salts of silver are the chloride, nitrate, oxide, cyanide, and sulphide; the sulphite, hyposulphite, acetate, and other salts have however been

tried for electro-plating purposes, but the one which has best stood the test of experience and time is the double cyanide of silver and potassium.

Most of the salts of silver are formed from the nitrate. The nitrate is produced by adding pure grain silver, in small quantities at a time, to a warm mixture of one measure of distilled water and four measures of the purest and strongest nitric acid. If the liquid is too hot, or too much silver is added at a time, the action will be very strong, and loss of materials, by boiling over of the liquid, may be occasioned ; in such a case add a small quantity of cold distilled water. When the liquid ceases to dissolve more metal, it should be evaporated and crystallised, or else kept in a covered vessel, protected from the light until required to be used.

The oxide is obtained by adding a solution of either caustic potash, caustic soda, or clear lime-water, to one of nitrate of silver as long as it produces a precipitate, filtering and washing the oxide, which is a brown powder. The wash-water contains a little oxide of silver in a dissolved state. The chloride may be obtained by adding dilute hydrochloric acid or a solution of common salt, in slight excess, to one of the nitrate in a similar manner, filtering and washing the white flocculent precipitate, which must be kept in the dark. In this case the wash-waters do not contain silver, and may therefore be thrown away.

Fluoride of silver may be obtained by saturating pure dilute hydrofluoric acid in a platinum vessel with argentic oxide or carbonate, and evaporating the clear solution to perfect dryness. It is extremely soluble in water and easily fusible. The electrical relations of substances in the fused salt I found to be as follows, the first-named substance being the most positive—silver, platinum, charcoal of lignum-vitæ, palladium, gold ; and in a dilute aqueous solution of the salt—aluminium, magnesium, silicon, iridium, rhodium, and carbon of lignum-vitæ, platinum, silver, palladium, tellurium, gold ('Chemical News,' vol. xxi p. 28).

Nearly all the salts of silver (except the sulphide) become converted into the cyanide by immersion in a solution of cyanide of potassium, and therefore the nitrate and chloride are sometimes used instead of the cyanide, to form with cyanide of potassium the plating liquid. But this is a bad mode of procedure, because the same amount of materials is used, and nitrate or chloride of potassium are thereby introduced into the plating liquid, and these substances are objectionable.

Cyanide of silver is generally prepared by adding a solution of cyanide of potassium to one of nitrate of silver just as long as a precipitate occurs ; the white precipitate, which is cyanide of silver, is insoluble in water, and is not perceptibly soluble in commercial hydrocyanic acid ; it dissolves very freely in a solution of cyanide of ammonium, potassium, or sodium, and in hyposulphite of soda solution ; it is also soluble in solutions of ammonia, carbonate of ammonia, sal-ammoniac, nitrate of ammonia, and ferro-cyanide of potassium.

Chemical characters of cyanide of silver.—According to Messrs. Glassford and Napier, cyanogen in the presence of cyanide of potassium, possesses a greater affinity than any other substance for silver; decomposing every salt of that metal except the sulphide, and forming cyanide of silver.

In dissolving the oxide, carbonate, chloride, or ferrocyanide of silver, in a solution of cyanide of potassium, they are all decomposed, and cyanide of silver is always formed. Cyanide of silver should be dried below 260° Fahr. Hydrochloric acid decomposes it with evolution of hydrocyanic acid gas; cold nitric acid has no action upon it; a boiling mixture of sulphuric acid and water decomposes it, with escape of hydrocyanic acid gas, and formation of sulphate of silver ; it is soluble in solutions of the alkaline chlorides, but its best solvent is an aqueous solution of cyanide of potassium, of which salt it requires one equivalent (sixty-five parts), to dissolve one equivalent (134 parts). The resulting solution, when evaporated, yields crystals of double cyanide of silver and potassium, which are soluble in eight parts of cold and in one part of boiling water. The solution of this double salt, which is nearly the same as the ordinary plating solution, may be boiled for any length of time without being decomposed, and it is very little affected by light; it is decomposed by all acids, and they precipitate the silver as cyanide of silver; the hydro-acids—hydrochloric acid for example—decompose the cyanide of silver also; sulphuretted hydrogen precipitates the silver as sulphide of silver ('Philosophical Magazine,' vol. xxv., 1844, pp. 56-71). According to Baup, the double cyanide of silver and potassium (K Cy, Ag Cy) requires for solution four parts of water, or twenty-five parts of 85 per cent. alcohol, each at 20° C. It does not become discoloured by exposure to sunshine, nor make stains upon paper nor on the skin. The double cyanide of silver and sodium (Na Cy, Ag Cy) dissolves in five parts of water, or twenty-four parts of 85 per cent. alcohol, at 20° C.

Electrolysis of salts of silver.—A highly-concentrated solution of argentic nitrate, made as neutral as possible, is easily decomposed, and yields coherent metal by a sufficiently feeble current. An anode of silver is, however, indispensable (Becquerel, 'Chemical News,' vol. vi. p. 126). I have electrolysed a solution of argentic nitrate, and obtained a thick, black, insoluble crust upon the silver anode; the crust was probably a peroxide of silver, but it also contained a compound of nitrogen. Fused argentic nitrate yields silver at the cathode and oxygen gas at the anode. Fused chloride of silver is resolved into silver and chlorine.

The sulphate of silver is too sparingly soluble, and too bad a conductor to be of much value in electro-plating. The hyposulphite is much more soluble, and has been used for practical purposes, but abandoned. Solutions of nitrate, chloride, and carbonate of silver in excess of aqueous ammonia have also been tried. They are good conductors, but contain argentate of ammonia or fulminate of silver, which is an extremely dangerous substance when in a dry state, and detonates with fearful violence by the slightest friction or percussion. The iodide of silver and potassium, the acetate of silver, the sulphocyanide of silver and potassium, and the potassio-tartrate of silver, have all been tried for depositing purposes, but do not appear to equal the double cyanide of silver and potassium ; in the electrolysis of all of them there is occasionally observed a black crust, probably containing some peroxide of silver, upon the silver anode.

I have electrolysed artificially-chilled anhydrous hydrofluoric acid by means of a pure silver anode, and a current from ten Smee's cells. The current was conducted more freely than with an anode of palladium, and still more so than with one of gold ; the anode corroded rapidly, and became covered first with some black powder upon its edges, and then with a grey powder (probably metallic silver) which contained only a trace of soluble silver salt. I also electrolysed a saturated neutral aqueous solution of argentic fluoride with a current from six Grove's cells, a small platinum anode and a large platinum cathode. Free conduction occurred, no gas or odour was evolved, a thick, hard, and strongly adherent, black crust quickly formed upon the anode, and a rapid deposit of yellowish scales of silver upon the cathode ; the crystals soon extended upwards and united the electrodes. From the behaviour of the black crust with strong nitric, hydrochloric, and sulphuric acids, and also with aqueous ammonia ; and from its evolving, when heated to redness, a gas which re-inflamed a red-hot splint explosively, and losing nearly the theoretical proportion by weight in the process, and leaving metallic silver, I concluded the crust to be peroxide of silver ($Ag_2 O_2$). By the electrolysis of a more dilute solution a similar crust was formed, but gas was also evolved at the anode.

I also made a number of experiments of electrolysing

argentic fluoride in a fused state with platinum electrodes, in a covered platinum cup, with a current from six Smee's elements. In each case conduction commenced before the salt had fused; and when the salt was liquid the conduction was as perfect as if the electrodes were united by a metal wire. No signs of genuine electrolysis could be detected in either instance. I also electrolysed the fused salt with a rod of highly-ignited charcoal of lignum-vitæ, and a current from ten Smee's cells, but only a small amount of conduction occurred in consequence of the resistance of the carbon. The anode was corroded and evolved gas (See ' Phil. Trans. Royal Society,' 1870, p. 234; ' Chemical News,' vol. xxi. p. 28).

According to F. Wöhler, if a current from two Bunsen's cells is passed through a dilute solution of sulphuric acid or sodic sulphate by means of a silver anode, the anode becomes covered with black amorphous argentic peroxide, and this oxidation is due to ozone. With a solution of potassic nitrate similarly treated, flocculent light-brown argentic oxide is formed. In one of potassic ferro-cyanide, the anode acquires a film of white amorphous argentic ferro-cyanide. And in a solution of potassic dichromate it becomes covered with a reddish black film of crystallised argentic chromate ('Chemical News,' vol. xviii. p. 189).

The electrolysis of melted caustic soda with an anode of silver causes the silver to dissolve and be deposited upon the platinum cathode; but, on cleaning the cathode with nitric acid, a black powder of platinum is left (Brester, ' Chemical News,' vol. xviii. p. 145). I have repeatedly met with a similar effect with fluoride of silver melted in vessels of platinum.

Deposition of silver by simple immersion (see pp. 77, 78). —Aluminium throws down the silver from acid or neutral solutions of argentic nitrate; slowly in dilute solutions; also immediately from solutions of chloride or chromate of silver in aqueous ammonia. It rapidly decomposes chloride of silver in a state of fusion, evolving great heat (A. Cossa, Watts' ' Dictionary of Chemistry,' 2nd Supplement, p. 54).

A solution of sulphate of silver is not reduced, but one of argentic nitrate deposits its silver by contact with hydrogen, obtained by a variety of methods (Brester, ' Chemical News,' vol. xviii. p. 144). According to J. Spiller, electro deposited nickel does not deposit silver from a solution of argentic nitrate ('Chemical News,' vol. xxiv. p. 175).

I have observed that hydrogen deposits silver from semi-fused argentic fluoride ('Chemical News,' vol. xxi. p. 28) ; also that carbon and crystalline boron do not separate silver from that salt at a red heat, nor does boron separate it from an aqueous solution of that salt ; that crystals of silicon thrown upon melted argentic fluoride become red-hot, undergo rapid combustion, forming fluoride of silicon and depositing silver ; that those crystals also deposit crystals of silver slowly from an aqueous solution of that salt ; and from a mixture of solution of argentic fluoride, hydrofluoric acid, and nitric acid, they liberate bubbles of spontaneously inflammable siliciuretted hydrogen gas ('Chemical News,' vol. xxiv. p. 291). A fragment of stannous fluoride also deposited metallic silver from an aqueous solution of argentic fluoride in a platinum dish.

According to Raoult, gold in contact with silver in a cold or hot acid or neutral solution of a salt of silver receives no deposit of that metal (' Journal of Chemical Society,' vol. ii. p. 465).

Silvering by simple immersion.—This process is chiefly applicable to small articles, such as pins, buttons, buckles, coffin nails, hooks and eyes, &c. where only a very thin coating of silver is required. The following solutions, in the proportions indicated, are used by adding a small quantity of water sufficient to form the ingredients into a pasty liquid of the consistence of cream, stirring the articles thoroughly about in it, or rubbing them over with it, until they have acquired the desired degree of whiteness.

1st. Take equal parts of chloride of silver and bitartrate

of potash. 2nd. Take chloride of silver one part, alum two parts, common salt eight parts, and cream of tartar eight parts. 3rd. Take chloride of silver one part, prepared chalk one part, common salt one and a quarter part, and pearlash three parts. 4th. A 'novargent' solution for resilvering old plated goods consists of 100 parts of hyposulphite of soda, and chloride or any other salt of silver, fifteen parts. Compounds of this description are also used for silvering clock faces, thermometer and barometer plates, and many other articles of copper and brass. Or one part of chloride of silver mixed with eighty or 100 parts of cream of tartar (bi-tartrate of potassium), to which may be added or not about eighty or 100 parts of common salt, is dissolved in boiling water, and the articles contained in a basket are immersed, and stirred about in the hot liquid. An old solution is of a green colour from the presence of dissolved copper, and works better than a new one. If there is the least particle of metallic zinc or iron present, it causes a red deposit of copper upon the articles. Another solution employed for a similar purpose is composed of 1000 parts of water, sixty parts of cyanide of potassium, and ten parts of nitrate of silver ; it is used at a boiling temperature. The action of this bath depends upon the fact that copper is more electro-positive to silver in proportion to the excess of free cyanide of potassium present.

The following solution of the double sulphite of silver and sodium, according to Roseleur, may be used cold. It is prepared as follows : Dissolve four parts of crystals of washing soda in five parts of distilled water. Pass sulphurous anhydride gas (SO_2) through the liquid by bubbling it through mercury at the bottom of the vessel to prevent the exit tube becoming clogged with crystals, until the liquid redissolves the crystals of bicarbonate and slightly reddens blue litmus paper. Set it aside for twenty-four hours to allow some bisulphite of sodium to crystallise out. Take, the liquid part only, stir it well to remove carbonic acid.

Now add, if it is alkaline, some more sulphurous gas, or if it is acid, a little more carbonate, until the liquid after stirring renders blue litmus paper violet or slightly red. To the clear liquid add, with stirring, a solution of argentic nitrate, until the precipitate produced begins to be slow in dissolving ; the solution is then ready.

This liquid is said to be ' always ready to work,'and produces quite instantaneously a magnificent silvering upon copper, bronze, or brass articles, which have been thoroughly cleansed, and passed through a weak solution of nitrate of binoxide of mercury, although this last operation is not absolutely necessary.' The bath is renewed by addition of nitrate of silver, and sooner or later also some bisulphite of soda. Some of the silver of the solution deposits itself upon the sides of the vessel. Roseleur states that he has used this bath for five consecutive years, and has daily silvered in it ' as many articles as a man could conveniently carry, and at prices varying ten cents to two dollars per kilogramme,' and he ' does not doubt that it would eventually replace all the other known methods. He further states that the deposit, without the aid of electricity, may become nearly as thick as desired, and in direct ratio to the length of the immersion.'

This particular process differs from nearly all other ones of metallic deposition upon metals, in being only in part an electrical action. When a metallic article is immersed in a bath of another metal, in which it coats itself, a portion of the immersed metal dissolves and generates an electric current, which decomposes the liquid, and deposits an equivalent of the other metal as a coating upon the immersed article ; this deposited film arrests the action, and prevents a thick coating being formed. But in this particular liquid a spontaneous *chemical* change also takes place ; the sulphurous anhydride of the sulphite of silver takes oxygen to itself, to form sulphuric anhydride, and sets the silver free, and this silver adheres to the articles, and to the vessel containing the liquid. The sulphuric anhydride unites with some of the soda of the undecomposed portion of the sulphite and liberates sulphurous anhydride, and forms sulphate and bisulphite of sodium. The process is partly like that of coating looking-glasses with pure silver.

Silvering by contact with zinc (see p. 82).—Mr. Joseph Steele took out a patent, dated August 9th, 1850, for silvering articles by immersing them in a silver solution in contact with a piece of zinc of proper size. The process is as follows : Dissolve four ounces of pure silver in twenty ounces of nitric acid ; also dissolve separately one and a half pound of common salt in one and a half gallon of water ; mix the two solutions together ; allow the mixture to remain until clear, pour away the clear liquid, and wash the precipitate, which is chloride of silver ; next fuse together twenty-four ounces of ferro-cyanide of potassium and twelve ounces of carbonate of potash, and, when the mass is cold, add it, together with the chloride of silver, to one gallon and a half of water, boil the mixture and filter it ; it is then ready for use.

Solutions for plating by separate current (see p. 86).— These may be made either by chemical means or by the aid of an electric current. The best solution for general purposes is the double cyanide of silver and potassium, and when required in large quantities it is usually made by chemical means. Many solutions have been proposed and tried for depositing silver by the battery process, but none have stood the test of time and experience like the one composed of double cyanide of silver and potassium dissolved in water, and a little free cyanide of potassium added. It must, however, always be remembered, when making cyanide depositing solutions with the aid of potassic cyanide, that the composition of this salt as usually sold is extremely variable, and unless the depositor is aware of this he may be led quite astray in his calculations, and unwittingly introduce various impurities into his depositing liquids.

Ordinary cyanide plating solution may be made of various strengths, from half an ounce of silver to the gallon of water, to two, four, six, or more ounces, and from an ounce of cyanide of potassium to several pounds per gallon, and still be effective in working. One part of silver in 100 parts of solution is enough for ordinary purposes, but a good proportion is two or two and a half in 100 ; some platers use as much as ten in 100. The following proportions were employed by M. Roulz, viz. a solution of one part by weight of cyanide of silver and ten parts of cyanide of potassium, in 100 parts of water ; the mixture being diluted with water to the desired strength.

Making cyanide of silver plating solution by chemical means. —Take four parts of grain silver, add it, in small portions at a time, to a warm mixture of about six and a half parts by weight of pure and strong nitric acid, and one part of water, contained in a capacious glass or stoneware vessel. Gas will be evolved from the surfaces of the pieces of silver, and reddish-brown fumes of nitrous anhydride will arise from the mixture, and should be conveyed out of the apartment by means of a chimney. The action should be maintained moderate and uniform, and, if it should become too strong, a little cold distilled water should be added, and the mixture kept more cool ; when the whole of the metal is dissolved, evaporate the solution nearly to dryness, which will drive off any excess of acid that may be present ; the resulting salt, nitrate of silver, may then be dissolved in a large quantity of distilled water, in the proportion of half a gallon (more or less) to each ounce of the silver used. At the same time a solution should be made of from two to three parts (according to its quality) of cyanide of potassium in twenty or thirty parts of distilled water, which is to be added gradually to the solution of nitrate of silver as long as it produces a precipitate ; if too much be added, it will cause some of the precipitate to redissolve and be wasted ; it will also make the liquid appear clear and slightly brown where it passes ; in such a case the liquid should be stirred, then allowed to settle clear, and a small quantity of nitrate of silver dissolved in distilled water should be added as long as it produces a white cloud. By conducting the operation in a glass vessel, adding the liquid towards the latter period in small quantities at a time, and at intervals of a few minutes each, with gentle stirring immediately upon each addition, carefully observing when it ceases to produce a precipitate, the point of neutralisation may be very accurately arrived at. The liquid must now be allowed to remain undisturbed until quite clear, the clear portion poured steadily away from the precipitate of cyanide of silver, and the precipitate washed five or six times in a large quantity of water, by simply adding the water briskly to it, allowing it to settle, and then pouring away the clear portion. Next dissolve from six to eight parts (according to its quality) of cyanide of potassium in twenty parts of distilled water, adding it in portions at a time to the wet cyanide of silver, with free stirring, until barely the whole is dissolved; then add about three parts more of cyanide of potassium to form free cyanide, and sufficient distilled water to reduce the whole to the proportion of about one ounce of silver to the gallon ; finally, when all the free cyanide is dissolved, filter the solution through a piece of unglazed calico. On the small scale, distilled water is used in all the various parts of the process, except the washing; but, on the large scale, clean rain-water may be used in all the operations.

If either the nitric acid, or the water in which the nitrate of silver is dissolved, contains chlorides, a white residue of chloride of silver will be left on dissolving the nitrate ; and if a small amount of brown matter is left on dissolving the cyanide of silver in the cyanide of potassium solution, it arises from decomposition of some of the latter salt. The insoluble matters and the wash-waters should be reserved for the purpose of recovering from them any traces of silver they may contain. The numbers given above are based upon the assumption that the cyanide of potassium employed is of an average quality, and contains about 50 per cent. of the actual substance ; in proportion as the percentage is higher, so will the quantity required be less.

The following solution is said to be an excellent one ; water 1000 parts, pure cyanide of potassium fifty parts, silver, for converting into cyanide, twenty-five parts. The silver is converted into nitrate. The nitrate is dissolved in distilled water, and prussic acid added until it produces no more precipitate. The cyanide of silver is filtered and washed, and then added to the water and cyanide of potassium. This method is only suitable for making small quantities of solution, because the ordinary prussic acid is so dilute.

' Some operators have employed the plan of forming cyanide of silver by generating hydrocyanic acid gas, by heating a mixture of dilute sulphuric acid and coarsely-powdered ferro-cyanide of potassium in a glass flask, and passing the evolved hydrocyanic acid (prussic acid) vapour through a solution of argentic nitrate as long as a precipitate is formed, and this yields a slightly purer cyanide of silver than that prepared by precipitation, but it is a slow and not an economical process, and the prussic acid vapour is highly dangerous to inhale.

The cyanide of silver plating solution may be made by other modifications of the chemical method than the one described ; for instance, some depositors make the solutions by adding oxide, carbonate, or even chloride of silver to a solution of cyanide of potassium, as long as it will dissolve, and then adding an amount of free cyanide ; by this process the depositor is enabled to use caustic potash, carbonate of potash, hydrochloric acid, or common salt, instead of cyanide of potassium, for precipitating the nitrate of silver ; nevertheless it still requires two equivalents of cyanide of potassium to be used as before, viz. one to convert the salt of silver into cyanide, and the other to dissolve the cyanide of silver formed, because, in all such cases, according to the researches of Messrs. Glassford & Napier (' Philosophical Magazine,' 1844), when any salt of silver is added to a solution of cyanide of potassium, it is first converted into cyanide of silver at the expense of one portion of the cyanide of potassium, it then combines with the remaining cyanide to form double cyanide of silver and potassium, which dissolves in the water, therefore by this modification of the chemical method no cyanide of potassium is saved, and the carbonate of potash, hydrochloric acid, &c., are wasted. This modification has a still greater disadvantage; it introduces substances into the depositing liquid which are injurious. A good depositing solution should dissolve the anode freely, hold abundance of metal in solution, and not act chemically upon base metals, because it is such metals we generally wish to coat; now, if instead of cyanide of silver we add oxide of silver to the cyanide of potassium liquid, it converts part of the cyanide into caustic potash; if we add carbonate of silver, it converts it into carbonate of potash; and if chloride of silver, it converts it into. chloride of potassium; and each of these substances, especially the last, diminishes the action of the liquid upon the dissolving plate, decreases its solvent power for cyanide of silver, makes its particles less mobile, and causes it to act in some degree upon base metals, and thus endangers the adhesion of the deposits upon them. Some electro-platers think the presence of these salts not injurious, but most consider them highly detrimental. One hundred ounces of silver, converted into chloride, and dissolved in cyanide of potassium solution, produces sixty-nine ounces of chloride of potassium as an impurity in the liquid; or if converted into nitrate, and so dissolved, produces ninety-three and a half ounces of nitrate of potassium as impurity.

A good plating liquid should contain one equivalent (sixty-five parts) of pure cyanide of potassium, and one equivalent (134 parts) of cyanide of silver, besides about 20 to 50 per cent. of free cyanide, and sufficient water to form a thin liquid. It is necessary to have free cyanide, because in working the solution insoluble cyanide of silver is formed at

the anode, and requires free cyanide of potassium to combine with it and form the soluble double cyanide; at the
same time cyanogen and cyanide of potassium are set free
at the cathode, or receiving surface, by the deposition of the
silver; and as it requires some time for those substances to
mix with the liquid, and reach the dissolving plate, sufficient
free cyanide must be provided; the necessity of having
sufficient water to form a thin liquid arises from the double
cyanide formed at the dissolving plate being specifically
heavier than the solution, and thus having a tendency to
sink to the bottom, whilst the hydrocyanic acid and cyanide
of potassium, set free at the surface of the articles, being
specifically lighter, tend to rise to the surface; at the same
time each of them mixes more or less with the surrounding
liquid by capillary attraction or adhesion, and the more
dilute the liquid is, the more mobile are its particles, and
the more rapidly does this mixture take place. This
explains why strong silver solutions require more frequent
stirring than weak ones to keep them uniform. In some
manufactories, where they have steam power at command,
the articles are kept in constant motion by machinery
swinging them gently to and fro, but in small electro-plating
establishments the silver solutions are stirred every evening
instead.

Many electro-platers use a cyanide solution containing
about half an ounce of silver to the gallon, and add a very
large proportion of free cyanide to make it conduct freely;
such a solution has the advantage of being comparatively
inexpensive in its first formation, quick in working, and
yields metal of an average character; but it is rather
difficult to manage in hot weather, and dissolves the anode
very rapidly, on account of the large proportion of free
cyanide. In practice, the amount of silver to the gallon
varies from half an ounce to about four ounces, but ordinary
solutions contain about one or two ounces to the gallon;
the amount of free cyanide of potassium also varies from

about half the weight of the silver dissolved in the liquid, to five or ten times this quantity. A very good proportion, is about three-fourths of the weight of the dissolved silver, but there is no rule generally recognised in the trade upon this point ; some manufacturers use a very large, and others a very small proportion.

Mr. Alexander Parkes took out a patent, dated March 29th, 1871, for improvements in the solid deposition of silver. He converts an ounce of silver into oxide, by first dissolving it in nitric acid, and then precipitating it by caustic potash ; he then dissolves the oxide, together with sixteen ounces of cyanide of potassium, in two gallons of water, and uses the resulting liquid for depositing solid articles.

Solid articles of silver are occasionally made, by first forming a thin mould in copper by the electrotype process, and then depositing silver upon this mould in a cyanide solution (containing about eight ounces of silver per gallon), until the deposit is sufficiently thick. The article is then immersed in a boiling solution of dilute hydrochloric acid, or in a hot one of perchloride of iron, until all the copper is dissolved.

Mr. Edmund Tuck took out a patent, June 4, 1842, for 'improvements in depositing silver upon german-silver.' For plating the commoner quality of that alloy, he uses a liquid composed of sulphate of silver, dissolved in a solution of carbonate of ammonium, and for the best quality, he uses cyanide of silver dissolved in a similar liquid. The mixtures are formed, by dissolving seventy parts of the carbonate in distilled water, then adding 156 parts of sulphate of silver, or 134 of cyanide of silver, and boiling the liquid until the silver salt is dissolved. For coating common german-silver, he adds half an ounce of sulphate of silver, to a solution of 107 grains of bicarbonate of ammonium.

One plater recommends the use of two liquids, the first to ' whiten,' and the second to 'finish.' The whitening

M

one is composed of one gallon of distilled water, two and a half troy pounds of cyanide of potassium, eight ounces of carbonate of soda, and five of cyanide of silver; and the finishing solution, of one gallon of distilled water, four and a half troy ounces of cyanide of potassium, and one and a half of cyanide of silver; using a series of from three to ten Smee's cells with the first solution, and one large cell only for the second, and keeping the anode and articles in the second solution, as closely together as possible. By these means, the silver may be made to adhere firmly to all kinds of brass, bronze, type-metal, &c., without the use of mercury.

Copper, brass, and german-silver, are the best substances to deposit silver upon ; lead is a very bad metal for the purpose, because it is so soft. Articles formed of zinc, or iron, are usually coated with a film of copper, in a cyanide solution, before putting them into the plating liquid. Those formed of Britannia-metal, tin, or pewter, are not dipped into acid before plating, but into a strong and boiling-hot solution of pure caustic potash, and are then either 'scratch-brushed' or taken direct from the alkali, without rinsing in water, and immersed in a cyanide of silver solution (at about 190° Fahr.), containing a considerable proportion of free cyanide, with a large anode; and an electric current of considerable intensity, is passed through the vat for several minutes, until the articles receive a thin coating, they are then transferred to the ordinary plating solution, to receive the full amount of deposit : steel articles, after being cleaned in the hot potash, are dipped (without brushing) into a solution of one pound of cyanide of potassium in a gallon of water ; and then coated thinly with silver in a similar manner, before plating. Those of lead are first scraped, or otherwise made quite clean and bright, by mechanical means, and then treated in the same manner as those of Britannia-metal. Articles of copper, brass, or german-silver, after being properly cleansed, are dipped into the solution of nitrate of mercury (see p. 166), or a very dilute one of cyanide of mercury and potassium (pp. 95 and 323), then rinsed in a vessel of water, and immediately suspended in the depositing vat. The preparation of articles by immersion in a bath of cyanide of mercury was patented by Dr. H. B. Leeson, June 4, 1842, and is in use by the electro-platers of Birmingham. If the articles are immersed without this precaution, the deposited silver does not always adhere firmly. All articles are attached to the cathode, immediately *after* immersion in the plating liquid.

For preparing articles of tin, lead, zinc, Britannia-metal, &c., a cold solution is sometimes employed, containing from two to three pounds of cyanide of potassium, and only two to five pennyweights of silver per gallon ; and the current from the usual battery is passed into it by means of a small anode. But, for coating steel direct with a preparatory film of silver in this solution, a powerful battery is used, so as to evolve hydrogen from the steel surface, and the anode is composed of a large sheet of platinum together with a small sheet of silver.

For silvering cast-iron, Böttger recommends a bath composed of fifteen parts of argentic nitrate dissolved in 250 parts of water, to which thirty parts of cyanide of potassium have been added. After complete solution, pour the mixture into 750 parts of water containing fifteen of common salt. The cast-iron articles, after being well cleansed, should be placed for a few minutes in nitric acid of sp. gr. 1·2, then rinsed thoroughly, and placed in the electro-depositing liquid.

Brass, copper, or nickel articles, also those of iron and zinc which have been coppered, may after thorough cleansing, be silvered, by treatment with a solution of fourteen grammes of silver dissolved in twenty-six grammes of nitric acid, to which (after the silver is all dissolved) is added a solution of 120 grammes of cyanide of potassium in one litre of water, and also twenty-eight grammes of finely powdered chalk.

The original paper gives further directions for producing dif-ferent shades of colour on the silvered articles (C. Paul, ' Journal of Chemical Society,' vol. xi. p. 955).

To electro-plate over soft solder.—Clean the articles well with alkali, then dip them in red nitrous acid, and thoroughly rinse away all the traces of acid. Dip the soldered portion for a short time, in a very dilute solution of cyanide of mercury and potassium (see p. 321). Rinse them again, and then place them in the plating vat.

In addition to cyanide mixtures, other solutions have been employed for electro-plating by the separate current process; amongst which the most practical ones are those containing the sulphite, or the hyposulphite of silver. To form the first of these, the patentee (Mr. Woolrich) directs as follows :—' Take of the best pearlash of commerce, twenty-eight pounds, and add to it thirty pounds of water, and boil them in an iron vessel until the pearlash is dissolved ; the solution should then be poured into an earthenware or other suitable vessel, and suffered to stand until the liquor becomes cold. It should then be filtered, and fourteen pounds of distilled water added ; sulphurous acid gas (obtained by any of the known processes) should then be passed into the filtered liquor until it is saturated, taking care not to add the gas in excess. The liquor should be again filtered, and the liquid so filtered, is what I term the solvent, or sulphite of potash.'

' To make the silvering liquor, which I use in coating with silver the surface of articles formed of metal or metallic alloys : I dissolve twelve ounces of crystallised nitrate of silver in three pounds of distilled water, in a clean earthenware vessel, and add to the solution, by a little at a time, the before mentioned solvent, so long as a whitish-coloured deposit is produced ; care being taken not to add more of the solvent than is necessary. After the precipitate has subsided, I pour off the supernatant liquor, and wash the sediment with distilled water. To the precipitate I add as much of the before mentioned solvent as will dissolve it, and afterwards add about one-sixth part more, so that the solvent may be in excess ; I then stir them well together, and let them remain about twenty-four hours, and then filter the solution, when it will be ready for use. This is what I designate silvering liquor ' ('Repertory of Patent Inventions,' 5th series, 1843, p. 210). This liquid is a very good one, except that it gradually decomposes and deposits its silver, by the influence of light.

The simplest way of forming the hyposulphite plating liquid, is by dissolving chloride of silver in a solution of crystals of hyposulphite of sodium. The liquid yields its metal easily by means of the electric current, but under the influence of light it is decomposed, and its silver falls to the bottom in the form of sulphide.

Mr. Alexander Parkes also took out a patent, October 29, 1844, for depositing silver by means of a battery and a silver anode, from melted chloride or iodide of silver, with or without the addition of from half to one and a half times its weight of iodide of potassium, to increase the bulk of the liquid ; and for gilding in a fused mixture of two parts of iodide of gold and eight parts of iodide of potassium ; but neither process appears to have been much used.

Mixtures composed of cyanide of silver dissolved in solutions of ferro-cyanide of potassium, in the proportion of one ounce of silver, to three pounds of the cyanide, have also been employed for plating, and yield with a feeble current, an excellent deposit of soft silver, but the silver anode becomes covered with an insoluble white crust, which falls off, and the solution soon becomes exhausted of metal.

Making silver plating liquid by battery process.—The ordinary cyanide of silver plating solution may very conveniently be made by the battery, process, and by some platers this plan is preferred to all others. To make by this method, a solution containing one ounce of silver per gallon—first ascertain the percentage of actual cyanide in the salt of potassium to be used. If it contains about 50 per cent., dissolve about three ounces of it in each gallon (=160 ounces) of distilled water ; or if it contains more, add less, and if less, add more, in proportion. Suspend a large anode and a small cathode of silver, in the liquid; and pass a strong current of electricity, until about one ounce of silver for each gallon of liquid has dissolved from the anode, or until with a moderate current, and electrodes of average size, a bright silver or other suitable cathode, receives a good deposit. As this process produces some caustic potash in the liquid, some of the strongest hydrocyanic acid may now be added to form cyanide, and some more silver then dissolved in the mixture by the battery process.

Condensed outline of the silver plating process.—(According to the French method.) Immerse the articles of copper, brass, or german-silver, during a few minutes, in a boiling solution of one part of caustic potash in ten parts of water. Swill them thoroughly in clean water. Dip them into a liquid, composed of ten parts of water, and one of sulphuric acid. Rinse them again. Immerse them during a few seconds in a mixture of twenty parts of common salt, twenty of calcined soot, and 1000 of yellow nitric acid of specific gravity 1·332, and swill them as quickly as possible in plenty of water. Dip them also rapidly in a mixture (prepared some time beforehand) of one part of sulphuric acid, sp. gr. 1·846, forty of common salt, and 1000 of yellow nitric acid, of specific gravity 1·332 ; and instantly wash them well in clean water. Dip them at once for a few seconds, or until they are quite white, in a 'quicking' solution, composed of ten parts of nitrate of binoxide of mercury, and 1000 parts of water containing sufficient sulphuric acid to make the solution clear; and swill them again in the fresh water. Immerse them in the plating liquid, using a weak current, and if the deposit looks good, continue the process, but if it looks uneven or spotted, take them out, 'scratch-brush' them, dip them into a hot solution of cyanide of potassium, and then in fresh water; 'quick' them afresh, rinse them again, and then continue the process. When the plating is finished, stop the current a few minutes, before removing the articles, in order to remove subsalts of silver from the deposit, and prevent its turning yellow. Swill them in water, then in water slightly acidified by sulphuric acid, again finally in water, scratch-brush them if necessary, swill them again, dry them in hot sawdust of boxwood, and weigh them.

Bright silver deposition.—The history of the discovery of this kind of plating has already been given (see pp. 26, 27). The brightening effect is produced by attending to the following directions. Take one quart of ordinary cyanide of silver plating liquid, *old* liquid by preference, containing two pounds of cyanide of potassium per gallon, add to it four ounces of strong liquor ammonia, four of bisulphide of carbon, and two of ether, and shake it occasionally. After it has stood twenty-four hours, add two ounces of the supernatant liquid, to 20 gallons of ordinary silver plating solution, with gentle stirring, every alternate day. Or add it every day, during the morning in summer, and evening in winter. But in every case it is highly important, to add only *the least possible quantity necessary to produce the effect;* for a greater number of silver plating solutions have been spoiled, by addition of an excess of brightening liquid, than by all other causes put together. If too much 'bright' is added, the plated articles become of a brown colour, and frequently spotted. The bisulphide of carbon mixture gradually becomes nearly black ; it may stand an indefinite length of time, and as often as two ounces of the supernatant liquid is taken out, an equal volume of old plating solution strong in cyanide, or a strong solution of cyanide alone, should be added ; this gradually decreases its blackness, and also prevents its producing a precipitate when added to the solution in the vat.

Messrs. Lyons and W. Milward, the patentees of the pro-

cess, give in their patent of March 23, 1847, the following instructions for forming a 'bright solution.'   'Add to the usual
solution of silver in cyanide of potassium, bisulphide of
carbon, terchloride or other chloride of carbon, sesqui-
chloride of sulphur, or hyposulphite of either potash or soda.
The bisulphide of carbon may be used alone, or dissolved in
sulphuric ether ; or it may be used in conjunction with any
of the other substances mentioned above.   But the patentees
prefer using it as follows :—Six ounces of bisulphide of
carbon are put into a stoppered bottle, and one gallon of the
usual plating liquid added to it ; the mixture is then shaken
and set aside for twenty-four hours ; two ounces of the resulting solution are then added to every twenty gallons of the
ordinary plating solution in the vat, and the whole stirred
together ; this proportion must be added every day, on
account of the loss by evaporation ; but when the mixture
has been made several days, less than this may be used
at a time ; (when hydrocarbons are used instead of the
bisulphide, a much larger quantity must be added.)
This proportion gives a bright deposit, but by adding a
larger amount, a dead surface may be obtained very different to the ordinary dead surface.'   This substance is
generally employed throughout the trade.   Other compounds have also been used, but to a very limited extent ;
among these are sulphur and collodion.   A solution of
iodine and gutta-percha in chloroform is said to be more permanent in its effect than the bisulphide of carbon.   Also one
and a half ounces each, of the carbonate and acid carbonate
of potassium, added once in nine or ten days, to a plating
liquid, containing twelve ounces of cyanide of potassium,
and three and a half of silver per gallon, is said to produce
the same effect.   But these do not equal the bisulphide.
(M. Planté silvers brightly, by adding a little sulphide of
silver to the bath.)   The brightening liquid is added to the
ordinary silver cyanide plating solution, and the proportion
either of silver or of free cyanide, per gallon, is not a matter

of much importance. The liquid in a brightening vat is always slightly cloudy.

Articles in the brightening solution become plated more slowly than in the ordinary plating liquid. They commence to become bright first at their lower parts, and become wholly bright in about fifteen minutes. The 'bright' vat is only used to 'finish' articles in, because its only service is to impart a superficial appearance. The electric current required, is stronger than that for ordinary silver plating. Silver deposited from a brightening solution is not however pure silver. I have found sulphur in it, by dissolving it in pure dilute nitric acid, determining the amount of silver, and testing for sulphuric acid, in two separate portions. Bright silver is also harder than that deposited from the ordinary cyanide solutions, and has very much the appearance of fused metal. If there are very small holes in the surface of the bright articles, dull streaks appear above them. Silver deposited in bright vats, blackens quickly on removal from the liquid, unless immersed in boiling water for a short time.

Vats for containing silver solutions.—These are of various dimensions and proportions, but usually they are about six feet long, three feet wide, and nearly three feet deep ; and they often contain 200 or 300 gallons of the liquid. They are made of different materials ; some are composed of wood only, others of two thicknesses of wood with lead between ; but the use of wooden vats is nearly discontinued, because they absorb a large quantity of the solution, become saturated with it, and it soaks through to the outside. A lining of gutta-percha cannot be employed, because cyanide of potassium acts upon the joints of that substance. They are now made of wrought iron, sometimes with a thin layer of wood as a lining upon the sides to prevent the anodes touching them, or they are lined entirely with cement, but the cement yields up a little impurity (probably oxide of iron) to the liquid.

Each vat has a wooden rim securely fixed to its upper

edge all round it; upon this rim is fixed a rectangle of brass tubing (see annexed sketch, Fig. 23) an inch in diameter, to which is soldered a large binding-screw, for connection with the positive pole of the battery. Within this rectangle of tubing is also similarly fixed, but insulated from the first one, a smaller rectangle of brass-tubing, about half an inch in diameter, with a screw for connection with the negative pole. Cross tubes of brass, about half an inch in diameter and as

FIG. 23.

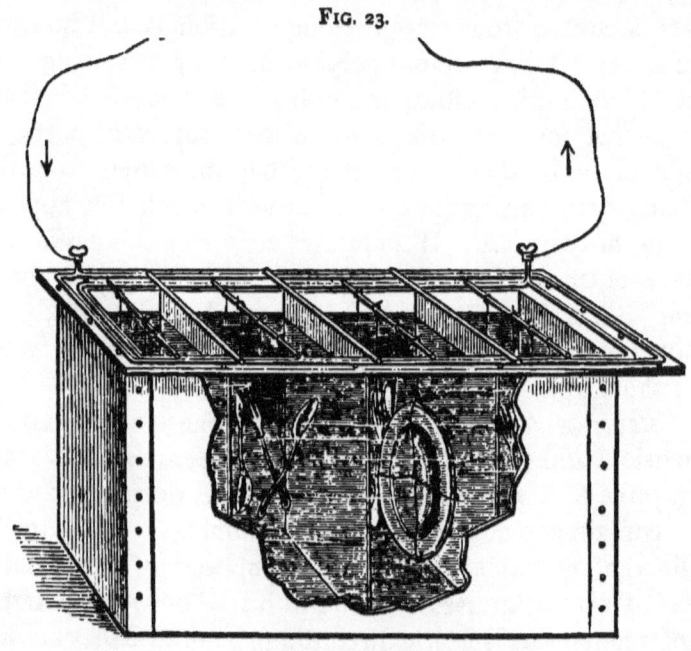

long as the vat is wide, are laid in clean metallic contact upon the larger rectangle, and these cross tubes support, and are metallically connected with the large and flat sheet silver anodes, by means of frames of iron, which extend downwards into the liquid. Similar, but shorter brass tubes, are laid across the vat, with their ends upon the inner rectangle, and these support, by means of wires, the articles to be coated. All the points of contact of the cross tubes with the rectangles, the supporting frames and wires with the cross tubes, and the other connections, are frequently examined, and kept scrupulously clean, by means of rubbing with emery cloth.

The wires for supporting the articles are usually formed of copper about the thickness of bell-wire, and are protected (excepting their ends and those parts which are not immersed in the liquid) from receiving a useless deposit of silver, by enclosing them in short tubes of glass, gutta-percha, or pure india-rubber ; and are bent at their lower ends into a sort of a loop, when required to support forks or spoons, so that those articles may be readily slipped into the loops and supported. (See Figs. 24, 25.)

FIG. 24.   FIG. 25.

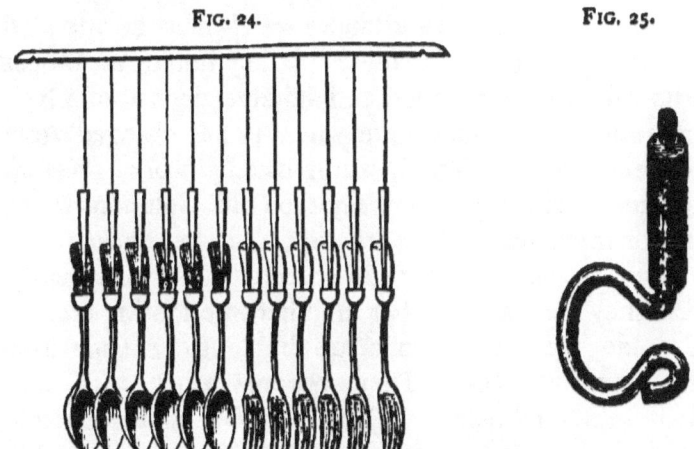

In vats where the articles are kept in continual motion, the cross rods supporting them are fixed to an iron frame, with four small wheels (about three inches diameter), which move backwards and forwards to an extent of three or four inches upon inclined rails fixed upon the edges of the vat, and impart to the articles a combined vertical and horizontal swinging motion, or they are suspended from a swinging-frame. (See Fig. 26, p. 174.)

Quality of the deposited silver.—The quality of the deposited silver, like that of all other metals, depends chiefly upon two circumstances, viz. the strength of the current in relation to the magnitude of surface of the article to be coated, and the composition and temperature of the plating solution. If the articles become grey or black in the solution, and evolve much gas, the current is too strong, and either the number of battery cells, or the depth of immersion of the battery-plates, must be diminished. If the deposit is good, but not sufficiently rapid, it is best increased by greater surface of immersion of the battery-plates.

Management of cyanide plating solutions.—Cyanide of silver plating liquid is more easily managed than cyanide gilding solution, and less easily than the ordinary sulphate solution employed in depositing copper.

There are various circumstances which must be attended to, in order to keep a cyanide plating liquid in proper condition for yielding a good and suitable deposit of silver, and to restore it to that state when it has changed from it. A new silver solution does not usually work as well as an old one, provided the latter is not too old. Solutions which are two or three years old work very well, but those which have been in use ten or twelve years, often work badly, because they generally by that time become too impure.

Cyanide plating solutions are liable to change from several circumstances. They become dirty with suspended solid particles. They become more concentrated, and of greater density, by evaporation of water, addition of cyanide of potassium, increased proportion of silver, &c. They either increase or decrease in their proportions of silver or cyanide by various causes; if that of cyanide of potassium to silver is large, or if the anode is large in relation to the receiving surfaces, the proportion of silver increases, and of free cyanide decreases; if the anode is relatively small, reverse effects take place. They acquire various metals besides silver, in solution, by corroding articles immersed in them to be plated, and also by dissolving impurities from the anode. They gradually decompose, become brown, and evolve ammonia by exposure to light, especially if they contain much free cyanide. They become spoiled by addition of too much brightening liquid ; and so on.

If the solution contains suspended solid particles, or sediment which is liable to rise by the motion of the liquid, the impurities settle upon the receiving surfaces, and produce a rough or uneven deposit, with vertical streaks ascending from them, especially if the articles are still, the liquid dense, and the deposition rapid. Solutions should therefore be filtered when necessary.

If the liquid is too dense, the articles still, and deposition is rapidly occurring, the goods are liable to become covered with vertical streaks, and to receive a much thicker deposit upon their lower parts than upon their upper ones. If the density is due to foreign salts, crystals of those salts are liable to form upon the articles in cold weather, and spoil the deposit. The specific gravity of a cyanide plating solution, may vary from 1·036 to 1·116 without detriment to the quality of the deposited metal, provided the greater density is not due to sparingly soluble substances ; if the density is greater than this, water may be added. The specific gravity of an old plating solution, analysed by me, was 1·1821 at 65° Fahr. ; it contained 16960 grains of solid matter (including 499 grains of silver), per gallon. If the solution contains too little water, but has the silver and cyanide in their proper relative proportions, it conducts freely, and gives a good and quick deposit; but is more difficult to manage than a weaker liquid, especially in hot weather, because, from the less mobility of its particles, it is more apt to settle into strata of different specific gravities ; its lower layers become nearly saturated with silver, and destitute of free cyanide, and its upper ones become exhausted of silver, and strongly charged with free cyanide. In consequence of this, the upper parts of the anode dissolve rapidly, whilst the upper parts of the articles receive very little deposit, and the lower parts of them are coated too rapidly, and neither receive a deposit of the best·

quality.   All these evil effects may be diminished in such a solution, by stirring it well every night, after having finished plating, or they may be entirely prevented by diluting the liquid to a proper degree, and stirring it every evening.   If the solution is too weak (i.e. contains too much water), it conducts sparingly, deposits slowly, and the deposit has a 'dead' white appearance.   This may be easily remedied by adding cyanide of silver and cyanide of potassium to it in proper proportions, and working it uniformly during a few

FIG. 26.

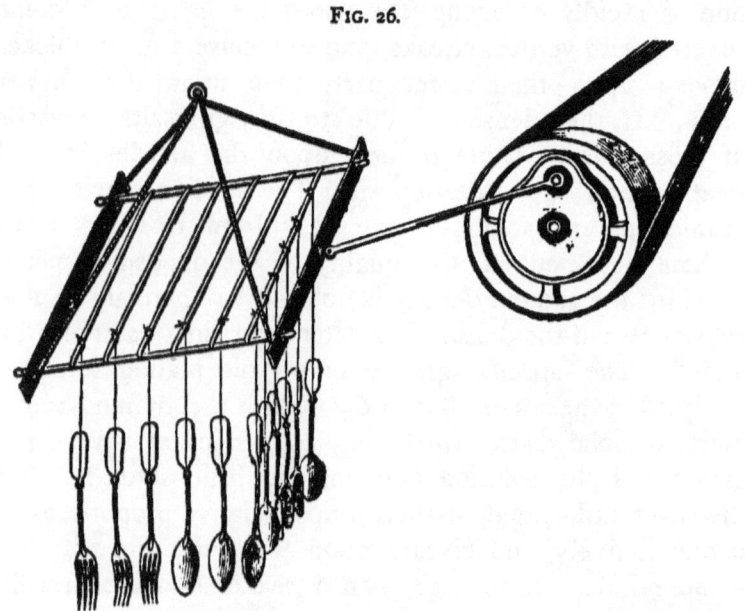

days.   The evil effects of improper degree of density of the solution upon the quality of the deposited metal, may be diminished, by keeping the articles in a state of gentle motion, by means of an apparatus (such as is shewn in Fig. 26) ; driven by power from a steam-engine.

If the anode is dirty, it indicates a deficiency of cyanide of potassium: this may be remedied by adding that substance ; but if the liquid contains much carbonate of potassium, as it usually does, either as an impurity of the cyanide employed, or by decomposition of the cyanide by light, as in a brown old solution, it may be remedied by addition of hydrocyanic acid, which will convert the carbonate into cyanide. The necessity of adding a little fresh cyanide is indicated, when the dissolving plate begins to change from its ordinary pure white appearance to a dull yellowish grey colour. It is best added in the evening after plating, about half an hour before stirring the solution.

If the anode appears grey during the passage of the current, and white whilst the current is stopped ; and if in addition to this, the deposit of metal is good, the solution is usually in proper condition. But if the anode is dark coloured or black whilst the current is passing, there is probably either too small a proportion of cyanide of potassium, or too large a one of silver; but if it is white, the reverse. If the cold solution coats bright copper rapidly with silver by simple immersion only, it probably contains too much free cyanide of potassium ; a bath should not produce this effect without the aid of a separate current. A solution containing much free cyanide, causes the anodes to corrode in holes, and fall to pieces, especially if the anodes touch the iron vat ; it also becomes rapidly richer in silver, and thus cures its own defect. The proportion of free cyanide may vary greatly without much injury to the solution, or to the quality of the deposited metal. One part of good cyanide of potassium is sufficient to dissolve one part of silver when converted into cyanide ; but unless there is, in addition to the quantity actually re- quisite to retain the silver in solution, a considerable excess of free cyanide of potassium, the conduction is defective, and the deposit granular and irregular.

To quickly ascertain if the plating liquid contains the proper proportion of silver to cyanide :—Put twenty-five parts by weight of the solution into a tall glass, and add to it, at first freely, but towards the last, drop by drop, with con- stant stirring, a solution of one part of crystallised argentic nitrate in ten parts of water. If the precipitate formed dissolves rapidly, with but little need of stirring, there is too little silver or too much cyanide. If it does not all dissolve, even after much stirring, there is too little cyanide, or too much silver, but if it wholly dissolves (the latter portions quite slowly), the proportion of silver to cyanide is about correct.

If a solution contains a great excess of free cyanide, and is also deficient in water, it becomes in hot weather very difficult to manage, and emits strongly the odours of ammonia and hydrocyanic acid. In such a solution, if from any cause the battery current becomes suddenly weak towards the evening, the silver deposited upon the articles will be re-dissolved, in consequence of the liquid about the anodes, having by the day's work, become saturated with silver, and that about the articles, become full of free cyanide ; the two electrodes (i.e. the dissolving plates and the articles) form a kind of voltaic battery, (of one metal in two liquids, see p. 69), which develops a current of electricity in an opposite direction to the original one, and thus re-dissolves the deposited silver. (See 'Polarization of electrodes,' p. 54.)

On some rare occasions, gas rises freely from the silver dissolving plate alone, and when this occurs, if the plate and articles are disconnected from the battery, and the ends of the wires brought into contact, and then suddenly separated, a minute spark (visible in the dark), is seen, produced by a current opposite in direction to that of the battery current, and probably produced by some polar conditions of the dissolving and receiving surfaces. This will even occur when the articles are kept in constant motion, and even after the dissolving plate has been taken out of the liquid, and re-immersed after the lapse of half an hour. On such occasions there is a tendency to a gritty deposit, and the solution is said to be out of order. (See p. 54.)

Peculiar phenomena often occur in the electro-deposition of silver, not only upon different metals, but also upon the same metals in different forms, or under other conditions of surface. For instance, if two perfectly similar pieces of thin sheet brass are taken (except, that one is perforated all over with small holes), and both be simultaneously immersed in the same solution to be silvered, and with the same battery power applied to each, the latter, although its amount of surface is reduced by the perforations, is said to become coated with silver much more slowly than the former. If a wire gauze cylinder of a Davy lamp, be suspended side by side with a piece of thin tubing of the same metal, and of the same dimensions, the latter will become coated much more rapidly than the former. If two pieces of the same metal, iron for instance, be immersed to be silvered in the ordinary cyanide solution, or to be coppered in the hot cyanide of copper and potassium liquid, each containing exactly the same amount of surface to be coated, but one being in the form of a thin sheet, and the other in that of a thick plate, or solid block of metal, the former will become coated much more rapidly than the latter. The edges and points of articles, whilst being plated, exhibit a greater tendency to a crystalline deposit than the flat parts, and this tendency is sometimes manifested in depositing silver upon table knives and forks. It is the knowledge of these, and many other peculiarities, of different metals and articles met with in practical working, and of the means of overcoming their attendant difficulties, which constitutes one of the chief differences between the practical operator, and the theoretical man.

Electro-deposited silver is sometimes of a yellow colour, and occasionally pinkish; these colours are due to impurities. The yellow appearance, which silver deposited from a cyanide solution, sometimes assumes after having been out of the liquid, is said to be due to subcyanide of silver in the deposit, and the pinkish tint is probably due to copper. Liquid ammonia, added to cyanide of silver plating solution not containing much free cyanide of potassium, frequently improves the colour of the deposit, and the condition of the liquid.

N

A bright solution is much more difficult to manage than the ordinary silver liquid; if it is not worked constantly, and in an uniform manner, it will lose its power of yielding bright metal. If any of the articles which are being plated in it are disturbed, or removed from the liquid and replaced, those will not now receive a bright deposit, and the disturbance of the liquid by their removal, will oftentimes cause all the neighbouring articles to lose their brightness. If too much 'brightening liquid' is added, the solution will be considerably injured; many silver solutions have been irretrievably damaged in this way. A bright solution requires a battery current of large quantity to work it, and the dissolving plates in it are generally of a darker colour than those in the ordinary silvering liquid.

Rapidity of silver deposition.—The rapidity with which metal of good quality can be deposited, depends largely upon the composition of the solution. To work rapidly, the solution should contain a rather large proportion of free cyanide of potassium, otherwise, the anode becomes covered with an insoluble film, before free cyanide from the articles can diffuse to it, and this film impedes the current. In a good silver solution, a dozen of ordinary table spoons or forks will acquire 1000 to 1500 grains of silver in twelve hours, but there are solutions used in Birmingham, in which it is said as much as one ounce of silver can be deposited upon a small table spoon in half a day.

Thickness of deposited silver.—Electro-plated articles vary greatly in quality, because any degree of thinness of silver may be put upon them. Great quantities of Britannia-metal articles are coated with only a few pennyweights of silver per square foot. 'The thickness of electro-deposited silver is in many cases from $\frac{1}{42}$nd to $\frac{1}{400}$th, or even $\frac{1}{8400}$th of a millimetre, or 1·24 grains upon a square metre of surface.' One ounce of silver per square foot of surface, is equal to a coating of about the thickness of thin writing-paper, and is considered an excellent coating. The prominent parts of a plated article usually receive the thickest coating, because electricity enters and leaves the projecting parts of bodies more readily than their hollow parts, and in deposition the prominences of the anode receive the most perfect supply of free cyanide, and those of the cathode receive chiefly the supply of saturated solution. ' M. M. Christofle, during the year 1865, deposited 33,600 kilogrammes of silver, upon 5,600,000 objects ; and covered a surface of 112,000 square metres, with three grammes of silver upon each square centi-metre.'

'In France, electro-plating is regulated by law, every manufacturer being required to weigh each article when ready for plating, in the presence of a comptroller appointed by the Government, and to report the same article for weighing again when the plating has been done. In this way, the comptroller knows to the fraction of a grain, the amount of precious metal that has been added, and puts his mark upon the wares accordingly, so that every pur-chaser, may know at a glance, what he is buying. As to the amount of silver consumed in ordinary plating—an ounce and a half of silver will give to a surface a foot square, a coat-ing as thick as common writing-paper, and, since silver is worth five shillings per ounce, the value of the silver covering a foot square, would be about seven-and-sixpence. At this rate, a tea-pot or coffee-pot is well plated, at a cost in silver, of not more than seven or eight shillings. The other expenses, including labour, would hardly be more than half that amount.

' The popular notion is, that genuine electro-gilding must necessarily add a good deal to the cost of the article coated. This is erroneous. A silver thimble may be handsomely coated, so as to have the appearance of being all gold, for threepence, a pencil case for tenpence, and a watch case for four shillings. An estimate of the relative value of electro-gilding as compared with silver plating, considering the cost of material alone, is about five to one. The quantity of silver used in plating the wares sent in such large quantities to the colonies, is about an ounce to the square mile ; one hard cleaning exposes the base metal, and your bargain from auction or cheap store, may be thrown on the dust heap' ('Technologist'). In Birmingham iron snuffers are sometimes 'silvered' wholesale, at as low a price as fourpence per pair ; and hooks and eyes, at one penny per pound.

In England also, the articles are weighed both before and after plating ; some French electro-platers use a 'plating balance,' the articles being suspended from one end of the beam, and a scale pan containing weights from the other, by means of which, when the articles have received any desired amount of deposit (determined beforehand), the circuit is automatically broken, and the deposition stopped.

The average cost of depositing silver has been estimated at twopence per ounce, but this would probably only include the cost of the battery power, and not of the numerous other incidental expenses. A very large amount of plating is done, at a cost of about eight shillings per ounce of deposited silver, by professed electro-platers ('electro-platers to the trade '), whose sole occupation is electro-deposition, for others, who are called 'electro-platers. Many manufacturers of spoons, forks, tea-pots, coffee-pots and other articles, have at different times commenced to plate for themselves, but have found, that the coated articles were so often required to be stripped and re-plated, and that the difficulties and incidental expenses were so great, that they have abandoned the actual performance of the process, and send the articles to the regular 'electro-platers to the trade' to be done.

Ornamenting silver articles. Dead silver; 'Oxidised silver.'—Frequently, for the purposes of ornamentation, and of producing a pleasing effect, the surface of deposited silver is treated in various ways. To obtain what is termed a 'dead' appearance, like frosted silver, deposit a mere trace of copper upon it in a sulphate of copper solution, and then a very thin layer of silver upon that. 'Oxidised' silver, is not silver coated with oxide, but with a film of platinum, or of sulphide of silver ; the former is produced, by applying a hot solution of perchloride of platinum, and allowing it to dry ; the colour varies from a light steel-grey to nearly black. The hotter the solution, and the greater proportion of platinum in it, the deeper black does it produce. The colour arises from finely divided black metallic platinum, deposited by simple immersion process ; and the silver is slightly corroded by the action, forming chloride of silver, which may be dissolved by means of diluted aqueous ammonia. A bluish black colour is produced by means of a freshly made, and hot solution, of sulphide of potassium ('liver of sulphur'). 'Nielled silvering' is a process of sulphurising engraved parts of silver surfaces, and consists in inlaying the surface with sulphide of silver which has been prepared beforehand, and causing it to adhere firmly to the metal, by heating the article in a muffle until the sulphide melts. This, however, is a separate art, closely allied to enamelling.

To produce a beautiful pink colour upon silver, dip the clean article for a few seconds in a hot and strong solution of cupric chloride, swill it in water, and then dry it, or dip it in spirit of wine, and ignite the spirit. (W. H. Fearn.)

A blackish colour is sometimes produced, by making a thin magma of plumbago and spirit of turpentine, and rubbing it upon the surface. After drying, the black is rubbed off from the prominent parts of the surface by means of a rag wetted with alcohol ; the process is called 'old or antique silvering.' To produce a brownish-black colour upon silver articles, a solution, composed of equal weights of salammoniac and sulphate of copper dissolved in vinegar, is applied.

Articles of silver are also ornamented, by depositing gold upon portions of their surfaces ; and those of gold are ornamented, by depositing gold of different tints of colour (sometimes as many as five or six) upon their different parts; this is effected by covering the parts which are not

to receive the coating, with a varnish which will resist the
solvent power of the hot alkaline gilding liquid.   The composition of such a varnish is as follows :—

Translucent resin	10 parts
Yellow beeswax	6 ,,
Extra fine red sealing-wax	4 ,,
The finest polishing rouge (i.e. impalpable per-oxide of iron)	3 ,,

Best quick drying copal varnish, with some peroxide of
iron, or ultramarine, mixed with it, is used for ' stopping off,'
in hot cyanide solutions, or mixed with chromate of lead, if
for use in cold liquids. It dries in about three or four hours.

Cleaning silver.—A yellow colour upon silver which has
been electro-deposited, is said to be due to the action of the
air upon a basic or subcyanide of silver in the pores of
the metal, and may be removed by immersing the article
either in a solution of cyanide of potassium, or in ordinary
plating solution containing free cyanide.   A weak solution
of cyanide of potassium is also a good substance to clean
discoloured silver, but it dissolves a little silver, and is also
a very dangerous liquid to be entrusted to ignorant persons,
being exceedingly poisonous.

The East Indian jewellers clean silver, by briskly rubbing
it with slices of a juicy lemon, and then covering the article
with the slices in a pan for a few hours.   They then swill
them two or three times in water ; stir them in nearly boiling
soap-suds, brush, rinse, and finally dry them on a metal
plate over hot water.   Green tamarind stems are more powerfully detergent than lemons, and are employed to remove
oxide and fire-marks from gold and silver ('Telegraphic
Journal,' vol. xi. p. 178).   Silver may be cleaned in water in
which potatoes have been boiled, and a superior polish is
thus imparted to them (Elsner, 'Chemical Society's Journal,'
vol. xi. 1072). A good polishing paste for silver, is composed
of washed whiting, precipitated carbonate of magnesia, and
precipitated peroxide of iron.

'*Stripping*' *articles.*—Sometimes, articles composed of copper, brass, or german-silver, which have been plated with a portion of silver, require to have the coating entirely removed, and a new deposit put upon them, because the first was defective. Occasionally also, the depositor has sent to him, to be re-plated, old articles of silvered copper, ('Sheffield plate,') the coating upon which has partly worn away; these require to have the remaining portions of silver removed, in order to obtain a uniform surface to deposit upon. The removal of the silver is termed '*stripping*.' To effect this, add a very little nitrate of soda ('Chili saltpetre,') to a quantity of strong and hot oil of vitriol, and immerse the articles in the mixture, until all the silver is dissolved. If the action becomes slow, apply more heat, and add more saltpetre at the moment of using. The copper will not be much acted upon, if the articles are not allowed to remain in too long. A number of such articles are generally '*stripped*' together, and are afterwards washed, and prepared in the usual manner for receiving a deposit; or the silver may be removed, but more slowly, without the aid of heat, by suspending the articles for a greater or less length of time, according to the thickness of the coating, in a bulky mixture of ten measures of strong sulphuric acid, and one measure of concentrated nitric acid, contained in a large stoneware vessel. The liquid must not be diluted, but be kept as free from water as possible, by not immersing wet articles in it, and by keeping it covered from the air; otherwise it will attack the copper, brass, bronze, or german-silver base of the articles. As the liquid becomes weaker, very small portions of strong nitric acid are added to it. Its action is less rapid than that of the hot mixture already described. In stripping an article for re-plating, the whole of the silver should be taken off, otherwise the deposit subsequently put upon it, is apt to shew lines, where silvered and unsilvered portions meet. Some depositers re-plate defectively coated articles without stripping them; they merely well 'scratch-brush,' 'buff,' and 'quick' them repeatedly, and then put on the second coating ; but this is a less satisfactory method.

If the base of the article is composed of iron, steel, zinc, or lead, the above mode of stripping by acids is not applicable, and the coating is best removed, by making the articles the anode, in an ordinary cyanide of silver plating solution, until the silver is dissolved.

Sometimes it is necessary to perform the converse operation, viz. to strip copper from the surface of silver, as when a thin copper mould has to be removed from a solid deposit of silver formed upon it by electro-plating process. To effect this, boil the article in dilute hydrochloric acid, or immerse it in a hot solution of perchloride of iron. This latter liquid may be made by digesting peroxide of iron (crocus, or jeweller's rouge) in warm hydrochloric acid, as long as it will dissolve ; the solution will remove tin, lead, or copper, from either gold or silver, without affecting those metals. A solution of chloride of zinc has been used for the same purpose. Copper may also be completely removed from silver or gold, by making it the anode in a sulphate of copper solution, until all the copper is dissolved ; the silver will remain unaffected if the current employed is feeble, and has not a greater intensity than one or two pairs.

Sometimes articles of silver which are gilded, require to have the gold removed from them; this is effected by heating them to redness, and throwing them into dilute sulphuric acid ; the gold scales off in spangles, and falls to the bottom. The process is repeated until all the gold is removed. With articles which are hollow, it is better to make them the anode in a solution of one part of cyanide of potassium and ten parts of water, with a cathode of platinum.

Analysis of cyanide of silver plating solutions.—The only points usually required to be known respecting the composition of a silver plating liquid, are the proportions of free cyanide of potassium, and of metallic silver.

The common method of ascertaining the percentage of free cyanide, is that of Glassford and Napier, and may be carried out as follows :—Dissolve a known weight, say fifty grains of crystallised nitrate of silver, in distilled water, and dilute it to a known weight or bulk. Take a known measure of the plating liquid, say one ounce, if it is not extremely strong in free cyanide. Add the solution of nitrate to the plating liquid, just as long as the precipitate formed dissolves completely on stirring, and ascertain how much of the silver salt has been consumed. Every 170 parts of nitrate, equal 130·2 parts of free cyanide.

The silver of an old cyanide plating solution, cannot be satisfactorily converted into pure chloride, by adding an excess of hydrochloric acid, and boiling the mixture ; the precipitate so obtained is of varied colours, and contains foreign substances. To ascertain the amount of silver, evaporate a known volume of the solution to dryness, fuse the residue at a red heat in a porcelain crucible, cool it, and dissolve the saline residue in water (testing the water for silver by means of hydrochloric acid), taking care to collect any chloride of silver present. Dissolve the metallic residue in warm dilute nitric acid ; and estimate the amount of silver in the solution, by ordinary volumetric or gravimetric methods. The silver may also be estimated as follows, by a method I have employed. Take 700 grains by measure of the solution ; evaporate it to dryness ; add sixty grains of sulphate, and fifteen of nitrate of ammonium, both in powder, mix, and dry thoroughly. Transfer the mixture to a capacious porcelain crucible, and heat very gradually to fusion at a low red heat. Dissolve the cooled residue in hot dilute nitric acid, taking care to collect any undissolved chloride of silver. Precipitate the dissolved silver by hydrochloric acid, filter, wash, ignite, and weigh the pure argentic chloride, every 143·5 parts of which equal 108 parts of silver. In this process, all the cyanogen of the solution, is expelled by the fusion with the sulphate and nitrate of ammonium, and the base metals and alkalies are converted into soluble sulphates. Frequently the fused saline mass is quite green with copper and other metals. A sample of plating solution, analysed by this method, gave 147·54 grains of metallic silver per gallon, and much copper ; another yielded 499·2 grains, and a great quantity of chlorides ; a third 1,720 grains, and much copper, iron, and chlorides ; seven others gave from 348·1 to 979·9 grains ; six, from 1,701 to 2,195 grains ; and eight yielded from 519 to 865 grains per gallon.

According to G. C. Wittstein, cyanide of silver plating solution may be completely analysed as follows :—1. Mix twenty cubic centimetres of the liquid with fourteen cc. of acetic acid of 20 per cent. ; evaporate ; extract the residue with absolute alcohol ; evaporate the alcoholic solution ; heat the residue with hydrochloric acid ; dry, and weigh as chloride of potassium (KCl). This gives the total amount of potassium (K) as free cyanide, as potassic silver cyanide, carbonate, and cyanate.

2. Take a second twenty cc. of the solution, add sulphide of ammonium—$(NH_4)_2S$—in slight excess, find amount of argentic sulphide, and calculate from it the amount of potassium in the double cyanide (KCy AgCy).

3. Precipitate a known portion of the plating liquid with calcic chloride, and weigh the carbonate of calcium (Ca CO_3) ; this gives by calculation the amount of potassic carbonate (K_2CO_3).

4. As the impure cyanide contains for every seven equivalents of cyanide of potassium, three equivalents of cyanate (7 KCy + 3 CyKO), calculate from this and the other data, the amount of potassic cyanate.

And, 5. Deducting from the entire amount of potassic chloride obtained in No. 1, the quantity of the same salt equivalent to the amount of double cyanide (KCyAgCy), the potassic carbonate and potassic cyanate ; the remainder, is the amount of potassic chloride, equivalent to the quantity of free cyanide of potassium in the solution ('Journal of the Chemical Society,' vol. xii. p. 1012).

Napier (commenting on Wittstein's process) recommends a simpler one as follows :—To find the amount of silver, evaporate one ounce of the solution to dryness, and fuse the dry residue in a small crucible. Dissolve the saline mass, obtain the little button of silver, and weigh it; or, precipitate one ounce of the solution, by hydrochloric acid in excess ; wash, dry, and weigh the product : three-fourths of the weight of which is silver. To one ounce of the liquid, add a solution of nitrate of silver as long as a cloud is produced ; wash the precipitate with water by decantation ; add to it an excess of a mixture of equal volumes of nitric acid and water, which, by digestion, dissolves all the silver salt not cyanide. Decant the liquid, add a little hydrochloric acid to it, wash, dry, and weigh any precipitated chloride of silver. One-half of the weight of this precipitate, approaches closely to the amount of cyanate or carbonate of potassium, the difference in their equivalents not being great. If there was any chloride in the original plating solution, the chloride of silver corresponding to it, will remain in the portion of precipitate insoluble in the dilute nitric acid.

Recovering silver and gold from residuary liquids, &c.[1]— Many silver plating solutions are spoiled by unsuccessful attempts to improve them ; by the accidental introduction of impurities; by adding excess of cyanide of potassium; by having been improperly made; and especially by addition of too much 'brightening liquid ;' and we wish to recover their metal. There are two general modes of recovering silver from wornout or spoiled solutions, or waste residuary liquids : one, by precipitating the silver as chloride by addition of sufficient hydrochloric acid, washing and drying the chloride of silver precipitate, and fusing it with carbonate of potassium containing a little saltpetre to oxidise the baser metals ; and the other, by evaporating the solution to dryness, and fusing

[1] See also p. 145.

the product without the addition of any oxidising substance, which would be very dangerous if cyanide of potassium is present. In both cases the silver and gold are set free.

To recover silver from 'stripping' solution in the form of cyanide, dilute the supernatant liquid with water, dissolve the precipitated yellow powder, or crystals of sulphate of silver, by addition of nitric acid to the water. Pour off the clear liquid, and add cyanide of potassium to it as long as a precipitate of cyanide of silver takes place, or until effervescence of hydrocyanic acid occurs by decomposition of the excess of cyanide of potassium by the free nitric and sulphuric acid present. Collect, wash, and dry the precipitated cyanide of silver ; and, after testing the liquid portion with hydrochloric acid to see if it contains any traces of silver, throw it away. Another plan is to add a solution of washing-soda to the 'stripping-liquid' until the latter is alkaline to red litmus paper (i.e. turns the paper blue), then acidify the mixture by addition of hydrochloric acid. After subsidence, throw away the clear liquid, dry the sediment, and fuse it with a mixture of the dried carbonates of potassium and sodium containing a little saltpetre. The nearly pure silver and gold remains.

As the recovery of silver and gold from silvering and gilding solutions, is an important matter in practical electrometallurgy, I extract the following valuable remarks from a paper by Elsner : 'I have undertaken a series of researches upon this object, and hasten to communicate the results to the public ; but, before proceeding to the communication, I think it necessary to mention the results of the experiments, upon which are based the methods given further on, for extracting both the silver and the gold of old cyanide of potassium liquids.

' 1. If we add hydrochloric acid to a solution of silver in cyanide of potassium until the liquid exhibits an acid reaction, we obtain a white precipitate of chloride of silver, which, when submitted to heat, melts into a yellow mass.

If this was cyanide of silver, the application of a red heat would have left a regulus of silver. The addition of the hydrochloric acid, precipitates all the silver present in the liquid in the form of chloride of silver.

' 2. If we evaporate a solution of silver in cyanide of potassium to dryness, and heat the residue to redness, until the mass is in a state of quiet fusion, and has assumed a brown colour, there remains, when we wash the mass with water, metallic and porous silver. The wash waters, when filtered, still contain a little silver in solution ; because, if hydrochloric acid is added to them, it produces a precipitate of chloride of silver. In evaporating and calcining a solution of gold in cyanide of potassium, the result is similar, i.e. we obtain metallic gold. The wash waters, acidulated with hydrochloric acid, give, when treated with sulphuretted hydrogen, a brown precipitate of sulphide of gold ; and, with the salt of tin, a violet precipitate (purple of Cassius),—a proof that these liquids still contain a little gold in solution.

' 3. If we pour upon finely-divided silver—for instance, silver leaf, or silver precipitated in the porous state by zinc, from a solution of silver—a concentrated solution of cyanide of potassium, at the ordinary temperature, and shake it frequently, the liquid, at the end of a certain time, exhibits silver in solution, and by adding hydrochloric acid to it, we produce an abundant precipitate of chloride of silver. This experiment explains why, in the wash waters of the various combinations of gold or silver with cyanide of potassium, we can still demonstrate the presence of gold and of silver after the most minute separation.

' 4. When hydrochloric acid, or ordinary sulphuric acid, is added to a solution of cyanide of copper and cyanide of potassium, until the liquid exhibits an acid reaction, the result is a reddish-white precipitate, which is a cyanide of copper in the anhydrous state. If the precipitate be well washed, and boiled in potash lye, oxide of copper is separated, of a beautiful red colour ; and if to the filtered alkaline liquid we add a solution of green copperas, a dirty-blue precipitate is obtained. A solution of carbonate of sodium furnishes the same results, and yields with the copperas, the same dirty-blue precipitate. If the reddish-white precipitate is dissolved in pure nitric acid, and a solution of nitrate of silver is added to it, an abundant white precipitate is produced, which, when washed, dried, and calcined, yields silver in the metallic state—a proof that the precipitate is cyanide of silver. The reddish-white precipitate is soluble in an excess of hydrochloric acid, in nitric acid, and in aqua regia ; it is also soluble in aqueous ammonia, and in a solution of cyanide of potassium.

' 5. When a solution of silver, prepared for silvering articles of bronze or of brass, has been employed a certain time for that purpose, the precipitate produced in it by the addition of hydrochloric acid is not pure white, but reddish, in consequence of the reddish-white cyanide of copper which is precipitated with it : for we know that those silvering liquids which have been used for some time, contain copper in solution. The same thing occurs with the solutions for gilding, in which articles of silver, copper, bronze, and brass, have been gilded for a long time ; the liquid contains after a certain time of service, not only gold, but also silver and copper. This case presents itself especially when gilded articles of silver, containing copper, or other alloys of silver, are in the solution of gold ; then the precipitate of cyanide of gold produced by the addition of hydrochloric acid, does not possess its proper pure yellow colour. It has happened to me to observe a precipitate of this kind, which, instead of being yellow, was green, and, in fact, articles of iron had been gilded in the solution, and the precipitate contained, besides cyanide of gold, Prussian blue, so as to be demonstrated in an examination, which consisted in boiling the green precipitate in aqua regia, filtering to separate the dirty-green residue, evaporating the filtered liquid to dryness, and dissolving the dry salt in water acidulated with hydrochloric acid ; the addition of sulphate of iron to this new liquid gave a brown precipitate, and the salts of tin a reddish-brown precipitate. In treating by aqua regia, the cyanide of gold was then decomposed, and converted into chloride of gold.

'Based upon the preceding facts, we may find several methods for recovering all the silver and gold of old cyanide of potassium solutions. The extraction of these precious metals may be effected, either by the wet or by the dry process.

'*Extraction of silver by the wet method.*—Adding hydrochloric acid until the liquid exhibits a strongly acid reaction. The precipitate of chloride of silver which is thus obtained, will be, as we have already said, of a reddish-white colour, because of the cyanide of copper, which is precipitated with it, when the solution has been used a long time for silvering objects containing copper. In this precipitation by hydrochloric acid, there is hydrocyanic acid gas set free, therefore the operation should only be performed in the open air, or in a place where there is good ventilation ; if the precipitate is very red, it must be treated with hot hydrochloric acid, which will dissolve the cyanide of copper. The chloride of silver, having been washed with water, must be dried, and then fused with pearlash in a Hessian crucible coated with borax, in the ordinary manner for obtaining metallic silver.

'This method is very simple in its application, and very economical, considering, that by the aid of the hydrochloric acid, all the silver contained in the solution of cyanide of potassium is precipitated, and there remains no trace of it in the liquid. But the quantity of hydrocyanic gas which is disengaged, is a circumstance which must be taken into serious consideration when operating on large quantities of silver solution, the vapour of which is most deleterious, and nothing but the most perfect ventilation, combined with arrangements for the escape of the poisonous gases, will admit of the process being carried on without danger to the workmen ; when, however, we have taken the precautions dictated by prudence, the method in question may be considered as perfectly practical. The liquid should be poured into very capacious vessels, because the addition of the acid produces a large amount of froth.

'*Extraction of silver by the dry method.*—The solution of cyanide of silver and potassium is evaporated to dryness, the residue fused at a red heat, and the resulting mass, when cold, is washed with water. The remainder is the silver in a porous metallic condition. There still remains in the wash waters a little silver, which may be precipitated by the addition of hydrochloric acid.

'*Extraction of gold by the wet method.*—A solution of gold and cyanide of potassium, which has long served for gilding articles of silver alloyed with copper, may still contain, as we have already remarked, independently of the gold, both silver and copper, and perhaps iron. In order to obtain these metals we operate in the following manner :—

'The liquid, the same as with the solution of silver, is acidulated with hydrochloric acid, in which case there is produced a disengagement of hydrocyanic acid gas, which requires the same careful ventilation. This addition of hydrochloric acid causes a precipitate, which may, according to circumstances, consist of cyanide of gold, cyanide of copper, and chloride of silver. The precipitate, washed and dried, is boiled in aqua regia, which dissolves the gold and copper in the form of metallic chlorides, and leaves the chloride of silver unaffected. The solution, containing the gold and the copper, is evaporated nearly to dryness, in order to drive off any excess of acid ; it is then dissolved in a small quantity of water, and the gold precipitated from it (by the addition of protosulphate of iron) in the state of a brown powder. The chloride of silver is reduced to the metallic state by the known means. The liquid from which we have precipitated the cyanide of gold, &c., by hydrochloric aci.l, may yet contain a little gold in solution. I refer to No. 5 for its further treatment.

This method is distinguished by the great simplicity of the

operation, and we may repeat for it all that we have already said respecting the extraction of silver by the wet method.

'*Extraction of gold by the dry method.*—The solution of cyanide of potassium which contains gold, silver, and copper, is evaporated to dryness; the residue fused at a red heat, cooled, and washed (the wash waters still contain a little gold and silver, and this occurs most often when the solution of gold or silver contains a very great excess of cyanide of potassium). The residue, after washing, consists of gold and silver in a metallic porous state, and carbide of copper resulting from the decomposition of cyanide of copper by the heat. The metallic residue is treated with aqua regia, which forms insoluble chloride of silver, and contains the chlorides of gold and copper in solution. In order to obtain these bodies in the metallic state, we must proceed in the manner previously indicated.

'If we operate according to the method of Professor Boettger, i.e. if we fuse the dried residue with its own volume of litharge, in a covered crucible, the regulus we obtain in this case, consists of gold, silver, and lead. In treating this alloy by nitric acid, of specific gravity 1·2, and applying heat, the gold remains in the form of a brown powder, whilst the lead and the silver are dissolved in the acid. This solution, after having been largely diluted with boiling distilled water, may have the silver separated from the lead, by the addition of hydrochloric acid.

'These methods of extracting the silver and gold, from old solutions of cyanide of potassium by the dry process, present this advantage, that the operator is not incommoded while working, by the disengagement of vapours of hydrocyanic acid. In these operations, the poisonous gases are not developed as they are in the processes for extracting the metals by the wet method.

'After the experiments here reported, those who are interested in the subject, may choose for themselves, which of these methods appears the most suitable to the circum-

o

stances in which they are placed, and the object which they wish to obtain.'

The process of recovering the silver from old plating solutions, depends largely upon the bulk of the liquid ; if that is small, the liquid may be evaporated to dryness, and the silver recovered by the dry method, taking care to ascertain that the fused salt above the button of silver, is free from precious metals, before throwing it away ; but if the bulk is great, the solution, contained in capacious vessels, must be precipitated by an excess of hydrochloric acid, added to it in the open air, taking extreme care not to breathe the vapour of hydrocyanic acid which is evolved.

Testing the purity of silver.—The silver recovered from solutions, strippings, and residues, is often not pure ; that employed as anodes is sometimes not perfectly pure, and electro-deposited silver, although not often alloyed, is not necessarily pure, because in some cases, other metals are deposited with it, in order to obtain the desired tint of colour.

According to Dr. Böttger, the purity of silver may be ascertained as follows :—Apply, by means of a glass rod, a drop of cold saturated solution of bichromate of potassium, in nitric acid of sp. gr. 1·2, to a clean part of the surface, and immediately wash it off with cold water ; if the silver is pure, a blood-red mark is left. All other metals behave differently from this ('Chemical News,' vol. xxiii. p. 119). Runge had also previously employed a somewhat similar test. He immersed the article in a cold mixture of thirty-two parts of water, four of sulphuric acid, and three of chromate of potash. The immersed part quickly assumed a purple colour, which was less deep and less lively, in proportion to the amount of alloy contained in the silver.

Silver may be purified, by dissolving it in warm dilute nitric acid, immersing clean sheets of copper in the liquid, until all the silver is precipitated, finally making the solution quite warm, to complete the precipitation, or until a little of it gives no white cloud on adding to it two or three drops of hydrochloric acid. The liquid may then be thrown away, the metallic silver precipitate washed, dried, and melted, with the addition of a little carbonate of potassium, and a trace of saltpetre put with it into the crucible; or the silver may be preserved for future use in making argentic nitrate. Silver may also be purified by dissolving it in nitric acid, then precipitating it by hydrochloric acid, and fusing the washed and dried chloride, with about one-half its weight of the anhydrous carbonates of sodium and potassium, and a little saltpetre, in an earthen crucible.

14. **Mercury.**—Elec. chem. eqt. $= \dfrac{200}{2} = 100$. The ordinary compounds of this metal, are the dioxide (red precipitate), nitrate, mercurous chloride (calomel), bichloride or mercuric chloride (corrosive sublimate), bisulphide (vermilion), and bicyanide. The most soluble of these salts are the nitrate, bichloride, and bicyanide.

Mercuric nitrate is made, by dissolving mercury in diluted nitric acid; corrosive sublimate may be more conveniently purchased than made. The bicyanide is prepared, by adding eight parts of Prussian blue, and sixteen parts of mercuric oxide (red precipitate), (both in a state of fine powder,) to thirty parts of water; boiling the mixture about fifteen minutes, filtering, evaporating, and crystallising the solution. It may also be made, by digesting an excess of red oxide of mercury, in the strongest aqueous hydrocyanic acid, until the odour of the acid has disappeared, and then evaporating the clear liquid. Also by dissolving one part of ferro-cyanide of potassium, in fifteen of boiling water, adding two parts of mercuric sulphate, digesting it hot for fifteen minutes, and then filtering the mixture, and crystallising the clear solution. Lielegg prepares it, by passing vapour of hydrocyanic acid through water in which finely divided oxide of mercury is suspended ('Chemical News,' vol. xxvi. p. 264). About fifty-five grains of cyanide of mercury, dissolve in one ounce of water at 60° Fahr. I have found that two and a half ounces of corrosive sublimate, converted into oxide of mercury, required about nine ounces of hydrocyanic acid of maximum strength (i.e. 'Scheele's strength,' see p. 197) to dissolve it with the assistance of heat. The double cyanide of mercury and potassium may be made, by dissolving an equivalent of bicyanide of mercury in water containing an equivalent of cyanide of potassium in solution, and evaporating the clear liquid. A very dilute solution of this salt, is used for ' quicking' the surfaces of articles of copper, brass, and german-silver, which are to be silver-plated. (See also p. 323.)

Deposition of mercury by simple immersion process (see also p. 78).—From solutions of mercuric chloride, cyanide, or nitrate, aluminium deposits mercury ; forming an amalgam which decomposes water at 60° Fahr., and rapidly oxidises and becomes heated in the air. Aluminium deposits mercury from a solution of mercuric chloride in alcohol ; also from one of mercuric iodide in potassic iodide (A. Cossa, 'Watts' Dictionary of Chemistry,' 2nd Supplement, p. 54). From a solution of mercuric chloride, magnesium deposits calomel and mercuric oxide (Commaille, ' Chemical News,' vol. xiv. p. 188). Nearly all the base metals coat themselves with mercury by simple immersion in solutions of mercuric salts.

Electrolysis of salts of mercury (see also p. 89).—Comparatively little as yet has been done in the electrolysis of salts of mercury. A solution of any such salt may be electrolysed, by putting some mercury as an anode in the bowl of a tobacco pipe, thrusting a platinum wire down the pipe into the mercury, and immersing the bowl in the liquid ; or the mercury may be placed in the vessel which contains the solution, and be connected with the positive pole of the battery by a platinum wire passing through a tube of india-rubber to protect it from the liquid. Gladstone and Tribe have observed, that on passing a weak electric current through a solution of mercuric chloride, into a cathode of platinum, a film of mercurous chloride is deposited ; but with a strong electric current, mercury itself is set free. I have noticed, that with a mercury anode and platinum cathode, in dilute sulphuric acid, the cathode soon receives a coating of mercury on passing a current.

Electrolytic vibrations and sounds.—As long ago as the year 1801, Gerboin observed, that mercury exhibited peculiar movements whilst acting as an electrode in electrolysis, and this phenomenon has been since investigated by Sir J. Herschel, Sir H. Davy, and others. According to G. Lippmann, the contraction of a globule of mercury, (whilst acting as a cathode in dilute sulphuric acid,) on the passage of the current, is due to a change in the capillary constant ('Journal of the Chemical Society,' vol. xi. p. 1094). It was whilst investigating these peculiar movements, and searching for thermic changes in electrolysis, by passing an electric current through a solution of double cyanide of mercury and potassium, with mercury electrodes, that I first heard a faint sound, and then observed the surface of the mercury covered with waves ; and by further research, was led to the discovery of *electrolytic sounds*; the dancing motion, and musical sound, being due to the alternate formation and destruction, of films upon the mercury, by electrolytic action. A paper on the subject in the 'Proceedings of the Royal Society' 1862, contains a full account of the phenomenon, and of the influence of various circumstances upon it.

The best liquid for producing the sounds, consists of ten grains of cyanide of mercury, and 100 of pure hydrate of potash, dissolved in 2¾ ounces of aqueous hydrocyanic acid, containing 5 per cent. of the anhydrous acid ('Scheele's strength') ; the liquid should be filtered. The phenomena usually occur only at the negative electrode, and out of a large number of solutions examined, the only ones in which *phonetic* vibrations occurred, were those of alkaline cyanides containing dissolved mercury. The quicksilver may be contained in two very small watch glasses submerged in the solution ; and the current, from either two Grove's, or five

Smee's cells, conveyed to the electrodes by platinum wires, protected, except at the ends, from contact with the liquid by means of tubes of glass or india-rubber.   During the FIG. 27.   occurrence of the sounds, the current itself is rendered imperfectly intermittent, and the arrangement may to a certain extent be employed for similar uses to those of a break-hammer.   A more perfect arrangement, consists of a glass vessel of the annexed form (see fig. 27), placed upon a sounding-board. One portion of mercury is poured into the centre, the other into the annular space, the wires immersed the two portions, and the liquid is above.

CLASS IV. BASE METALS.

COPPER—NICKEL — COBALT — IRON— MANGANESE— CHRO-
MIUM —URANIUM — TUNGSTEN—MOLYBDENUM — VANA-
DIUM—LEAD—THALLIUM—INDIUM — TIN—CADMIUM —
ZINC.

15. **Copper.**— Elec.-chem. eqt. $= \dfrac{63\cdot5}{2} = 31\cdot75.$ The commonest salts of copper, are the suboxide, or cupreous oxide (red oxide of copper) ; the protoxide, or cupric oxide (black oxide of copper) ; the nitrate, chloride, sulphate (blue vitriol), cyanide, and acetate.   The black oxide is made by heating the nitrate to full redness in a crucible, washing and drying the product.   The sulphate is most conveniently purchased (its price is about sixpence per pound) ; or it may be made by heating copper filings in oil of vitriol until the product is nearly dry, washing the residue in boiling water, evaporating and crystallising the filtered solution. The chloride and nitrate may be formed, the first by dissolving copper in aqua regia, or by saturating hydrochloric acid with protoxide of copper, and evaporating and crystallising the liquids ; and the second, by dissolving copper in nitric

acid, evaporating and crystallising the solution. Acetate of copper is most conveniently purchased ; its commercial name is crystallised verdigris. Cyanide of copper may be made, by adding a solution of cyanide of potassium to one of sulphate of copper (each liquid being cold), just as long as a precipitate is produced, but no longer ; filtering and washing the precipitate, which is the required compound ; it is a fine powder of a pale green colour. In the operation, a large quantity of cyanogen gas is evolved, which if inhaled is dangerous to health. The powder contains two equivalents of cyanogen, for every three equivalents of copper ; it is freely soluble in a solution óf cyanide of potassium ; it is also soluble in aqueous ammonia, and in a solution of carbonate of ammonium. The liquid, after precipitation, is invariably greenish blue, and contains much dissolved copper, but no use has hitherto been made of this remainder. Carbonate of copper may be prepared, by precipitating a cold solution of cupric sulphate, by one of washing-soda ; it is a green powder.

Electrical relations of copper.—Copper is electro-positive to iron in the following liquids at 60° Fahr. : powerfully in a solution of sulphide of ammonium ; feebly in a saturated aqueous one of ammonia ; in a solution of oxide of copper in liquid ammonia, in aqueous ammonia, or in a saturated solution of ferro-cyanide of potassium, each but for a short time—it then becomes negative ; in a saturated solution of bichromate of potassium ; in a strong aqueous one of sulphide of potassium, it is increasingly positive up to the boiling point of the liquid. This last liquid has a similar effect on brass.

Deposition of copper by simple immersion process (see also p. 78).—From a solution of cupric sulphate, magnesium deposits the metal, its hydrated protoxide, and a green subsalt; but from one of cupric chloride, it precipitates Brunswick-green and no metal (Commaille, 'Chemical News,' vol. xiv. p. 188). Metallic aluminium immersed in a similar solution, very slowly deposits copper; and, in one of the nitrate, it slowly separates (after several days) a basic salt and metallic copper; the reduction is quicker if an alkaline chloride is added. From a solution of the chloride, it quickly deposits copper, but from one of the acetate, the copper is more slowly separated (A. Cossa, 'Watts' Dictionary of Chemistry,' 2nd Supplement, p. 54). By adding crystals of silicon to melted black oxide of copper, I observed a sudden incandescence, which raised the temperature to a full white heat; copper was also deposited and melted to a red metallic bead, and could be hammered into a thin sheet. By heating to redness also, one part of fragments of magnesium and six of cupric fluoride, copper was separated. I have also observed, that crystals of silicon immersed in a solution of fluoride of copper containing free hydrofluoric acid, instantly coat themselves with bright copper, and evolve gas. According to Smee, iron does not decompose a neutral solution of acetate of copper, nor alkaline ones of the ammoniuret, ammonio-nitrate, or ammonio-sulphate; but decomposes a solution of oxide of copper in nitric acid. Raoult states, that gold in contact with copper, in either a cold or boiling, acid or neutral, solution of a salt of copper, receives no deposit of copper ('Journal of the Chemical Society,' vol. xi. p. 465).

Electrolysis of salts of copper.—I have electrolysed fluoride of copper (fused at a bright red heat in a deep copper vessel), by means of a platinum wire helix as the anode, and a copper wire helix as the cathode, and a current from six Smee's cells. The conduction was copious, as if the fused salt conducted like a metal, and an acid vapour was evolved. No copper was deposited, and the anode was unaltered; the cathode lost 3·35 grains in weight, by corrosion near the surface of the fused salt, and the copper vessel was similarly acted upon in several experiments, and caused to leak. It was evidently an instance of conduction by a liquid without electrolysis, as with fused argentic fluoride. A solution of fluoride of copper in pure dilute hydrofluoric acid, with copper electrodes, conducted well, and yielded a good deposit of copper, with a current from a single Smee's cell. Fluoride of copper is insoluble in anhydrous hydrofluoric acid, but dissolves quickly in aqueous ammonia.

If a feeble current be passed by means of copper electrodes, through a solution of chloride of copper dissolved in dilute hydrochloric acid, the anode becomes covered with snow-white crystals of cupreous chloride, and the cathode receives a thick deposit of loose, spongy copper ('Chemical News,' vol. xxii. p. 167). Gladstone and Tribe observed, that if a strip of platinum was connected with one of copper, and both were immersed in a solution of cupric chloride, the platinum became covered with a layer of insoluble cupreous chloride, also, that if such a solution was electrolysed by a feeble current with platinum electrodes, chlorine appeared at the anode, and cupreous chloride at the cathode ; but if the current was strong, metallic copper was also deposited upon the edges of the cathode. Smee states, that a solution of cupric chloride is less readily decomposed by an electric current than the nitrate, but more readily than the sulphate, and that it is one of the worst liquids for the reduction of copper, and the metal is apt to assume a very peculiar appearance. He also states, that the ammoniuret, acetate, and hyposulphite, offer no advantages ; that they are difficult to decompose, and require a current from several cells; also the ammonio-chloride is a bad solution, having a tendency to evolve hydrogen, and yield a spongy copper deposit ; that iodide of copper, dissolved in a solution of iodide of potassium, cannot be employed, because it liberates iodine ; that a copper anode is but little corroded in a solution of sulpho-cyanide of potassium, and the liquid does not contain much metal ; and the anode is very slightly acted upon in a solution of the potassio-tartrate. (Compare Elsner's statement, p. 209.)

Applications of electro-deposition of copper.—The purposes
to which the electro-deposition of copper has been applied,
are much more numerous than those of any other metal. It
is sometimes used for protecting iron and steel from oxidation, also to form a basis for silvering and gilding upon zinc,
iron, steel, tin, lead, Britannia-metal, &c. ; to the production
of medallions, busts, and even colossal statues, and many
other works of art; to the refining of crude copper on the
large scale, the separation of the metal from cupriferous
solutions ; to making copies of engraved, stereotype, and
daguerreotype plates ; to the coating of leaves, flowers, fruits,
insects, &c., with copper, for the purpose of ornamentation ;
to the arts of glyphography, galvanography, and electro-tint
printing ; corrosion of the anode by electrolysis, has also
been applied in voltaic etching.   Bank-notes ; postage
stamps ; playing cards ; maps of the Ordnance Survey ; and
the illustrated papers in which grocers and others wrap their
goods, are printed from electrotype copper plates.

Coating articles with copper by simple immersion process.—
A vast number of articles of a common kind, to which it
is desired to impart merely the appearance of copper, such
as steel pens, iron and steel wire, &c., are coated with a film
of copper, by simply immersing them in an acidulated and
dilute solution of the sulphate, washing them thoroughly, and
drying them quickly by rubbing them with hot sawdust. To
effect this object, mix together one measure of hydrochloric
acid, three of water, and a few drops of a solution of sulphate of copper ; clean and immerse the iron, wash it, rub
it with the cupreous liquid, and re-immerse it repeatedly,
adding a few drops of the copper solution occasionally.

According to O. Gaudain, articles of cast-iron, wrought
iron, and steel, may be coated with copper, by dipping them
into a melted mixture of chloride and fluoride of copper, with
five or six parts of cryolite, and a little chloride of barium,
contained in a plumbago crucible ('Journal of the Chemical Society,' vol. xi. p. 955). To coat brass with copper,

Dr. C. Puscher directs us to dissolve ten parts of sulphate of copper, and five of salammoniac, in 150 of water. Immerse the previously cleaned articles in this liquid for one minute, drain them, and then heat them over a charcoal fire until the ammoniacal salt is expelled, and the coating of copper appears perfect, then wash them with water, and dry them ('Chemical News,' vol. xxiii. p. 215).

M. Weis Kopp, deposits copper upon cast iron, by the simple immersion process, in a bath composed of ten parts of nitric acid, ten of chloride of copper, and eighty of hydrochloric acid of specific gravity 1·105. He immerses them several times, until a sufficient deposit is obtained, rubbing them with a woollen cloth between each immersion. The articles are first cleaned in a mixture of one part of nitric acid, and fifty of hydrochloric acid of specific gravity 1·105 ('Chemical News,' vol. xxi. p. 47). For coating iron wire with a film of copper by this method, Roseleur uses a mixture, composed of fifty to a hundred parts of water, one of sulphate of copper, and one of sulphuric acid; and, if the deposit is not sufficiently adherent, the copper is compressed by drawing the wire through a hole in a steel plate. He coats small articles, by burying them in sawdust wetted with this solution, diluted with three or four times its volume of water, and tossing the articles and sawdust about.

Separation of copper from cupriferous liquids.—Extremely large deposits of sulphide of iron, containing a greater or less percentage of cupric sulphide, exist in mineral strata. These sulphides, by exposure to air and water, become more or less oxidised into sulphates, and are rendered soluble. In this way, the water of certain mines becomes impregnated with sulphate of copper, and from very ancient times, copper has been separated in the metallic state from such waters, by immersing fragments of iron in them. Of late years, these immense deposits, (notably those of the Tharsis and Rio Tinto mines in Spain,) have been worked by scientific processes, and all their constituents utilised. The sulphur contained in them is burned and converted into oil of vitriol, and has become one of the chief sources of supply of that acid. The sulphide of iron, converted into oxide by that process, is next mixed with common salt, and the mixture roasted, to render the copper soluble, a quantity of hydrochloric acid being at the same time produced from the common salt, and collected. By washing the roasted mixture with the dilute hydrochloric acid, and with water, all the copper is extracted as a greenish-blue liquid. This liquid is run into large vats filled with scraps of iron, and kept hot by passing steam into it. In a short time all the copper is reduced in the form of feathery crystals upon the iron, and falls to the bottom of the liquid as a red powder ; which is separated from the iron, and refined in the usual manner. The washed oxide of iron is employed as a source of metallic iron, and for ' fettling' furnaces. In this way, great quantities of copper are yearly obtained, and many hundreds of thousands of tons of a substance, for which there previously existed no uses, are beneficially utilised. The process usually employed, is the one patented by Mr. Henderson.

Deposition of copper by contact with another metal (see also p. 82).—The electro-deposition of copper upon cast-iron fountains, lamp posts, &c. is very common in Paris. The process of M. Fred Wiels, is as follows :—Dissolve in 1,000 parts of water, 150 of sodio-potassic tartrate, eighty of soda-lime containing from 50 to 60 per cent of free soda, and thirty-five of sulphate of copper. Iron and steel, and the metals whose oxides are insoluble in alkalies, are not corroded in this solution. The iron or steel articles, are cleaned with dilute sulphuric acid, of specific gravity 1·014, by immersing them in that liquid from five to twenty minutes, then washed with water, and finally with water made alkaline by soda, then cleaned with the scratch-brush, again washed, and then immersed in the cupreous bath, in contact with a piece of zinc or lead, or suspended by means of zinc wires; the latter is the most economical way. The articles must not be in contact with each other. They thus receive a strongly-adherent coating of copper, which increases in thickness (within certain limits) with the duration of immersion. Pure tin does not become coppered, by contact with zinc in this solution'; it oxidises, and its oxide decomposes the solution, and precipitates red suboxide of copper, and by prolonged action, all the copper is thus removed from the liquid. The iron articles require to be immersed from three to seventy-two hours according to the colour, quality, and thickness of the required deposit. The copper solution is then run out of the vat, and the coated articles washed in water, then cleaned with a scratch-brush, washed, dried in hot saw-dust, and then in a stove. To keep the bath of uniform strength, the liquid is renewed from below, and flows away in a small stream at the top. After much use, the exhausted liquid is renewed, by precipitating the zinc by means of sulphide of sodium, (not in excess), and re-charging the solution with cupric sulphate. He also supplies to the bath, hydrated oxide of copper ('Chemical News,' vol. xiii. p. 1).

By simple cell process. Coppering cast-iron cylinders for calico printing.—M. Schlumberger's process. The cylinder (having been perfectly cleaned by the usual methods,) is made the cathode (with a current from four or six elements) during twenty-four hours, in a mixture of two liquids, composed of

	Parts			Parts
Water . . .	12	Water . . .		16
Cyanide of potassium	3	Sodic carbonate .	.	4
		,, sulphate . .	.	2
		Cupric ,, . .	.	1

It is then well washed, rubbed with pumice powder, then washed again with an aqueous solution of cupric sulphate, of specific gravity 1·161, containing $\frac{1}{300}$ part of its volume of sulphuric acid, scraps of copper being kept in the bath to supply the loss of copper, and prevent the liquid becoming too acid. It is immersed again in the above alkaline solution, or else in a mixture composed of two liquids, viz. :—

	Parts			Parts
Water	10	Water		16
Cyanide of potassium	3	Sodic carbonate	.	4
Aqueous ammonia .	3	„ sulphate . .	.	2
		Cupric acetate . .	.	2

In these mixtures, at a temperature of 15° to 18° C., it is surrounded by porous cells containing zinc rods and dilute sulphuric acid, and connected with the zinc by copper wires. The cylinders are turned partly round once a day, in order to render the deposit uniform, and the action is continued during three to four weeks, until the deposit is $\frac{1}{25}$th of an inch thick (G. Schaeffer, 'Chemical News,' vol. xxx. p. 219 ; 'Journal of the Chemical Society,' vol. xiii. p. 196).[1]

Deposition of copper by separate current process.—In depositing copper by the single cell method, a nearly saturated solution of sulphate of copper answers very well, but for the battery process, an excellent solution may be made, by dissolving four parts, by weight, of finely powdered sulphate of copper (best quality), and one of sulphuric acid, in about eighteen or twenty of water, and then filtering it ; neither of these solutions however, are fit to deposit copper upon steel, iron, or zinc, because the electrical relations are unsuitable ; these metals decompose such liquids rapidly, and deposit the copper upon themselves by simple immersion. Some persons use a solution containing a smaller proportion of acid, a greater one of copper salt, and less water ; and others add a small quantity of sulphate of zinc or sulphate of potassium to the liquid ; the latter is very good.

Deposition of copper upon metals.—The sulphate solution is used for coating all metals and alloys, such as brass and german-silver, which do not decompose that liquid ; but zinc, iron, steel, tin, lead, Britannia-metal, type-

---

[1] See also H. Wildes, patent No. 4515, Dec. 28, 1875.

metal, &c., which precipitate the copper from such a liquid by simple immersion, are coated in the cyanide or other alkaline solution; and as the deposition of copper from an alkaline liquid is more expensive than that from the sulphate, if a greater thickness of metal is required, the additional thickness is put on the articles in the sulphate solution.

Deposition of copper upon zinc, iron, &c.—Various solutions are employed for this purpose, but they are all alkaline ones, and mostly contain cyanide of copper dissolved in cyanide of potassium. A very good one may be formed thus :—Dissolve cyanide of copper to saturation in water containing about two pounds of cyanide of potassium to the gallon, and then add about four ounces more of the potassic salt per gallon, to form free cyanide ; the liquid is then ready, and should be used at a temperature of about 150° Fahr. Cyanide of copper is not very soluble in cyanide of potassium solution, the liquid formed, does not readily dissolve the anode, nor does it conduct well ; it also has a strong tendency to evolve hydrogen at the cathode, but this may be lessened, or wholly prevented, by avoiding the use of any free cyanide of potassium, employing a weaker current, and adding some aqueous ammonia and oxide of copper.

Watt recommends a solution, composed of one gallon of water, six ounces of cyanide of potassium, four of carbonate of potassium, two of liquid ammonia, and two of cupric sulphate. Dissolve the sulphate of copper in rain water, and, when cold, add the carbonate of potassium, and the ammonia. When the precipitate is re-dissolved, add the cyanide. Decant the clear liquid for use. Another, recommended by Roseleur, is, to reduce twenty parts of crystallised cupric acetate to powder, and rub it to a paste with a little water, add to it 200 parts of water containing twenty parts of dissolved washing-soda, and stir the mixture ; a light green precipitate is formed ; twenty parts of bisulphite of sodium are now dissolved in 200 parts of water, and the solution mixed with the former one ; the precipitate becomes dirty yellow. And, lastly, dissolve twenty parts of perfectly pure cyanide of potassium, in 600 of water, and add it to the previous mixture. If the solution is not quite colourless, add more cyanide until it is so. This liquid may be used either hot or cold, and requires a current of moderate strength. Another very good liquid, which may be employed either hot or cold, consists of 2,500 parts of water, fifty of potassic cyanide of 70 per cent., thirty-five of acetate of copper, thirty of bisulphite of sodium, and twenty of aqueous ammonia. The cyanide and bisulphite, are to be dissolved in one part of the water, and the ammonia and acetate of copper in the other, and the two solutions mixed together. If the blue solution of acetate of copper in aqueous ammonia, does not then become quite colourless, a little more cyanide must be added. If these liquids are used hot, the deposition is more rapid. If they become green, or blue, by working, it is from an excess of copper dissolved, and either the anode should be reduced in size, or some cyanide of potassium added. And if the anode acquires a brown or white insoluble coating, the liquid is deficient in copper, and some of the solution of acetate of copper in ammonia, must be added.

A good depositing solution, for coating iron and steel by the battery process, may be made, by dissolving ammoniuret of copper in a solution of cyanide of potassium. Nine hundred or 1,000 parts of water, containing eighty parts of cyanide of potassium, dissolve forty parts of the blue ammoniuret, and form a colourless liquid.

W. H. Walenn's solution for coppering iron, consists of cyanide of copper dissolved to saturation, in a solution of equal parts of potassic cyanide, and ammonium tartrate, and sufficient oxide of copper, and ammoniuret of copper added, to prevent evolution of hydrogen at the surface of the articles. The solution is used at 80° C. (=176° Fahr.). In this process, one Smee's cell may be employed. The cost has been found to be about 2*s*. 6*d*. per pound of metal deposited. One ounce and a half of copper per square foot will protect iron from rust ('Chemical News,' vol. xxi. p. 247; vol. xxii. p. 181).

Dr. Elsner coats base metals with copper, in a solution made as follows :—One part of bitartrate of potassium in powder, is boiled in ten parts of water, and as much freshly prepared, and wet hydrated carbonate of copper (which has been washed with *cold* water), stirred with it, as the liquid will dissolve. The beautiful dark blue alkaline liquid is filtered, and rendered still more alkaline by addition of a small quantity of carbonate of potassium. A copper anode dissolves readily in the liquid (compare Smee's statement, p. 201), and the solution may be employed to coat objects composed of tin, cast iron, and zinc (' Chemist,' vol. vii. p. 124).

Management of copper depositing liquids.—Sulphate of copper depositing solutions are the most easy of any to manage, because the range within which the density of the current may be varied, without causing a bad deposit, is greater than with those of any other metal, not even excepting antimony ; much however, depends with copper, as with other metals, upon the kind of liquid employed ; many cupreous solutions are difficult to manage, and none are as easy as the sulphate ; the alkaline ones, employed for coppering iron, etc., are much more difficult to obtain a thick deposit from, than the sulphate.

The general rules which determine and regulate the quality of other deposited metals, operate also with copper ; if the current is too great in relation to the amount of receiving surface, the metal is set free as a brown or nearly black metallic powder, and hydrogen gas may even be deposited with it and evolved. In the sulphate solution, if the liquid is too dense, streaks are apt to be formed upon the receiving surface, and the article (especially if a tall one) will receive a thick deposit at its lower part, and a thin one at the upper portion, or even have the deposit on the upper end re-dissolved. If there is too little water, crystals of sulphate of copper form upon the anode, and sometimes even upon the cathode, at its lower part, and also at the bottom of the vessel. If there is too much acid, the anode is corroded whilst the current is not passing. The presence of a trace of bisulphide of carbon in the sulphate solution will make the deposit brittle, and this continues for some time, although the solution is continually depositing copper ; in the presence of this substance, the anode becomes black, but if there is also a great excess of acid, it becomes extremely bright. Solutions of cupric sulphate, containing sulphate of potassium, and the bisulphide of carbon applied to them, are sometimes employed for depositing copper in a bright condition. The copper obtained from the usual double cyanide of copper and potassium solution, by a weak current, is of a dull aspect, but with a strong current it is bright.

Rapidity and cost of depositing copper.—If a good sulphate solution, and a strong current are employed, a thickness of ⅛ of an inch of firm copper can be deposited, either by the single cell, or the separate current process, in about seven to ten days. The cost of copper deposited by a current from a battery, is usually estimated at 2s. or 2s. 6d. per pound, but upon the large scale, by means of currents from magneto-electric machines driven by steam power, obtained under the most economical circumstances —as, for instance, by waste heat from other processes—it is probably less than 2d. per pound, exclusive of the value of the metal.

Composition of the dirt upon copper anodes.—Everyone who has deposited copper from the sulphate by electrolysis, has observed how black and dirty the anodes become, because he has been obliged to frequently wash them. Max, Duke of Leuchtenberg, analysed this black matter and found in it—

Tin .	33·50
Oxygen	24·82
Copper	9·24
Antimony	9·22
Arsenic	7·20
Silver	4·45
Sulphur	2·46
Nickel	2·26
Silica	1·90
Selenium	1·27
Gold	·98
Cobalt	·86
Vanadium	·64
Platinum	·44
Iron	·30
Lead	·15
	99·69

(See Erdmann's 'Journal of Practical Chemistry,' vol. xlv., 1848, pp. 460–468.)

The following analyses of similar residues have been kindly supplied to me by a friend :—

No. I.

Copper	85·850
Water and oxygen	4·950
Arsenic	2·480
Silver .	1·815
Sulphuric acid	1·150
Insoluble earthy matter	0·950
Antimony	·750
Iron .	·750
Bismuth	·650
Alumina	·250
Chlorine	·250
Gold .	·085
Lead .	·050
Loss .	·020
	100·000

Per ton of 20 cwt.

ozs. dwts. grs.
Silver = 623 2 8
Gold = 27 15 8

No. II.

Lead .	27·70
Water and oxygen	21·05
Copper	19·40
Antimony	7·35
Sulphur	6·35
Silver .	5·61
Arsenic	5·20
Earthy matter	4·35
Bismuth	1·25
Chlorine	·70
Iron .	·60
Nickel	·20
Organic matter	·20
Gold .	·01
Loss .	·03
	100·00

Per ton of 20 cwt.
Silver = 1835 ozs.
Gold = 3 ozs.

P 2

No. III.

Copper	67·90
Sulphur	18·10
Iron	5·55
Insoluble earthy matter	3·40
Organic matter	2·25
Lead	2·05
Silver	0·55
Loss . ·	·20
	100·00

Refining crude copper by means of electrolysis.—The fore-going analyses throw great light upon the question, why it is, that electro-deposited copper (and also that found in the metallic state in fissures in rocks at Lake Superior, and other places,[1]) is so very pure ; and also upon the process of re-fining crude copper by electrolysis. Several hundreds of tons of copper, are yearly refined by Mr. James Elkington's patent process ('Chemical News,' vol. xxi. p. 264). This simply consists in making large slabs of the crude metal (ob-tained from the ores by the usual melting process), the anodes in the ordinary sulphate of copper depositing liquid, and passing electric currents from numerous magneto-electric machines through them and the liquid ; a ' compound depo-siting vessel ' (see pp. 24, 90) and a series of electrodes being employed. The copper dissolves, and nearly all the impurities are separated, and fall to the bottom as a dirty powder. The impurities vary of course, with the composition of the crude metal, and the success of the process depends largely, upon the selection of such crude metal as contains no such substances, as would be electro-deposited along with the copper.·

In this process, the metalloids, such as oxygen, sul-phur, selenium, phosphorus (and perhaps arsenic), carbon,

---

[1] A mass of copper, estimated to weigh between 75 and 100 tons, was discovered at Eagle Harbour, Lake Superior.

boron, and silicon, are not liberated at the cathode. Any silver present, will not dissolve, because the sulphuric acid employed, contains a little hydrochloric acid, which converts it into insoluble chloride. Gold, if traces of it are present, is also insoluble. Lead is converted into sulphate, which is its most insoluble compound, and the small portion which does dissolve, is not deposited, because lead is electro-positive to copper, and therefore copper is deposited first. Carbon, together with any sulphides present, also falls to the bottom as an insoluble powder. Iron dissolves, but being highly electro-positive to copper in such a liquid, is not deposited, even when present in very great amount. Zinc is probably not present, but, if it is, behaves like iron, and is more highly electro-positive. Tin is probably not dissolved, and if it were, being positive to copper, would not be deposited. Cobalt behaves like iron, but the quantity present is very minute. Nickel dissolves, and if in very large quantity, would be deposited to a small extent along with the copper, but its proportion is also very small. Antimony (and perhaps also arsenic to a less extent) would be the most likely metal to contaminate the deposit.

Analytical estimation of copper in alloys and ores by means of electrolysis.—In consequence of the possibility, by attending to proper precautions, of depositing copper alone from a solution of its sulphate containing other metals, attempts have been made to separate the whole of the copper from those metals by such a process, and thus determine its amount, and a premium of 300 thalers, or 45*l.*, was offered by the directors of the Mansfield copper mines, in Germany, for a satisfactory method. The successful competitor was Mr. C. Luckow, chief chemist to the Cologne-Minden Railway Company. The process is extensively used, and a full description of it may be found in the 'Chemical News,' vol. xix., 1869, p. 221. (See also Watts's 'Chemical Dictionary,' 2nd supplement, p. 384.) J. M. Merrick has also electrolysed known weights of pure sulphate of copper in aqueous solution, in a covered platinum crucible, with a platinum wire for the anode, and the crucible for the cathode, with a current from two or three Grove's cells, until all the metal was deposited, and then weighed the deposits. In two experiments, the percentages of copper obtained, were 25·44, and 25·46, theory requiring 25·46 per cent. The deposits were washed with alcohol and cautiously dried ('Chemical News,' vol. xxiv. pp. 100–172).

Preventing adhesion of deposit.—In all cases where a perfect copy is required of a metal plate, or other metallic object, the surface to receive the deposit, must be sufficiently clean to enable the electricity to freely enter, but not so perfectly unstained as to cause the deposit to adhere so firmly that it cannot be separated. To prevent adhesion, the metal, after having been cleaned and dried, should remain exposed some time to the air before being put into the depositing liquid; it should also not be heated immediately before immersion; and if a wire has to be soldered to it, that should be done beforehand. It should also not be put into the solution without first making all the connections of the circuit complete, and attaching it to the battery, otherwise the momentary immersion may corrode and clean its surface. To make perfectly certain of preventing adhesion, without preventing a deposit, the dry surface of the metal may be brushed with a thick camel-hair brush with short hairs, and some fine black-lead, before immersion, in addition to taking the above precautions. In some cases, the articles to be copied, are rubbed with cotton-wool slightly moistened with a very weak solution of beeswax, prepared by dissolving a fragment of wax of the size of a pea in a quarter of a pint of spirit of wine. In other cases they are rubbed over with a little olive-oil, and then wiped as clean as possible by rubbing with cotton-wool. To prevent adhesion of deposits on copper, or on steel, vapour of iodine is also employed.

Copying engraved copper-plates, medallions, &c.—En-

graved steel plates, are copied by coating ('stopping off') the back and edges with copal varnish, allowing the varnish to become perfectly dry; immersing the plate in the cyanide coppering liquid, and depositing a thin film of metal upon it, then washing it well, and at once suspending it in the sulphate of copper solution, and depositing the desired thickness; this will require from twenty-four to forty-eight hours. The surface of the steel should be previously prepared for a non-adhesive deposit, otherwise the two metals cannot be separated. If the plate to be copied, is a copper, brass, silver, or gold one, it may receive the entire deposit in the sulphate solution. Medallions and other forms of metal, may also be copied in a similar manner. As the metal in all such cases, has a great tendency to be deposited upon the edges, the deposit creeps round those parts, and this superfluous copper has to be filed off before the two can be separated. When they are taken apart, their surfaces are found to be perfectly dry.

Copying Daguerreotype pictures in copper.—First solder a wire to the back of the plate near the edge; varnish the back and edges, and allow it to dry; hang it (taking the above precautions) in a clean sulphate of copper solution, which is perfectly free from dust or grease on its surface, and in the course of twenty or thirty hours (if about two pairs of Smee's batteries have been used), the deposit will be sufficiently thick to be removed; it should then be taken out, well washed, wiped perfectly dry, and a narrow strip cut off its edges with a strong pair of scissors or shears; the two may then be easily separated by inserting the blade of a knife, or the end of a thin wedge of hard wood, between them at the edges. If the process has been carefully managed, and the original picture is a strong one, a most beautiful and vivid copy will be obtained; and if the picture is not only a strong one, but has been fixed by Fizeau's process, a number of successive copies may be taken from it, but their intensity, as well as that of the original, diminishes in each succeeding trial. With a vivid original picture, clear solution, very regular and undisturbed action of the battery, and a fine deposit, there may be observed a most strange phenomenon; it will be found that the picture has not entirely disappeared, even in twenty hours, although the coating of copper has constantly increased in thickness; the image has penetrated quite through the deposited metal, and appeared upon the back, even with deposits as thick as an address card. In some cases the figure is optically positive, and in others negative.

Coppering cloth.—Mr. J. Schottlaender took out a patent, Dec. 8, 1843, for depositing either plain or figured copper upon felted fabrics. He passes the cloth under either a plain or engraved copper roller, horizontally immersed in a sulphate of copper solution, (not containing much free acid,) and a deposit takes place upon the roller as it slowly revolves; the meshes of the cloth are thus filled with metal, and the design of the roller copied upon it; the coppered cloth is slowly rolled off, and passed through a second and closely contiguous vessel filled with clean water. The roller is previously prepared for a non-adhesive deposit.

Deposition of copper upon non-metallic surfaces.—A sufficiently large conducting wire of copper is first attached to the object, and, if necessary, a number of ' guiding wires ' formed of very fine brass wire, are attached to the main wire, and their free ends stuck into the surface of the article, in those parts only where deposition is the most difficult to effect, such as in recesses or deeply undercut parts of the mould, which are the most distant from the anode, or into which exhausted portions of the solution, would ascend and collect. All light moulds and articles, require to be weighted, in order to make them sink in the solution ; lead is usually employed for this purpose.

To deposit copper (or other metal) upon non-conductors, their surfaces must be rendered conductive. To effect this object, there are two methods in use: first, to cover them with a thin film of black-lead, or the finest quality of metallic bronze-powder, by brushing; or, second, to coat them with a minute film of gold or silver, by chemical means. The first of these methods is generally used for moulds composed of gutta-percha, wax, resinous composition, or plaster saturated with oil, where the parts are not much undercut; the second for elastic moulds, because they will not bear the friction of black-leading, and because the black-lead cannot be readily applied to all their recesses.

In employing black-lead, care should be taken to select a kind which conducts electricity freely; and this can only be found by actual trial. Some specimens conduct very badly, and others very well. It should be applied to the surface, by persistent brushing with a camel-hair brush, having a large and thick body of short hairs; breathing upon the face of the article occasionally, to facilitate the adhesion of the black-lead, and when the surface is perfectly black and bright, blowing off the superfluous powder. The whole operation occupies about ten or fifteen minutes with a small object the first time of preparing it, but less in subsequent operations with the same surface. A small quantity of black-lead is sufficient for a very large surface. The conducting power of black-lead is greatly improved by gilding or silvering it. It may be gilded as follows :—Dissolve one part of chloride of gold in 100 parts of sulphuric ether in a bottle, add fifty parts of the plumbago, mix them thoroughly, and expose the mixture in the open bottle to sunlight, stirring it frequently until it is perfectly dry; apply it by brushing.

A plan which I have devised, and which is cheaper than this, is to mix with the black-lead, one-third of its weight of the finest white bronze-powder. The particles of this powder, being composed almost wholly of tin, when immersed in a solution of sulphate of copper, dissolve, and coat themselves with copper by the simple immersion process, and also cause those of black-lead in contact with them to become coated, and thus a thin deposit of copper is very quickly formed all over the bronzed surface. This effect will of course take place without connecting the mould with the battery, but they may be immediately connected together, and a deposit will spread over the whole of the bronzed surface by the ordinary battery process, through the medium of the bronze and the thin deposit already mentioned, and it may be continued to any required thickness in the usual way. By this plan, gutta-percha medallions were repeatedly covered with a deposit of copper, in from two to five minutes, which would occupy from twenty to forty-five minutes when prepared by black-lead in the usual manner. The addition of white or tin bronze, causes the deposit to spread as rapidly as when the surface is prepared by the phosphorus solution, but without the disadvantage which occurs in using the latter, of making the deposited metal brittle. Silver may be deposited nearly as easily as copper upon black-leaded surfaces, but it must be remembered that wax moulds are injured in a cyanide of potassium solution, and should be protected by a layer of copal varnish on the parts not to be coated.

For moulds of elastic composition, the depositor may employ the following liquids, patented by Mr. Alexander Parkes : A, the phosphorus solution : to make nearly three ounces of this, melt sixty-four grains of beeswax or tallow ; then dissolve eight grains of india-rubber (cut up very small), in 160 grains of bisulphide of carbon, and when it is dissolved, add to it very carefully (as it is highly inflammable) the melted wax, and stir the mixture thoroughly ; then dissolve sixty-four grains of phosphorus, in 960 grains (about two and a quarter ounces) of bisulphide of carbon, and add to it eighty grains of spirit of turpentine, and sixty-four grains of asphalte in fine powder ; when they are dissolved, add this solution to the previous one of india-rubber and wax, and thoroughly mix them by shaking. B, the silver solution : dissolve thirty grains of nitrate of silver in a pint (twenty ounces) of distilled water. And, C, the gold solution—to make twenty ounces of which, dissolve five or six grains of pure gold, in about twenty or twenty-five grains of a hot mixture, of one measure of nitric acid, and two or three of hydrochloric acid, and when dissolved, dilute the solution with twenty ounces of distilled water.

In making surfaces conductive by this plan, the article, after the conducting and guiding wires are attached to it, is either dipped into the phosphorus solution, or its surface is covered with that liquid; and after it has been drained, it is allowed to remain until perfectly dry; the silver solution is next applied to it, and is drained away in like manner for several minutes, until the surface acquires a metallic lustre like black china; it is then gently rinsed with distilled water, and the gold solution applied in the same way, which gives it a yellowish aspect: after another rinsing in distilled water, it is ready for receiving a deposit.

The same patentee, includes in his patent a phosphorus moulding composition, by the use of which the immersion in the phosphorus liquid, is dispensed with, the moulds themselves containing the required amount of phosphorus. To make about one pound of this composition, melt together half a pound each of wax and deer's fat, then dissolve nineteen or twenty grains of phosphorus in about 300 grains of bisulphide of carbon; keep the wax mixture barely melted, and add the phosphorus solution slowly to it, with brisk stirring of the fat, pouring it in at the bottom of the melted mixture by means of a vessel with a long spout, to prevent its inflaming. It is highly dangerous to have spilt portions of the phosphorus composition or solution in contact with wood, paper, rags, etc., or other fibrous or porous substances, as after a lapse of some time (even hours), they will often burst into flame.

Another method of rendering the surface of the article conductive, is to wet it with a solution of nitrate of silver, and then expose it to sulphuretted hydrogen gas; this converts the film of silver salt into a conducting substance, viz., sulphide of silver; the liberated nitric acid should then be removed, by dipping the article in distilled water. Or, wet it either with nitrate of silver or chloride of gold solution, and then expose it to hydrogen gas; this reduces the salts to metals. The article must then be rinsed. The film of silver salt may also be reduced to metal, by placing the object in a well closed box, containing at its lower part a porcelain dish containing a small quantity of a strong solution of phosphorus in bisulphide of carbon. In a few hours the vapour will reduce the salt to a film of black silver.

Non-conducting surfaces may also be rendered conductive by washing them, first, with a mixture composed of equal parts of white of egg, and a saturated solution of common salt; second, with a strong solution of argentic nitrate, and exposing them to sunlight until they are quite black; and, third, with a saturated solution of green vitriol. (R. Piffard, 'Chem. News,' vol. ii. p. 323.) Or by coating the surface with a film of gum-water, and drying; then with a solution of the nitrate, and drying; and then exposing it to sulphuretted hydrogen gas. (R. Piffard, 'Chem. News,' vol. iii. p. 110.)

Hockin recommended, for metallising the surfaces of non-metallic bodies, to plunge them into iodised collodion, then immerse them in a nitrate of silver solution, expose them to the light for a few seconds, and then precipitate the silver in a metallic state by means of a bath of protosulphate of iron acidulated with nitric acid, and finally deposit copper upon them in a nearly neutral solution of cupric sulphate. ('The Chemist,' New Series, vol. i. part 4, January 1854, p. 196.) Liquids used for dyeing hair black, composed of a solution of ammonio-nitrate of silver, followed by one of pyrogallic acid, might be similarly employed.

In preparing a gypsum mould, Professor Heeren soaks it in wax, then covers it thickly with a mixture of a solution of one gramme of argentine nitrate, dissolved in two grammes of water, to which two and a half grammes of aqueous ammonia is next added; and then also three grammes of absolute alcohol. The mould is then exposed to sulphuretted hydrogen gas. By employing four or five Daniell's cells, the copper spreads quickly. ('Journal of Chemical Society,' vol. x. p. 1133.)

Berland prepares non-conducting surfaces for receiving a deposit thus :—Wet the article with spirit of wine, wash it with distilled water, and whilst wet, pour over it a solution of one part of nitrate of silver in four parts of distilled water. After draining it a few minutes, a solution of one part of pure green vitriol in three parts of distilled water is poured upon it. After five minutes, repeat with the silver solution, and then with the green vitriol, three or four times, till the surface of reduced silver has a whitish-grey colour. Then wash it with pure water, and it is ready to receive the deposit. At the first moment of immersion, the entire surface is covered with a thin layer of copper ('Philosophical Magazine,' fourth series, vol. xxx. p. 451. See also Alexander Jones's patent, 1841).

Coppering lamp-posts, &c.—M. Oudry electro-deposits copper upon gas-lamps, pillars, candelabras, fountains, and ornamental ironwork generally. He first coats the articles with a kind of red paint containing benzine, then black-leads the dried paint, and deposits copper upon it to the thickness of one millimetre, during four and a half days, by the battery process. The copper is afterwards bronzed by applying a solution of ammonio-acetate of copper.

Coppering fruit, flowers, insects, reptiles, &c.—Objects of this kind, some of which will scarcely bear handling, are first coated with silver by means of a saturated solution of argentic nitrate in hot alcohol. The nitrate is reduced to fine powder in a mortar, and an excess of it digested with alcohol in a flask placed in warm water, with occasional shaking, until the liquid is saturated. One hundred parts of alcohol, dissolve about two and a quarter parts of the salt. The articles are then dipped for a moment in the warm solution, and the liquid, being volatile, soon evaporates. The film of salt left upon the objects, is reduced to metal by either of the means already described; the articles then coated in the solution of cupric sulphate, and either silvered or gilded as may be desired.

E. T. Noualhier and J. B. Prevost, in their patent of January 1, 1857, propose to 'metallise soft surfaces—a human corpse, for instance—by the following process :—All the apertures are stopped with modeller's wax, the body is placed in a suitable attitude, and pulverised nitrate of silver spread over it by means of a brush or otherwise ; it is then electro-coppered in a bath of sulphate of copper ;' the 'result being a metallic mummy.'

Coating plaster models and clay figures with copper.—Busts, and other similar objects, may be coated by saturating them with linseed-oil (or better, with beeswax), then well black-leading, or treating them with the phosphorus, silver, and gold solutions, attaching a number of 'guiding wires,' connected with all the most hollow and distant parts, and then immersing them in the sulphate of copper solution, and causing just sufficient copper to be deposited upon them by the battery process to protect them, but not to obliterate the fine lines or features.

Copying wood engravings in copper.—This process is largely used. In cases where a great number of impressions of a particular woodcut is required, the plan of taking copies of the engraved wooden block in copper by the electro process, and using those copies instead of the original block to print from, has attained a considerable degree of importance ; the vignette at the head of the title-page of the 'Illustrated News,' the title-page of 'Punch,' many of the large engravings in the 'Illustrated News,' and even the illustrations of some of the penny periodicals, are regularly produced in this way. To copy an engraved wooden block, the surface is first either black-leaded, or moistened with water, and firmly surrounded by a shallow frame of metal; a thick piece of gutta-percha, more than sufficient to fill the enclosed space, and made quite soft by heat, is then laid upon it, commencing its contact at the centre of the engraving, and proceeding outwards, so as to exclude all air-bubbles ; a plate of cold iron is then laid upon the gutta-percha, and the whole subjected to gradually increasing pressure as the substance cools. The block and its copy are then separated, and the figured surface of the copy (with the main connecting wire previously attached) is treated in the usual manner, with black-lead (or with the phosphorus, silver, and gold solutions); 'guiding-wires' are then affixed, and copper deposited upon the mould in a solution of sulphate of copper, until a moderate thickness of deposit is obtained, which will occupy at least twelve or eighteen hours; when sufficiently thick, the deposit is removed, its back made rigid by a layer of solder or type metal (the surface being previously moistened with a solution of chloride of zinc, to make the solder adhere), the back is planed flat, and mounted upon a block of wood to the height of the type. In London this process is employed upon a large scale, some of the copies being upwards of two feet square. Engravings upon steel are copied in an exactly similar manner. In some instances, successful deposits of large 'Illustrated News' engravings, have been formed and taken off in eight hours; but this can only have been effected by the most perfect black-leading, keeping the solution in excellent condition, and working it with the maximum of battery power. A mould of electrotype copper of the 'Times' newspaper is said to have furnished as many as twenty millions of impressions before it was quite worn out.

Copying set-up type in copper.—The process of electrotyping has been gradually encroaching upon that of stereotyping, and has, we are informed, almost superseded that process in America. The plan adopted, is similar to that of copying woodcuts, viz., to lay a sheet of softened gutta-percha upon the surface of the page of type (which is previously black-leaded), and subjecting it to increasing pressure until it is cold; the gutta-percha copy is then removed, and treated as in copying wood engravings. The advantages of electrotyping over stereotyping are numerous: the metal is harder, takes a sharper impression of the mould, and delivers the ink much more rapidly than type metal, besides being a cleaner process ; it also takes up less ink, and consequently the printed pages dry more quickly.   Both woodcuts and letter-press, have also been copied in plaster of Paris, and the deposit of copper formed upon that ; but this material is much inferior to gutta-percha for the purpose.   In deposit-ing copper upon moulded surfaces of set-up type, the deposit is thin, and easily broken, where there are lines ; to prevent this, the lines are wetted with a solution of nitrate of mercury, or other ' quicking' liquid (see pp. 195, 323), at those parts, and then deposited upon again.

Moulding, and copying coins, &c. — Some electro-de-positors confine themselves to multiplying printing surfaces, some to plating with nickel, others to plating with silver and gilding, to which latter process in other establishments, is added the multiplication of works of art, the production of busts, statues, &c.   The electro-depositor, therefore, who includes in his business the multiplication of works of art, as well as the simple plating of metal articles, will require a knowledge of the art of moulding.

Ability to reproduce works of art on a large scale, requires very considerable experience, and amateurs should first acquire ability to copy smaller ones, such as medals, coins, etc.   The moulding materials commonly used for small objects, are fusible alloy, wax, stearine, gutta-percha, plaster of Paris ; a mixture of gutta-percha and marine glue, a composition of spermaceti, etc., etc.

Fusible alloy, consists of a melted mixture of eight parts bismuth, five of lead, and three of tin, and fuses at about the temperature of boiling water.   A much more fusible mixture, which melts at 151° Fahr., consists of seven and a half parts of bismuth, four of lead, two of tin, and one and a half of cadmium.   The ingredients should be thoroughly mixed in each case.   The melted alloy should be poured upon a slab of stone, its surface skimmed by means of a card ; and the medal or coin to be copied dropped upon it. As soon as the alloy has solidified, the coin may be removed ; the end of a clean copper wire attached to the mould by means of heat, the back of the alloy varnished, and the copy hung in the solution of sulphate of copper to receive the deposit.

To copy a medal in wax, the medal (slightly oiled) should be surrounded by a rim of stiff paper about one inch deep, fastened by means of sealing-wax; then made quite warm, and the white wax in a melted state (but not too hot), poured upon it. When the wax has become solid, put it in a cold place for several hours, and then separate the coin and its copy; but if the medal be a large one, the cooling process must be gradual, otherwise the wax may split.

A good composition for copying coins, consists of two parts of gutta-percha, and one of Jeffery's marine glue. The two substances are cut up very small, heated very gradually, with constant stirring, until most thoroughly mixed. To copy both sides of a medal in this mixture, take a strip of thin sheet copper, brass, or tinned iron, about an inch wide, wind it closely round the edge of the medal, and solder its ends together ; wipe the medal, and take two balls of the composition, quite hot and soft, and press them simultaneously against the two faces of the medal, working the material from the centre towards the circumference, to exclude bubbles of air ; place two thick plates of cold metal, one on each side, and gradually screw up the whole in a vice or press, gently at first, but increasing the pressure to a high degree as the materials become hard. When it is quite cold, which will be in about two hours, the two copies may be easily removed from the original, by inserting the ends of gimlets in their backs and drawing them out ; they are easily removed, because the composition slightly contracts in cooling. They will present fine impressions of the original, and be perfectly free from air-bubbles, if the operation has been carefully performed. A slight disadvantage attending the use of gutta-percha

Q

(and mixtures containing it) is, that it shrinks a little, in course of a long period, but unless the surface is a large one, this defect is too small to be perceived

All these mixtures require, of course, to have suitable conducting wires attached to them, and their surfaces black-leaded, or otherwise prepared, to render them conductive ; and those parts to which the conducting material has accidentally adhered, and which are not to be coated, must be ' stopped off' by means of a suitable varnish. Quick-drying spirit varnish is very suitable for the purpose, especially if some superfine red sealing-wax is dissolved in it, to render the coating more visible. It must also not be forgotten, that in all cases, the copy taken of an original object is not a fac-simile of the object, but its reverse, and that to obtain a fac-simile we must take a copy of the copy. For instance, if we take a mould of a coin in copper, either by means of electrolysis or in any other way, we must take a copy from this mould in order to obtain a real fac-simile.

Copies of coins may also be taken in plaster of Paris. To copy a coin or medal in plaster it should be slightly oiled, and surrounded by a paper rim one inch in depth. Take the finest and freshest plaster (which has been kept in a well-closed bottle), mix it with water in a lipped vessel, to the consistency of treacle, then without delay brush a little of it over the surface of the coin with a camel-hair brush, and at once pour on the remainder. The plaster quickly sets to a solid state, and soon afterwards may be removed from the medal. The mould may be either itself copied in wax, etc., or be thoroughly dried, and then saturated with wax or tallow, by standing whilst still hot in a shallow layer of the melted substance, until the latter has spread throughout its mass, and then at once remove it, and prepare its surface for receiving a deposit. Or it may be prepared for black-leading by saturating it with skimmed milk, and then drying it.

Copying Busts, Statuettes, Statues, &c.—There are, however, many objects which cannot be copied by any fo the

methods above described, such as medals which are 'undercut'; busts, statues, and figures of various kinds, because
the mould formed upon the object cannot be removed without breaking either the original or its copy. In such cases
either the mould or its copy, or both, are formed in pieces,
so arranged that each piece may be removed ; or the copy
of the object is taken in an elastic moulding material, which
by allowing itself to be stretched, may be removed from
overhanging, projecting, and undercut parts of the object,
and then returns to its original form and dimensions. The
best substance of this kind, and almost the only one used, is
composed of four parts of the best thin glue and one of
treacle; the glue is broken into small pieces, and soaked
several hours, or until it is quite soft, in sufficient cold
water to cover it. The superfluous water is then thrown
away, and the gelatine together with the treacle, is heated in
a glue-pot (i.e. by immersing the vessel in boiling water), to
nearly 100° C., and stirred until the two substances are
thoroughly mixed. The use of the treacle is to prevent the
mould drying and shrinking. Some operators add half an
ounce of beeswax for each pound of glue.

The great disadvantage of such moulds is their tendency to absorb water, to swell, and to be partly dissolved in
the solution of sulphate of copper. These difficulties are overcome, by using a depositing liquid containing the minimum
proportion of water, and covering the mould as quickly as
possible with the metallic deposit, any portions of it not
requiring a deposit being previously well coated with a
quickly-drying varnish, best, a solution of india-rubber in
bisulphide of carbon. Various attempts have been made to
enable the gelatine to resist more perfectly the action of the
water, one of the most effectual of which is to dissolve in
the mixture, two parts of tannic acid for each 100 parts of
the dry glue, or to immerse the mould a few seconds in a
solution of ninety parts water and ten of bichromate of
potassium, and then expose it to the sun.

If the object to be copied is a medallion with undercut parts, it is treated thus :—First well oil the medallion ; then encircle its edge by a strip of stout paper, and pour the mixture (quite hot, and of the consistency of treacle), upon its surface, to the depth of half an inch or more, according to the size of the medal, and the depth of its hollow parts, brushing its surface beneath the liquid with a brush having fine and long hairs, to remove air-bubbles. Allow the mixture to remain until it is quite firm, which will be from two to twenty-four hours, according to its bulk ; take off the paper, and remove the mould very gently, carefully stretching and drawing it at the same time in the direction of the overhanging parts, to prevent injury.

Should the object to be copied be a hollow metallic bust, proceed as follows :—First oil it, then partly fill it with sand, to make it heavy, and thus prevent its rising in the liquid, and cover its opening by fixing a piece of millboard strongly over it ; then place the bust in the centre of a circular and *taper* vessel, a few inches deeper and wider than itself, and pour the melted composition in steadily, until it is a few inches above the top of the head, tapping the bust, and inclining the outer vessel, to facilitate the escape of air-bubbles. The composition will become firm in about twenty hours, and may be easily removed from the vessel by shaking, if the latter has been previously well oiled ; the mould may then be extracted from the bust, by previously marking on its lower end the position of the face, passing a knife carefully up the back of the bust nearly to the crown of the head, and opening the elastic mould with the hands whilst a second person lifts out the bust. If the original bust is composed of plaster, it must be previously saturated with oil, to prevent the melted composition adhering to it. In all cases, after fixing the necessary conducting and guiding wires to an elastic mould, it is rendered conducting by means of the silver or gold solutions, reduced to metal by means of phosphorus or hydrogen, &c.

Notwithstanding the greatest possible care having been taken in making and copying an elastic mould, failures are not infrequent; either the coating of deposited metal is imperfect in the inmost parts of the mould, or the latter swells so greatly as to alter the figure of the object. Some objects are first copied in elastic composition, then the elastic mould re-copied in the phosphorus and wax mixture at the lowest possible temperature, so as not to melt the gelatine; the mould removed, and the other deposited upon.

Bubbles of air often adhere to moulds immersed in depositing solutions; they may be prevented by previously dipping the object in spirit of wine; or be removed by the aid of a soft brush, or by directing a powerful upward current of the liquid against them by means of a vulcanized india-rubber bladder, with a long and curved glass tube attached to it; but the liquid should be free from sediment.

With large objects, such as statues, a different plan from any of those already described is resorted to; in this case, instead of employing elastic moulds, the copy itself is sacrificed. The original figure, formed of plaster of Paris, and obtained from a modeller or sculptor, is saturated all over its surface with boiled linseed oil. It is then coated with extreme care in all parts with a shining film of black-lead, by prolonged brushing, or with a film of silver by means of the phosphorus and silver solutions; but the presence of phosphorus is apt to make the deposited copper brittle. It is then immersed in a large cistern of the sulphate of copper solution (see p. 206), and coated entirely with copper to a thickness of about $\frac{1}{18}$th of an inch, or sufficiently to retain its form when the inner figure is removed. It is now lifted out of the vat, washed, the copper cut through at suitable places, the plaster figure broken away with great care, and the whole of it extracted. The outer surfaces of the copper forms, (with wires attached) are now thoroughly varnished all over, to prevent any deposit being formed thereon; the forms exposed to sulphuretted hydrogen, or dipped into a weak solution of sulphide of potassium, to prevent adhesion of the deposit ; the parts immersed in the depositing vat again, and filled with copper solution. A dissolving plate of pure electrotype copper is suspended within each portion, and a deposit of copper thus formed all over its interior, until a considerable thickness, varying from ⅛ to ⅓ of an inch, is deposited, which requires a period of three or four weeks. Each piece is now removed from the liquid, washed, and the outer shell torn off, when all the parts of the figure remain nearly complete and ready for fixing together. Some of the objects made by this process by Messrs. Elkington are colossal ; that of the Earl of Eglinton is 13½ feet high, and weighs two tons ; and the vat in which it was formed, is 15 feet long, 9 feet deep, and 8 feet wide, and is capable of containing 6,680 gallons of liquid. Messrs Christople of Paris, made a statue of 9 metres (=29 feet 6 in.) high, weighing 3,500 kilogrammes (=about 3 tons 9 cwt.), and of a thickness of 4½ millimetres. It occupied about ten weeks in depositing. As it is difficult in practice to deposit the figures of a man, horse, &c. all in one piece, this plan of dividing the first copper figure in suitable parts, usually at the lower edge of the vest, the shoulders and wrists, and depositing upon the interiors of these, and fastening the separate deposits together to form the complete figure, is nearly always adopted.

Lenoir employs a different process, which may be briefly stated as follows :—An external copy of the figure is made of gutta-percha in several parts, so as to be capable of being put together and form the complete figure ; and the internal surfaces of these pieces are black-leaded. An outline figure of the object, but of somewhat smaller dimensions, is formed of platinum wire, to act as an anode, and the pieces of gutta-percha are fixed together to form the complete figure around it. The mould is placed in a vertical position, the platinum outline figure being suspended in it by means of its connecting wires, and prevented from touching the mould by partly covering the wire with a spiral of india-rubber thread. The mould and its anode (previously weighted) are now immersed in the same vertical position in the copper solution, the battery connected, and a current of the liquid caused to continually enter the mould by a hole at the top of the head and escape by two holes at the feet. After a sufficient deposit is formed, the flexible anode of wire is drawn out through the hole in the head, the parts of the gutta-percha mould are taken asunder, and the seams in the copper at the junctions of the model are then removed by filing, &c.[1]

Glyphography.—The process of glyphography consists in coating a plate of copper with two thin layers of engraver's wax composition, the first one white, and the second black, engraving the design through the wax to the copper beneath, then black-leading the entire surface of the wax, varnishing the back of the copper plate, and depositing copper upon the entire front surface until a stout plate of metal is formed. The plate is then removed, strengthened with solder, mounted like a stereotype plate, and employed for printing in the usual manner. Before black-leading, it is sometimes necessary to thicken the wax coating over large white spaces, the middle portions of which might otherwise print black.

Etching copper plates.—In etching a copper plate by galvanism, we first solder a wire to it, then varnish the back, and cover the front with a thin layer of engraver's etching-ground ; draw the design upon the front surface with an etching needle, cutting through this material to the clean surface of the copper. Having completed the etching, hang the plate as an anode in the ordinary sulphate of copper solution, opposite a suitable cathode of copper. The current of electricity in passing out of the engraved lines into the liquid, causes the copper in them to dissolve, and thus etches the design on the plate. The various gradations of light and shade are produced by suspending cathodes of different forms and sizes opposite the plate to

[1] M. Planté employs, instead of the platinum wire outline, a thin hollow anode of lead pierced with holes.

be etched, in varied positions, and at different distances from
it, thus causing the plate to be corroded to unequal depths
in different parts, the deepest action being always at those
portions of the electrodes which are nearest together.

Depositing copper upon glass, &c.—The only effectual
way of obtaining an adhesive deposit upon glass or porcelain, is to send the article to a glass and porcelain gilder,
and have gold burnt into its surface, and then depositing
upon the gold coating in the usual manner.

16. **Nickel.**—Elec. chem. eqt. $= \frac{59}{2} = 29\cdot5$. The commonest salts of nickel, are the oxide, nitrate, chloride, carbonate, and sulphate. The oxide is a black powder,
soluble in nitric, hydrochloric, and sulphuric acids; the
nitrate and chloride are green salts freely soluble in water;
the carbonate is a pale green powder readily soluble in most
acids; the sulphate is a freely soluble salt. The oxide may
be made by heating the carbonate or nitrate to redness, or
by precipitating a solution of a salt of nickel with caustic
potash or soda. The nitrate may be made by digesting
the metal, its oxide or carbonate, in dilute nitric acid, and
evaporating the solution; the chloride may be formed with
hydrochloric acid in a similar manner, or by adding an
excess of that acid to a solution of the nitrate, and evaporating the mixture to dryness. The carbonate may be formed
by adding a solution of carbonate of sodium to one of any
salt of nickel, and washing and drying the precipitate. The
sulphate may be made by digesting either the oxide, nitrate,
chloride, or carbonate in an excess of dilute sulphuric acid,
and evaporating the solution nearly to dryness. A solution
of the nitrate, chloride, or sulphate, may also be obtained by
making a bar of nickel (or fragments of nickel, suspended
on platinum wire gauze) the anode in dilute nitric, hydrochloric, or sulphuric acid, and passing the current until the
acid is sufficiently saturated with metal.

Electrolysis of salts of nickel.—I have electrolysed dilute hydrochloric acid by means of one Smee's element, an anode of nickel and a cathode of copper. The conduction was very feeble, and a film of iron grey metal was deposited upon the cathode in twelve hours. A dilute solution of nitrate of nickel did not yield its metal freely. By making a strong solution of salammoniac, or of sulphate of ammonium, and passing a strong current through it, by means of an anode of nickel during several hours, until the liquid acquired a pale greenish blue colour, I obtained a deposit of coherent white metal. A good solution for depositing nickel, is the double cyanide of nickel and potassium, to which some common salt has been added.

Nickel has also been deposited from a liquid formed by precipitating a solution of nitrate of nickel with carbonate or cyanide of potassium, washing the precipitate and dissolving it nearly to saturation in a solution of potassic cyanide, and operating upon the liquid by the battery process with an anode of nickel. The metal deposited from this solution is said to be nearly equal in whiteness to silver. According to Smee, a solution of chloride of nickel is an excellent one for deposition, because of its small tendency to evolve hydrogen at the cathode. The nickel deposited from it 'has a peculiar white brilliant lustre, looking almost like glass ; this deposit is so very beautiful, though brittle when removed from the negative pole, that its examination would amply repay any person taking the trouble to precipitate it.' He also states that a solution of the acetate yields a black powder deposit, and is a bad one for obtaining reguline metal.

Merrick has electrolysed a number of solutions of salts of nickel by means of a current from two Grove's cells, an anode of nickel and a cathode of platinum, placing a voltameter for mixed gases in the circuit to measure the strength of the current, and a rheostat to keep the current uniform. He weighed the metallic deposits. The nitrate yielded a thick greenish non-metallic deposit (probably a basic nitrate), with a metallic layer beneath. A solution of sp. gr. 1·0503 of the pure chloride, yielded an adherent deposit of metal with a non-adherent layer of black powder upon it. The quantity of metal obtained equalled 83·6 per cent of the theoretical amount. With a solution of commercial sulphate of nickel, there was a great evolution of gas from the cathode, and much from the anode; and two layers of deposit, the outer one greenish and non-adherent, the lower one speckled and blotched metal. A solution of sp. gr. 1·0223 of the pure sulphate, gave a blackish deposit of metal, equalling 52 per cent of the theoretical amount, streaked with a non-adherent greenish deposit of subsulphate above it. With a solution of sp. gr. 1·0232 of the acetate, much of the deposit was black oxide in powder. The coherent metallic layer beneath, equalled 10 per cent of the theoretical amount. A solution of the cyanide of nickel and potassium, evolved much gas at the cathode, and gave a dull blackish-grey metallic deposit equal to 14 per cent of the theoretical quantity. One of the ammonio-nitrate, of sp. gr. 1·016, yielded a variously coloured metallic deposit amounting to 97·4 per cent of the theoretical amount, beneath a layer of greenish subsalt. The double chloride of nickel and ammonium, gave a pulverulent deposit, with a metallic one beneath, equal to 47 per cent of the theoretical quantity. The ammonio-chloride yielded 96 per cent of the required quantity of metal, partly bright, and partly dull. The double sulphate of nickel and ammonium, gave a good metallic deposit, equalling 93·5 per cent of the theoretical amount. The ammonio-sulphate gave 96 per cent in the form of greyish-brown metal; the solution was formed by adding an excess of ammonia, and then alcohol, to a strong solution of nickel sulphate, and dissolving the precipitated salt in water. A solution of the pure sulphate of nickel and potassium, evolved much gas from the cathode, and gave a blackish-green deposit, with a dull metallic layer beneath. The layer of metal contained

37 per cent of the theoretical amount of nickel. ('Chem. News,' vol. xxvi. p. 209; also 'Journal of the Chemical Society,' vol. xi. p. 204.) A solution of hydrated oxide of nickel, in a mixture of cream of tartar, and a little soda and water, electrolysed by means of a current from two Daniell's cells and platinum electrodes, yields a layer of solid hydrated peroxide of nickel upon the anode (W. Wernicke, 'Journal of the Chemical Society,' vol. ix. p. 307).

Deposition of nickel by simple immersion.—Nickel is not usually deposited by simple immersion process in aqueous solutions; it is too electro-positive. Crystalline silicon separates nickel from its anhydrous fluoride by the assistance of heat. I mixed 1·5 grain of crystals of silicon with ten of dry fluoride of nickel, and heated the mixture. At a gentle red heat in a porcelain crucible, vivid incandescence occurred, and metallic nickel was deposited and melted by the evolved heat. The globules were grey, looked like nickel, and were feebly attracted by a magnet, I also immersed crystals of silicon in an aqueous solution of fluoride of nickel containing free hydrofluoric acid ; the crystals did not coat themselves with metal. According to I. C. Davies, nickel is scarcely precipitated at all from acid solutions by means of zinc; but if ammonia be added to the liquids, precipitation occurs. Zinc throws down the metal perfectly from a solution of nickel chloride rendered ammoniacal. A. Merry obtained similar results with sulphate solutions ('Journal of the Chemical Society,' vol. xiii. p. 311). According to M. Becquerel, the simple immersion of copper in a solution of the double chloride of nickel and sodium, is sufficient to deposit the metal ('The Chemist,' vol. v. p. 408). Magnesium deposits hydrated protoxide of nickel from a solution of nickel sulphate (Commaille, 'Chemical News,' vol. xiv. p. 188). But from slightly acid solutions of salts of protoxide of nickel, magnesium deposits hydrogen and metallic nickel (Roussin, 'Chemical News,' vol. xiv. p. 27).

Depositing nickel by contact with another metal (see also p. 82).—C. Mène coats articles of iron, steel, copper, brass, zinc, and lead, by immersing them in contact with zinc, in a boiling neutral solution of chloride of zinc, containing metallic nickel in fragments or plate. If the solution is acid, the coating will be dull ('Chemical News,' vol. xxv. p. 214). Stolba adds two measures of water to one of concentrated solution of chloride of zinc in a copper vessel, boils the mixture, and re-dissolves any precipitate by the least possible quantity (a few drops) of hydrochloric acid. A few particles of powdered zinc are thrown into the liquid; and this causes a deposit of zinc upon the vessel. Sufficient chloride or sulphate of nickel is then added until the liquid is distinctly green; and the previously cleaned articles are at once immersed in contact with zinc in the boiling liquid during fifteen minutes. For thick coatings the operation is repeated. Articles of zinc, cast iron, wrought iron, steel, brass, and copper, are coated by this process. According to Raoult, gold in contact with nickel, either in a cold or boiling, acid or neutral, solution of salts of nickel, receives no metallic deposit ('Journal of the Chemical Society,' vol. xi. p. 465).

Depositing nickel by separate current process (see also p. 89).—Nickel is not usually deposited by the single cell method, because that process robs the liquid of metal, and sets free its acid, and an acid solution of nickel is difficult to manage. Various solutions have been tried for practical use by the separate current method, but the most successful ones have been those composed of the double salts of nickel and ammonium. In the year 1855 I employed the double salts of nickel and ammonium for depositing the metal, and published the results of the use of them. In August 1869 Dr. Isaac Adams patented those salts for nickel plating purposes, and nearly all the deposition of nickel has been done by their aid. The liquids contain various proportions of the salts to the water, but are usually strong.

The credit of depositing nickel upon a large scale, and

coating other metals extensively with it, has been given to
Dr. Adams (see 'Chemical News,' vol. xxi. p. 69). According
to the terms of his patent he claims that the solution must
be free from potash, soda, lime, alumina, and nitric acid;
according also to M. Gaiffe it should not contain a trace of
salt of potash or soda; but Becquerel disproves this, and
shows that the double sulphates of potassium and nickel,
may be used with equal success to those of ammonium and
nickel; and that the bath must not be allowed to become
acid ('Chemical News,' vol. xxi. p. 57). H. Bouillet also
denies the necessity of absence of the fixed alkalies in
depositing nickel, and says: ' I have deposited good nickel
from the double sulphate of nickel and magnesium'
('Chemical News,' vol. xxii. p. 22).

Becquerel employed a solution of sulphate of nickel, with
its free acid neutralized by ammonia, and kept the sulphuric
acid which was liberated by the electrolysis, saturated with
metal by means of oxide of nickel placed in the liquid, or
neutralized it by occasionally adding ammonia. When the
oxide is employed for replenishing, the liquid remains of the
same degree of concentration; but with ammonia, it deposits clear green crystals of double sulphate of nickel and
ammonium; these are very slightly soluble in water alone,
but more freely in that containing ammonia. The deposited nickel is brilliantly white, and may be formed into
bars, &c., by electrolysis, by having proper moulds to receive
it; the bars possess magnetic polarity. The solution of
double sulphate of nickel and ammonium, whether containing
free ammonia or not, yields metal by electrolysis (Becquerel,
' Chemical News,' vol. vi. p. 126).

Böttger states that he has tried many nickel solutions, but
the best was made by adding to dry crystals of protosulphate
of nickel, as much liquid ammonia as was necessary to dissolve them. The dark blue fluid was then ready to use by
the battery process (' Pharmaceutical Journal,' vol. iii.
p. 358).

M. Nägel adds one part of aqueous ammonia to thirty parts of water, then dissolves in it two parts of crystals of sulphate of nickel, and adds six parts of aqueous ammonia of sp. gr. ·909. He uses the solution at a temperature of about 100° Fahr. with a platinum anode, and a moderate current.

Another solution is formed as follows :—Take 150 parts of water, add twelve and a half of nitric acid, five of chloride or sulphate of ammonium, five of nitrate of ammonium, heat the mixture to 80 C., and saturate it with freshly precipitated hydrate of nickel, made by precipitating a solution of chloride or sulphate of nickel, by one of caustic potash or soda; cool the mixture, add twenty-five parts of aqueous ammonia, dilute the whole with water to 250 parts, dissolve in it five parts of carbonate of ammonium, filter the liquid and use it at a temperature of 50 C. (' Chemical Society's Journal,' vol. xii. p. 928).

Another solution is composed of 100 parts of sulphate of nickel, fifty-three of tartaric acid (dissolved in water), and fourteen of caustic potash added: it is said to yield a deposit of very great beauty, having a bright silver lustre, without scratch-brushing.

According to Roseleur, nickel may be deposited as a dull grey metal, from a solution made by dissolving nitrate of nickel in its own weight of aqueous ammonia, and diluting the mixture with twenty or thirty times its volume of aqueous bisulphite of soda solution of sp. gr. 1·199.

Thomas and Tilley, according to their patent of Dec. 26, 1854, precipitate a solution of chloride of nickel with ferrocyanide of potassium, and dissolve the washed precipitate in a solution of cyanide of potassium, to form a depositing liquid.

Management of nickel plating solutions.—The solutions are generally contained in vats of wood, lined with asphaltum (see p. 313). As metallic nickel is much like cast iron, and cannot be rolled, the anodes are composed of plates of the cast metal, usually about 12 inches deep, 9 inches wide, and half an inch or more in thickness; and should have a much larger surface facing the articles, than that of the articles themselves. Cast nickel contains a variable proportion of copper, carbon, and silicon ; and when it dissolves, the carbon and silicon are thrown out upon the surface in the form of a black powder, which falls to the bottom; the copper dissolves and is deposited. If plates of nickel cannot be obtained, fragments of metallic nickel should be suspended in baskets of platinum-wire gauze, but this is not a very practical plan, because the current becomes impeded by the impurities ; or a quantity of freshly precipitated hydrated oxide of nickel in a wet state should be added to the solution, and the liquid stirred up each evening.

Nickel solutions are less easy to manage than those of silver. Some of those employed, contain a large quantity of the double salt, others contain about four ounces per gallon. The chief point to be attended to, is to keep the solution neutral or slightly alkaline; a few operators, however, prefer a slight degree of acidity. They should be frequently tested with neutral tint litmus-paper ; and ammonia should be added if necessary. It is also desirable to employ a much larger surface of anode than that of the articles, and to keep the anodes in the liquid whilst the current is not passing ; by these means liberation of much free acid is avoided. The current should also be maintained very uniform, and the articles kept in motion. During the electrolysis, more or less hydrogen is usually evolved at the cathode, but this of course depends upon the composition of the solution, and the 'density' of the current ; in consequence of this, it is difficult to obtain thick deposits, the hydrogen is set free in the metal, and causes it to split off in films. The articles to be coated should be very clean, and also free from scratches; the latter cause an irregular deposit. A current from one to three cells of large surface is usually sufficient.

Properties, uses, &c., of electro-deposited nickel.—Electro-deposited nickel is hard, too hard to be burnished, and therefore resists rough usage, and is very much more durable than an equal thickness of silver, but the deposits are usually very thin. ' Laminæ of electro-deposited nickel sometimes contain forty times their volume of hydrogen '(MM. L. Troost and P. Hautefeuille ('Chemical News,' vol. xxxi. p. 196).

Nickel has sometimes a dull appearance when deposited, but in consequence of its hardness a high degree of polish may be imparted to it by mechanical means. It must not be 'scratch-brushed' with brass brushes, because they make it yellow. Its colour when polished, is more blue than that of silver, and changes to a slightly yellowish tint by lapse of time. It does not readily oxidise in the air, even when wet; but it is easily corroded by acids. Whilst silver is rapidly blackened by sulphuretted hydrogen, nickel is not affected. In consequence of its hardness, and indifference to sulphuretted gases, it retains its polish a very long time. Nickel should not be employed for coating the interior of cooking utensils, because of its being corroded by acids, and having poisonous properties. Although metallic nickel is much less expensive than silver, the cost of nickel-plating is not proportionately less, because the value of the metal is not the greatest part of the expense. In America, spoons and forks are said to be coated at twenty-five cents per dozen, and saddlery trimmings at thirty cents. It is very useful for harness furniture, carriage fittings, scales and weights, points of lightning conductors, and for many other purposes, and the electro-deposition of it has greatly extended. There are establishments in which it is carried on, in Birmingham, Sheffield, London, New York, Philadephia, and other places ; at the Star Nickel-plating works, Phila-delphia, they operate in accordance with the patents of Dr. Isaac Adams.

Estimation of nickel by means of the battery.—A solution containing one gramme of the pure double sulphate of nickel and ammonium was analysed; with a current from two Grove's cells, it required two hours for its reduction, and yielded in three experiments 14·79, 14·78, and 14·77 per cent of nickel; the theoretical amount is 14·72 per cent. Samples of the commercial sulphate, gave 22·07, 21·87, 21·84, 21·43, and 22·05 per cent ; theory required 20·71 per cent. The double sulphate of nickel and potassium gave 13·33, 13·37, and 13·24 per cent; theory required 13·29 per cent. With a larger amount of the same salt, 13·24 per cent was obtained. The double phosphate of nickel and potassium, yielded 13·27 per cent, theory requiring 13·29. The metallic deposits were washed with alcohol, and cautiously dried (I. M. Merrick, ' Chemical News,' vol. xxiv. pp. 100, 172).

17. **Cobalt.**—Elec. chem. eqt. $= \dfrac{59}{2} = 29 \cdot 6$. Metallic cobalt, especially in the form of bars or plates, can rarely be obtained, because it is not only extremely difficult to melt, but there is very great loss by oxidation in the process. The commonest salts of cobalt, are the oxide, nitrate, and chloride. The oxide may best be purchased : there are two varieties of it, one a black powder containing more oxygen, and the other brownish black, and containing less, having been more strongly heated. The chloride may be made, by digesting the oxide in hot and strong hydrochloric acid, and evaporating the deep blue liquid ; it yields purplish red crystals, freely soluble in water. The nitrate is prepared by digesting the oxide in nitric acid, and evaporating the solution ; it is in the form of deliquescent red crystals ; freely soluble in water.

According to Becquerel, copper immersed in a solution of double chloride of cobalt and sodium, acquires a coating of cobalt (' The Chemist,' vol. v. p. 408). Magnesium slowly deposits hydrated oxide of cobalt from a solution of the sulphate (Commaille, 'Chemical News,' vol. xiv. p. 188). But from slightly acid solutions of protoxide of cobalt, magnesium deposits metallic cobalt and hydrogen gas (Roussin, 'Chemical News,' vol. xiv. p. 27).

Electrolysis of salts of cobalt.—Very little has been done to ascertain the behaviour of these salts by electrolysis. I electrolysed the fluoride dissolved in pure dilute hydrofluoric acid, with a current from a single Smee's element, an anode of cobalt, and a cathode of copper. The conduction was very sparing, and only a film of black powder appeared on the cathode in twelve hours. According to Smee, this metal may be reduced from its chloride (to which an excess of ammonia has been added) by using a cobalt anode, a series of battery cells, and a cathode of copper. The reduced metal is white, but is not deposited freely. It may also be reduced from a solution, formed by digesting oxide of cobalt in cyanide of potassium ; but only in small quantity, hydrogen being freely evolved.

W. Wernicke says that the hydrated oxide, dissolved in water containing cream of tartar and a little caustic soda, yields, when electrolysed with electrodes of platinum, solid hydrated peroxide of cobalt upon the anode. The deposit exhibits magnificent interference colours, which are valuable for the purposes of metallo-chromy, and are readily produced, and permanent ('Chemical News,' vol. xxii. p. 240 ; 'Journal of the Chemical Society,' vol. ix. p. 307).

Deposition of cobalt by contact with another metal.—C. Méne deposits the metal upon articles of zinc, lead, iron, brass, or copper, by immersing them in contact with zinc, in a boiling hot and neutral solution of chloride of zinc, containing fragments of metallic cobalt ('Chemical News,' vol. xxv. p. 214). According to Stolba, salts of cobalt treated in a similar manner to those of nickel (see p. 234), yield a metallic deposit of a steel-grey colour, less lustrous than nickel, and more liable to tarnish.

Deposition of cobalt by separate current process.—According to Becquerel, a concentrated solution of the chloride, with its excess of acid neutralised by addition of caustic potash or ammonia, and electrolysed by a very weak current, deposits its metal in coherent tubercles, or in uniform layers, according to the strength of the current. The deposited metal is brilliantly white, hard, and brittle, and may be obtained in cylinders, bars, and medals, by using proper moulds to receive it. The deposited rods are magnetic, and possess polarity. With an anode of cobalt, it is unnecessary to alter the solution after its first preparation. Part of the chlorine of the solution is disengaged during the electrolysis. If the liquid contains iron, the greater portion of it is not deposited with cobalt ('Chemical News,' vol. vi. p. 126).

To deposit the metal, dissolve five ounces of the dry chloride in a gallon of distilled water, and make the solution slightly alkaline with ammonia. Pass the current through the liquid, either by using a plate of cobalt as anode, or a bar of gas-carbon in contact with a heap of fragments of cobalt contained in a gutta-percha basket. From two to five Smee's cells are required. The solution must be kept slightly alkaline ('Telegraphic Journal,' vol. ii., p. 246). 'Laminæ of electro-deposited cobalt, sometimes contain thirty-five times their volume of hydrogen' (Troost and Hautefeuille, 'Chemical News,' vol. 31, p. 196).

18. **Iron.**—Elec.-chem. eqt. $= \frac{56}{2} = 28$. A plate of iron 15 centimetres square, and two millimetres thick, was deposited on copper by Herr Bockbushmann, in the year 1846. In 1857, M. Feuquieres exhibited specimens of electro-deposited iron at the Paris Exhibition. In 1858, M. Garnier patented his process, termed acierage (steeling), for protecting the surfaces of engraved copper plates; and in the same year, Klein produced his beautiful specimens of electro-deposited iron.

Iron is a very impure metal, and liable to contain carbon, silicon, &c.; its least impure form is wrought iron. Its commonest salts are the peroxide, (called also sesqui-oxide of iron, jewellers' rouge, &c.); the protosulphate, (called

also green vitriol); the carbonate; the protochloride; and
the perchloride.   The peroxide may be readily obtained,
by adding some nitric acid to a solution of green vitriol,
boiling the mixture, adding an excess of aqueous ammonia,
and washing and drying the precipitate; it is a brick-red
powder. The protosulphate is most conveniently purchased,
it is very cheap ; it may however be prepared, by digesting fine iron wire in partly diluted sulphuric acid, in a nearly
filled and covered glass vessel, until no more will dissolve,
and then evaporating the solution, as much as possible out
of contact with air; it is a green crystalline salt, freely soluble
in water. The protochloride may be similarly prepared,
by using hydrochloric acid instead of sulphuric; it is also
of a green colour, and very soluble. The solutions of all
protosalts of iron, rapidly absorb oxygen from the atmosphere, and become persalts. The perchloride is made, by
adding some nitric acid to a solution of the protochloride,
and then boiling the mixture down to the crystallising point
in a wide porcelain dish; it is a brick-red salt, very soluble
in water, and liable to produce a cloudy solution.

Deposition of iron by simple immersion (see p. 79).—Magnesium deposits from a neutral solution of ferrous sulphate,
hydrated ferrous oxide, but from an acidified solution it
deposits metallic iron (Commaille, 'Chemical News,' vol.
xiv. p. 188). From slightly acid solutions of proto and
sesqui salts of the metal, magnesium deposits pure iron and
hydrogen gas (Roussin, 'Chemical News,' vol. xiv. p. 27).
Gold in contact with iron, and immersed in cold or boiling,
acid or neutral, solutions of salts of iron, does not receive
a metallic deposit (Raoult, 'Chemical News,' vol. xxvi.
p. 240, vol. xxvii, p. 59 ; 'Journal of Chemical Society,' vol.
xi. p. 465).

Electrolysis of salts of iron.—Iron may be reduced from a
solution of its protosulphate (green copperas), or from its
protochloride, which is preferable. I have deposited it in
the state of reguline white metal, by passing a current of

considerable intensity (from fifteen or twenty Smee's cells), for one hour, through an anode of iron immersed in a saturated aqueous solution of salammoniac ; its appearance when deposited from this liquid is rather white, very similar to that of freshly broken cast iron. By similar means it may also be deposited, using a saturated solution, either of carbonate of ammonium, acetate of ammonium, or acetate of potassium. Good metal may also be deposited from a saturated aqueous solution, of a mixture of two parts of protosulphate of iron and one of salammoniac. I have deposited it from an aqueous solution of ferrate of potassium, formed, either by igniting peroxide of iron very strongly for some minutes with caustic potash and saltpetre, and dissolving the product in water ; or by making a very strong solution of caustic potash, immersing in it a large iron or steel anode, and a small copper or platinum cathode, and passing a strong current from fifteen or twenty Smee's cells through it, until it acquires a deep amethystine or purple colour ; by that time, the cathode will have obtained a coating of iron, which will be in the state of a dark powder if the power has been too great, or it will have the appearance of white cast iron (or intermediate between that and the appearance of reguline deposited zinc), if the power has been sufficiently weak. The solution rapidly decomposes, becomes colourless, and deposits all its metal in the state of peroxide at the bottom of the vessel. Iron may be very easily deposited from its sulphate thus : dissolve a little of the salt in water, and add a few drops of sulphuric acid to the solution ; one Smee's cell may be used to deposit it upon copper or brass. The metal in its pure coherent state, has a very bright and beautiful silvery appearance. An aqueous solution of cyanide of potassium is a very bad conductor with an iron anode, even if it be maintained hot. Solutions of persalts of iron yield no metallic deposit, but are reduced to protosalts, by the passage of an electric current through them.

Walenn deposited reguline, white, silvery-looking iron,

(attended by the evolution of much hydrogen) from a cold and slightly acid solution, composed of one part of crystallised ferrous sulphate, dissolved in five of water; employing a current from three Smee's elements of about ten times the amount of surface as that of the electrodes. The addition of sulphate of ammonium increased the conducting power, and formed a very good depositing solution ('Chemical News,' vol. xvii. p. 170).

One of the best liquids for depositing iron, is that of M. Klein, prepared as follows :—Precipitate a solution of ferrous sulphate (green vitriol) by means of carbonate of ammonium, and dissolve the washed precipitate in sulphuric acid, taking care to avoid any free acid. Use the bath as concentrated as possible. Good reguline metal may be obtained from it, by means of a current from four weak Meidinger's cells, with an anode of iron and a cathode of copper. It is highly important to prevent the bath becoming acid by the working, &c., and this is effected by using an anode eight times the size of the receiving surfaces, and by attaching a plate of copper or platinum to it, so that the two form a voltaic pair in the liquid, and thus cause the iron to dissolve whilst the battery current is not passing. The metal obtained from it, is as hard as tempered steel, and very brittle, but after annealing, it is malleable, and may be engraved as easily as soft steel. He also employs a bath, composed of the double sulphate of iron and magnesium, of sp. gr. 1·155; the liquid must be neutral, and the electric current very feeble. The iron obtained from it has a sp. gr. of 8·139; it occludes thirteen times its volume of hydrogen, and possesses a higher electric conductivity than any commercial iron ; it does not warp or contract when heated, but slightly expands, and is not porous ('Chemical News,' vol. xviii. p. 133, and vol. xxxi. p. 137 ; 'Telegraphic Journal,' vol. ii. p. 128).

' *Copying engraved metal plates with copper and giving them a surface of iron.*—If the engraved plate is of steel, boil it one hour in caustic potash solution. Brush and wash it well. Wipe it dry with a rag, and then with one moistened with benzine.

'Melt six pounds of the best gutta-percha very slowly indeed, the gum being previously cut up into very small pieces. Add to it three pounds of refined lard, and thoroughly incorporate the mixture. Pour the melted substance upon the centre of the plate. Allow it to stand twelve hours, and then take the copy off.

'*Phosphorus solution.*—Dissolve a fragment of phosphorus half an inch in diameter, in one teaspoonful of bisulphide of carbon, add a similar measure of pure benzine, three drops of sulphuric ether, and half a pint of spirits of wine. Wash the mould twice with this solution, allowing it to dry each time.

'*Silver solution.*—Dissolve one-sixth of an ounce of nitrate of silver, in a mixture of half a pint of strong alcohol, and half a teaspoonful of acetic acid, wash the mould once with this liquid, and allow it to dry.

'*Copper solutions.*—Dissolve fifty-six pounds of sulphate of copper, in nineteen gallons of water, and add one gallon of oil of vitriol. Deposit a plate of copper upon the mould in this solution.

'*Iron solution.*—To coat the copper plate with a surface of iron, dissolve fifty-six pounds of carbonate of ammonium, in thirty-five gallons of water. Dissolve iron into the liquid, by means of a clean anode of charcoal iron, and a current from a battery. Clean the anode frequently, and add one pound of carbonate of ammonium once a week. The copper plate before receiving the deposit, should be cleansed with pure benzine, then with caustic potash, and thoroughly with water. Immerse the cathode in the iron solution for four minutes, take it out, wash, scrub, re-place in the vat, remove and brush it every five minutes, until there is a sufficient deposit. Then wash it thoroughly, well dry, oil and rub it, and clean with benzine. If it is not to be used at once, coat it with a film of wax.

Meidinger coats engraved copper plates with iron, in a solution of sulphate of iron and chloride of ammonium ; and the plates serve for as many as from 5,000 to 15,000 impressions (Wagner's ' Technology,' p. 116).

Management of iron depositing solutions.—Hydrogen is very apt to be evolved at the cathode in solutions of iron. Such liquids also soon spoil for the purpose of electro deposition, because they absorb oxygen from the air, and become persalts ; and the energy of the electric current, instead of being expended in liberating metal, is consumed in de-oxidising the solution at the surface of the articles. Solutions of ferrous chloride become turbid, and continually deposit a slimy precipitate upon the electrodes. Iron solutions should, therefore, be covered as much as possible from the air, but this cannot usually be conveniently done. Klein adopted the expedient, of adding glycerine to the liquid, in order to retard the change. His solution keeps pretty clear, but has upon its surface a slimy foam, which sometimes falls upon the articles.

When iron is deposited from some liquids—for instance, one composed of the double chloride of iron and ammonium, to which is added a small quantity of glycerine,—after the deposit has attained a certain thickness, its surface cracks, and brittle spangles are detached and fall to the bottom. Deposits of iron should be immersed in boiling water, in order to remove dissolved salts, and prevent rusting.

Properties and uses of electro-deposited iron.—Voltaic iron receives magnetism like soft steel. According to W. Beetz, iron electro-deposited from a solution containing salammoniac, is, in a very eminent degree, capable of permanent magnetism ; and if deposited under the influence of power-ful magnets, is itself strongly magnetic. The salammoniac is essential to the formation of good electrolytic magnets ('Telegraphic Journal,' vol. ii. p. 399).

The earlier formed deposits of iron, were full of holes, and quite spongy in texture, in consequence of bubbles of hydrogen, but by keeping the solutions free from uncombined acid, and employing a sufficiently feeble current, that defect has been almost entirely obviated. Iron also, in common with many other metals in the act of electro-deposition, occludes hydrogen (see p. 97). According to R. Leng ('Chemical News,' vol. xxi. p. 179), it contains 185 times its volume, chiefly in the first layers deposited. According to Troost and Hautefeuille ('Chemical News,' vol. xxxi. p. 196), it sometimes contains as much as 260 times its bulk. L. Cailletet states, that by decomposing by an electric current, a neutral solution of ferrous chloride, to which salammoniac has been added, the iron has been deposited in the form of mammillary masses, brittle, brilliant, and hard enough to scratch glass; and when the deposit is plunged into water, or other liquid, numerous bubbles of pure hydrogen are given off. One volume of iron absorbs about 240 volumes of hydrogen, and by contact with a flame, the gas ignites, surrounding the metal with a pale colour ('Chemical News,' vol. xxxi. p. 119; 'Journal of the Chemical Society,' vol. xiii. p. 425).

The advantages of facing printing-type, engraved copper plates, those used for printing bank-notes, &c. with iron, are very considerable. The iron being very much harder than the other metals, is so much the more durable, and is not so readily injured by knots in the paper, or by other substances accidentally present. It also takes the ink readily, and, unlike copper, is not injured by ink containing vermilion; also, when the iron facing has lost its sharpness of impression, it may be dissolved off by means of dilute sulphuric acid, without affecting the copper, and a new coating may be put on, possessing all the original sharpness; and this may be repeated a very great number of times.

19. **Manganese.**—Elec.-chem. eqt. $= \dfrac{55}{6} = 9\text{·}16$. Very few investigations have yet been made, in the electro-deposition of the more intractable metals, manganese, chromium, uranium, tungsten, molybdenum, vanadium, &c. Junot, in

December 1852, took out a patent for 'preparing silicium, titanium, tungsten, chromium, and molybdenum, by causing them to be deposited from their solutions, by means of electric currents, upon metals and other substances ;' but nothing has since been heard of the working of this patent.

The commonest salts of manganese, are the black oxide, chloride, and carbonate. The black oxide (pyrolusite), is found abundantly as a mineral; contaminated however with iron and various earthy matters. The chloride may be formed in an impure state, by digesting the black oxide in hot and strong hydrochloric acid, and evaporating the solution; or pure, by saturating dilute hydrochloric acid with the carbonate ; using pure materials. The sulphate may be formed by similar means, using however boiling hot dilute sulphuric acid. The chloride and sulphate are pink salts, freely soluble in water.

Deposition of manganese by simple immersion.—Manganese is deposited, by the simple immersion of sodium-amalgam, in an acidulated solution of a salt of manganese, it then alloys with the mercury (Roussin, 'Chemical News,' vol. xiv. p. 27). Giles deposited manganese upon mercury, by simple immersion of an amalgam of sodium in a saturated solution of proto-chloride of manganese ('Philosophical Magazine,' 4th series, vol. xxiv. p. 328). According to Phipson, magnesium deposits manganese as a black powder, from a neutral solution of a proto-salt of that metal. ('Proceedings of the Royal Society,' 1864, vol. xiii. p. 217 ; 'Chemical News,' vol. ix. p. 219). Magnesium deposits hydrated manganous oxide from a neutral solution of manganous sulphate, but from the same solution acidified, it deposits metallic manganese (Commaille, 'Chemical News,' vol. xiv. p. 188).

Electrolysis of salts of manganese.—Bunsen filled a porous cell, with a hot and saturated aqueous solution of chloride of manganese, placed it in a charcoal crucible containing hydrochloric acid to the same level, put the latter vessel in a sand-bath (to keep the liquids hot), immersed a platinum wire cathode in the centre of the chloride solution, and connected the charcoal crucible with the positive pole of a four-cell Bunsen's battery. Metallic manganese was separated with the greatest facility; but if the density of the current at the cathode was reduced, either by enlarging the cathode, diminishing the anode, or weakening the current; or if the degree of concentration of the solution was diminished, black manganoso-manganic oxide was obtained. ('The Chemist,' No. 11, August 1854, p. 685 ; Watts' 'Dictionary of Chemistry,' vol. ii. p. 438).

I melted some fluoride of manganese in a platinum crucible, and employed two spirals of platinum wire as electrodes, and a current from six large Smee's cells. The conduction was moderate, and gas was evolved from the anode. In a few minutes, both the cathode and the crucible became quite rotten, by the union of the deposited manganese with the platinum. The anode was not corroded. I also melted the same salt in a crucible of copper, and passed the current by means of a sheet platinum anode, and sheet copper cathode, during half an hour. The conduction was free ; abundance of gas was evolved from the anode, but none from the cathode, and it ceased on stopping the current. The deposit on the cathode was black, and did not evolve hydrogen with dilute hydrochloric acid, and was therefore not metallic manganese. The crucible was much corroded at the line of surface of the liquid.

I also electrolysed a dilute solution of fluoride of manganese, by a current from six Grove's cells, and electrodes of platinum. Much heat was evolved, gas was set free at the anode, and a film of black deposit formed upon the cathode. By similar treatment of a saturated solution of the salt, not containing any free hydrofluoric acid, a film of purple colour was instantly formed upon the anode, but it dissolved quickly, and did not colour the liquid. Gas came from both electrodes freely; the liquid also became heated. No solid deposit was obtained.

Salts of manganese yield peroxide at the anode. A solution composed of one part of chloride of manganese and eight of water, yields very beautiful alternating rings of purple-green, golden yellow, and blue, surrounded by a broad belt of golden yellow. With a solution composed of one part of acetate of manganese and fifteen of water, one uniform tint is invariably produced, first golden yellow, then purple, then green (B. Böttger, 'Poggendorff's Annalen,' vol. 1. p. 45).

According to W. Wernicke, solutions of acetate and nitrate of manganese, with a feeble current from two Daniell's cells, and platinum electrodes, yield a deposit of hydrated peroxide of manganese upon the anode ('Journal of the Chemical Society,' vol. ix. p. 307).

20. **Chromium.**—Elec. chem. eqt.$= \dfrac{52\cdot5}{6} = 8\cdot75.$ The ordinary compounds containing chromium, are the sesquioxide, chromic acid, the two chromates of potash, and chrome-alum, *i.e.* the double sulphate of chromium and potassium. The sesquioxide is a substance particularly insoluble in water and acids. The other salts are freely soluble in water, and are more conveniently purchased than prepared.

Deposition of chromium by simple immersion.—Magnesium deposits oxide of chromium from a solution of a salt of that metal; but sodium amalgam shaken up with an acid solution of such a salt, becomes an amalgam, from which the metal itself may be obtained as a spongy mass, by distilling away the mercury in a current of hydrogen. If the amalgam of chromium is heated in the air, the particles of chromium scintillate singularly, and the amalgam then suddenly becomes incandescent (Roussin, 'Chemical News,' vol. xiv. p. 27). Magnesium precipitates hydrated sesquioxide of chromium, from a mixed solution of chromous and chromic chloride (Commaille, 'Chemical News,' vol. xiv. p. 188). Vincent deposited chromium upon mercury, by simple immersion of sodium amalgam in a solution of chloride of chromium ; the latter metal was then obtained in a finely divided state, by distilling away the mercury in a retort filled with vapour of naphtha ('Philosophical Magazine,' fourth series, vol. xxiv. p. 328).

Deposition of chromium by separate current.—Bunsen, by operating in a similar manner upon a concentrated solution of chloride of chromium, as upon one of manganese (see p. 250), deposited chromium readily; the deposit presented the appearance of iron, but was less affected by damp air ; it resisted the action of boiling nitric acid, but dissolved in hydrochloric or dilute sulphuric acid. It was friable, and presented a high polish on the side next the cathode. On diminishing the current, a black powder was deposited, containing more oxygen in proportion as the density of the current was lessened ; adding protochloride of chromium to the solution had a reverse effect,—it caused metallic chromium to be deposited ('The Chemist,' No. 11, August 1854, p. 686).

I melted some acid chromate of potassium, and passed through it the current from five Smee's elements, by means of electrodes of platinum. A deposit slowly formed upon the cathode. I also electrolysed a strong solution of fluoride of chromium, containing some free hydrofluoric acid, and a little hydrochloric acid, by a current from six Grove's cells, and platinum electrodes. The liquid soon became hot ; no gas was evolved from the cathode, but chlorine and ozone were set free at the anode, which was not corroded.

21. **Uranium.**—Elect.-chem. eqt. $= \dfrac{120}{4} = 30$. The commonest salt of uranium is the sesquioxide, which is a yellow powder, insoluble in water, but soluble in several

mineral acids. The nitrate, chloride, bromide, and sulphate of uranium may be formed, by digesting an excess of the sesquioxide in the corresponding diluted acid, until it is saturated, and evaporating the solution; they are each of a pale yellow colour, and soluble in water.

Magnesium deposits golden-coloured hydrated sesquioxide of uranium, from a solution of oxalate of uranium (Commaille, 'Chemical News,' vol. xiv. p. 188). I fused some fluoride of uranium in a platinum crucible, and added to it some crystals of silicon; the salt was not decomposed.

Electrolysis of salts of uranium.—I also fused some fluoride of uranium in a copper crucible, and passed a current from six Smee's cells through it, by means of a platinum wire anode, using the crucible as a cathode; a little gas was set free at the anode, and the crucible melted. A second trial was made, using a platinum crucible, and two spirals of platinum wire as electrodes, and the current continued during one hour. Conduction was very free, much gas was evolved from the anode, but none from the cathode; a bulky deposit quickly formed upon the negative spiral, especially on the side towards the anode. The deposit weighed 43·66 grains, and consisted of hard jet-black crystals. The anode was not corroded. In a third trial, four Grove's cells were employed, and a special apparatus devised and employed to collect the evolved gas, and about five cubic inches were obtained. The crystals were not metallic uranium; they were insoluble in boiling water, but soluble in cold dilute hydrofluoric acid, without evolving gas. About one-fourth of the deposit, consisted of a fine crystalline powder, nearly of the colour of copper, but darker, and was composed of the crystals, with a film of less reduced fluoride upon them; they evolved gas in cold nitric acid, or in hot dilute nitric acid. They were not fused by heating alone to redness upon platinum foil; but if caustic potash was added, they oxidised. I also electrolysed a fused mixture of the pure fluorides of uranium and potassium, with platinum electrodes; the results were very similar, except that the deposit upon the cathode fell off as fast as it was formed; and the crystals had to be extracted by dissolving the cooled saline mass in slightly diluted and hot hydrochloric acid; they were very much like those of silicon; their form was that of a short pyramid with a square base. The anode was very slightly corroded and made bright by the action; and twenty cubic inches of gas were collected from it.

I also electrolysed a strong aqueous solution of fluoride of uranium, with a current from six Grove's cells, and platinum electrodes. Much gas, having the odour of ozone, was evolved from the anode, and the liquid became hot. I then added some aqueous hydrofluoric acid; the conduction was very free, and abundance of gas evolved from each electrode, but no solid deposit was formed.

22. **Tungsten.**—Elec. chem. eqt. $= \dfrac{184}{6} = 30\text{-}66$. The only readily available salts of this metal, are tungstic acid, and tungstate of sodium; the former is a yellow powder, insoluble in water and in acids; the latter is a colourless salt, soluble in water.

I fused some tungstate of sodium to a clear liquid in a porcelain vessel, and electrolysed it by means of a current from five Smee's cells, a gas-carbon anode, and a platinum wire cathode. The conduction was moderately free, gas was evolved from the anode, and at the cathode, black matter was set free, which floated, and became diffused in the liquid, and partly re-dissolved.

23. **Molybdenum.**—Elec.-chem. eqt. $= \dfrac{96}{6} = 16$. The only common salts of this metal, are molybdic acid, sulphide of molybdenum, and molybdate of ammonium. Molybdic acid may be prepared, by digesting sulphide of molybdenum in strong nitric acid; it is a pale yellow powder, insoluble in

water. The sulphide is a mineral substance, looking like black-lead, and insoluble in water. The mol)bdate of ammonium is white, and sparingly soluble in water.

Electrolysis of molybdic acid.—I fused some molybdic acid in a porcelain crucible, and passed the current from five Smee's cells through it, by means of a gas-carbon anode, and platinum cathode. It conducted freely. The action was rather strong at the anode, but little gas was evolved. Black crystals quickly collected round the cathode, but no gas was set free with them ; the entire liquid soon became full of the crystals, which spread quickly to the anode. The carbon was not dissolved, or disintegrated. The cooled residue was a solid black mass of crystals, which dissolved sparingly in water, and formed a blue liquid. In a second experiment, using twelve large Smee's elements, and platinum anode and cathode, there was free action, and much gas evolved ; and the bluish-black deposit quickly formed upon the cathode. Most of the gas came from the anode. A large quantity of crystalline needles ($\frac{1}{8}$th to $\frac{1}{4}$th of an inch long), formed upon the cathode, and stood out at right angles to its surface in the fused substance. The deposit imparted a transient green colour to water.

Molybdic acid dissolved freely in pure hydrofluoric acid, evolving a little heat. The solution was electrolysed, both with a carbon and with a platinum anode. The colourless liquid conducted freely a current from ten large Smee's cells, became instantly blue, and almost black, at a platinum cathode. Gas was evolved at each electrode, that from the gas-carbon anode was the most abundant, and had a slightly chlorous odour. On stopping the current, the deep blue film on the cathode quickly dissolved, and the liquid soon became colourless. During the action, the cathode was several times removed from the electrolyte, and dipped into water ; much blue matter dissolved, but the water became nearly colour-less in half a minute, even without stirring, and however large the quantity of blue matter was, which dissolved in it.

24. **Vanadium.**—Elec.-chem. eqt. $= \dfrac{137}{6} = 22\cdot8$. The
only readily attainable compounds of this element, are
vanadic acid, and vanadate of ammonium. I electrolysed
a solution composed of vanadic acid dissolved in pure
dilute hydrofluoric acid, with a gas-carbon anode and platinum cathode, and ten Smee's cells. Gas having an odour
of ozone was evolved from the anode. I also saturated
dilute sulphuric acid with pure vanadate of ammonium, and
electrolysed it with platinum electrodes, and a current from
four platinum and zinc elements. The conduction was very
sparing; the solution became gradually of a very intense
bluish-black colour from action at the cathode, and a jet-
black powdery deposit of some thickness formed upon that
electrode.

25. **Lead.**—Elec.-chem. eqt. $= \dfrac{207}{2} = 103\cdot5$. The commonest salts of lead, besides its three oxides, viz. litharge,
red-lead, and peroxide of lead, are the nitrate, chloride, carbonate, sulphate, and acetate. All of these may be readily
purchased in a comparatively pure state. The nitrate and
acetate, are the two common soluble salts of the metal; the
others may be readily made, by dissolving litharge, red-lead,
or carbonate of lead, in the particular acids, to saturation,
and evaporating the solutions.

Deposition of lead by simple immersion.—(See also p. 79.)
—The old experiment of producing a lead tree, by suspending
a spiral of zinc wire in a solution of nitrate or acetate of
lead, is well known. According to A. Cossa, aluminium
slowly deposits lead in crystals, from a solution of plumbic
nitrate or acetate, and immediately from a solution of the
chloride. An alkaline solution of plumbic chromate, is also at
once decomposed by that metal, with separation of metallic
lead, and formation of chromic oxide (Watts' 'Dictionary of
Chemistry,' vol. vii. p. 54). Magnesium deposits lead and
oxychloride of lead, together with much hydrogen, from a

S

neutral solution of plumbic chloride (Commaille, 'Chemical News,' vol. xiv. p. 188).

I melted some plumbic fluoride in a platinum crucible, and added some crystals of boron ; metallic lead was separated with vivid incandescence, and made a hole in the crucible.   Crystals of silicon also exhibited incandescence, and set free metallic lead.   Metallic antimony or copper, did not liberate lead from the fused fluoride.   By stirring the melted fluoride with an iron rod, the latter was rapidly corroded, heat being evolved, and lead deposited.   Metallic aluminium behaved similarly, but more rapidly.   Zinc exploded, and magnesium detonated, under similar circumstances.   The latter is a dangerous experiment.

Articles of zinc or tin, but not of iron, become coated with lead by simple immersion, in a liquid formed by boiling litharge in a solution of caustic potash.   Those of iron, but not of copper, coat themselves with lead by mere immersion, in a solution of plumbic acetate, *i.e.*, sugar of lead.

Deposition of lead by contact with a second metal.—(See also p. 83.)—F. Weil coats articles with lead, by a similar process to that he employs for tin (see p. 266), using a salt of lead instead of one of tin.   And to produce a deposit of lead free from zinc, he uses similar means to those described for tin ('Chemical News,' vol. xiii. p. 2).   According to Becquerel, if a piece of bright copper in contact with zinc, be immersed in a solution of chloride of lead and sodium, the copper becomes covered with lead ('Chemist,' vol. v. p. 408).   I connected together a wire of zinc, and one of platinum, and immersed them in a solution of litharge in aqueous ammonia; both became coated with a black powder in a few minutes.   The deposit, moist with the liquid, became yellow by contact with the air, and was apparently re-converted into litharge.

Haeffelly coats copper or brass with lead, by immersing it in contact with a bar of tin, in a hot alkaline solution of oxide of lead.   The tin dissolves in the form of an alkaline stannate, but the lead is precipitated in a spongy state ('Chemical News,' vol. vi. p. 163).

Deposition of lead by separate current.—(See also p. 89.) —According to Faraday, fused protoxide of lead yields metal at the cathode, and oxygen at the anode ; the chloride gives lead at the cathode, and chlorine at the anode; and the borate liberates metal at the cathode, and oxygen and boracic acid at the anode. Beetz electrolysed fused plumbic fluoride, and observed that a colourless gas was evolved from the positive pole, and lead set free at the negative pole ('Poggendorf's 'Annalen ;' also 'The Chemist,' new series, vol. i. p. 253) Fremy also electrolysed it in a platinum vessel, and found it easily decomposed, the lead was set free, and alloyed with the vessel ('Chemist,' new series, vol. ii. p. 548). I also electrolysed 400 grains of pure fluoride of lead (melted in a thick copper crucible), with a current from six Smee's cells, using a platinum wire as an anode, and one of copper as the cathode; conduction was copious, and a bulky crust quickly formed upon the cathode, and advanced towards the anode in lumpy projections. A little gas appeared at the latter, but during a short time only. The deposit upon the cathode was not lead, nor was there any metal contained in a free state in it, or in the saline mass, after action lasting one hour; it was a mass of lead-salt, brittle, and of a red-brown colour (like that of peroxide of lead), when cold. The conduction was very perfect, and the fused salt appeared to conduct without being decomposed. The anode was not corroded. I also electrolysed the fused salt in a deep, narrow, and thick copper cup, with an anode of gas-carbon, during one and a quarter hours ; the latter was corroded, and the metal liberated; action was copious, gas was evolved at the anode, and about seven or eight cubic inches were collected. G. J. Knox also electrolysed fused fluoride of lead with an anode of charcoal, a platinum wire cathode, and a current from sixty voltaic cells ('Philosophical Magazine,' third series, vol. xvi. p. 192).

Lead may be deposited from an aqueous solution, either of its nitrate or acetate, by means of a separate current, with an anode of lead; also from a liquid, formed by saturating a boiling solution of caustic potash with litharge; but it is difficult to obtain any considerable thickness of reguline metal from either of these liquids. The nitrate and acetate, yield peroxide of lead at the anode.

*Deposition of peroxide of lead.—Electro-chromy.—*According to W. Wernicke, an alkaline solution of the tartrate of lead and sodium, with platinum electrodes, and a current from two Daniell's cells, yields a black deposit of peroxide of lead upon the anode; and a solution of one part of plumbic nitrate and eight of water, gives a similar deposit by such treatment ('Journal of the Chemical Society,' vol. ix. p. 306; 'Chemical News,' vol. xxii. p. 240).

Nobili in the year 1826, discovered, that if a solution of acetate of lead be electrolysed, by means of a large sheet platinum anode, and a platinum wire cathode, a deposit is formed upon the positive plate; and that if a polished steel plate be employed as the anode, with a current from four or six Grove's cells, the deposit is in the form of a thin film, and exhibits all the colours of the spectrum; and by placing the positive plate horizontally beneath the vertical negative wire, the colours were in the form of rings, the centre of which was the wire, and were arranged in the order of the chromatic scale. These colours are known as 'Nobili's rings.' Becquerel, Gassiot, and others, have, by varying the strength of the battery, and of the solutions employed, and interposing non-conducting patterns between the anode and cathode, and by using cathodes of different shapes, obtained effects of great delicacy and beauty. Salts of other metals, such as manganese, bismuth, cobalt, nickel, &c., which yield deposits of peroxide of the anodes (see pp. 112, 235, 242, 252), may be employed instead of those of lead. Becquerel obtained films of peroxide of iron, by electrolysing in vacuo, a solution of protoxide of iron in liquid ammonia ('Chemist,' vol. iv. p. 457).

The colours occur, sometimes upon the anodes, and sometimes upon the cathodes, according to the liquid employed, and with a variety of metals in a number of different liquids. At other times, they arise wholly from deposits from the liquid, as with peroxides on anodes of platinum, or films of metal upon the cathodes ; and sometimes they consist of insoluble substances, formed by the union of the anode with an element of the liquid.

Becquerel prepared his plumbic solution as follows :— Dissolve 200 grammes of caustic potash in two quarts of distilled water, add 150 grammes of litharge, boil the mixture half an hour, allow it to become clear, take the clear portion, and dilute it with its own bulk of water (' The Chemist,' vol. iv. p. 457). The solution is used cold, and is rapidly deprived of its metal, because lead is deposited upon the cathode at the same time.

By this means may be imparted to polished surfaces of metals, all the richest colours of the rainbow. 'They commence with silver blonde, and progress onwards to fawn colour, and thence through various shades of violet to the indigo and blues ; then through pale blue to yellow and orange ; thence through lake and bluish lake to green and greenish orange, and rose orange ; thence through greenish violet and green, to reddish yellow and rose lake, which is the highest colour on the chromatic scale' (Walker's 'Electrotype Manipulation,' part ii. 16th edition, p. 40). Too great a strength of the current covers all the tints with a uniformly dark brown coating. The deposits, if properly prepared, resist friction well.

Metallo-chromy, effected by means of a solution of oxide of lead in caustic soda or potash, is largely employed in Nuremberg. to ornament metallic toys (Wagner's 'Technology,' p. 117). Bells are similarly coloured in France, and the hands and dials of watches in Switzerland.

26. **Thallium.**—Elec. -chem. eqt. = 204. Solutions of this metal are easily formed, by making a piece of it the anode, for a sufficient length of time, in the respective acids, sufficiently diluted to dissolve the compounds. The sulphate is one of the most soluble salts, and requires about twenty times its weight of water to dissolve it.

According to A. Cossa, aluminium deposits metallic thallium from a hot solution of its chloride ('Watts's Dictionary of Chemistry,' vol. vii. p. 54). Zinc coats itself with metal in solutions of salts of thallium, but tin does not. According to Lamy, zinc precipitates it from the solution of the sulphate and nitrate, in the form of brilliant crystalline laminæ. I found that crystals of silicon had no reducing effect, upon a solution of fluoride of thallium, containing free hydrofluoric acid.

Electrolysis of salts of thallium.—Solutions of the salts of thallium are easily decomposed by a feeble electric current; and the metal deposited in the form of beautiful crystalline plates upon the cathode. When one of the sulphate is electrolysed by a weak current, brown thallic peroxide is deposited upon the anode.

I electrolysed an aqueous solution of the fluoride, by means of one Smee's cell, a thallium anode, and a platinum cathode. It conducted freely, and quickly yielded a deposit of the metal, in long feathery crystals, like those of electro-deposited tin, but of a less white colour.

Thallium deposits metal from the sulphate, nitrate, and acetate of copper, nitrate of silver, solutions of gold, mercurous sulphate, and acetate of lead, but a basic salt from nitrate of cobalt (W. C. Reid, 'Chemical News,' vol. xii. p. 242).

27. **Indium.**—Elec.-chem. eqt. $= \dfrac{113 \cdot 4}{3} = 37 \cdot 8$. This metal being at present very costly, little has been done with it in electro-metallurgy. It is allied to thallium and aluminium. Its salts are generally freely soluble in water ; the double

sulphate of indium and potassium, called indium-alum, dissolves in half its weight of water at 16° C.

Indium is precipitated as metal from the solutions of its
salts, by means of zinc. A solution of the sulphate or chloride, may be used for this purpose.

28. **Tin.** Elec. chem. eqt. $= \dfrac{118}{4} = 29 \cdot 5$. The commonest salts of tin, are stannous and stannic oxides, the
two chlorides, the sulphides of tin, and the stannates of
sodium and potassium. Stannous chloride is the most useful salt, and should be freshly prepared, because it becomes
less soluble by being kept a long time exposed to the air; it
may be easily made, by adding abundance of fragments of
pure tin to strong hydrochloric acid, and keeping the acid hot,
until it has acquired an oily consistence, and gas ceases to
be evolved. Stannic oxide may be prepared, by pouring the
anhydrous bichloride very gradually into water, with stirring,
and then adding sufficient ammonia to precipitate the oxide.
Wash the precipitate.

A protochloride of tin depositing liquid, may be easily
formed, by dissolving the ordinary commercial salt in water,
and adding a little hydrochloric acid, to remove any cloudi-
ness which may appear ; a similar, but better liquid, may be
made by the battery process, by passing the current through
dilute hydrochloric acid, by means of a large tin anode, until
sufficient metal is dissolved. This (or the other chloride of
tin) is not a good solution to obtain reguline metal from ;
it has a very great tendency to deposit the tin in the form
of long crystalline needles, of a fernlike appearance, which
often project from the corners and edges of the cathode, to
a distance of upwards of half an inch. A solution composed
of eleven ounces of water, one of hydrochloric acid, and
eighty grains of protochloride of tin, admits of this effect
being produced in a striking manner. Nearly all the compounds of tin, and especially those formed with mineral
acids, exhibit this tendency in a greater or less degree, when

acted upon by electrolysis, rendering the deposition of tin in thick layers of fine white coherent metal, a matter of considerable difficulty.

The stannate of potash solution is made, either by dissolving the solid salt in water, or mixing freshly precipitated peroxide of tin (whilst still moist) with a boiling solution of caustic potash. It may also be easily formed by the battery process, by passing a strong current of electricity, by means of a large tin anode, through a strong and boiling solution of caustic potash, until the immersed cathode receives a free white deposit. This solution, if used at 150° Fahr., yields fine white tin ; but it decomposes by exposure to the atmosphere, and soon deposits its metal as oxide, at the bottom of the vessel. A solution of cyanide of potassium and of tin, has been proposed as a depositing liquid ; but it is a bad conductor with a tin anode, even if hot, and does not dissolve the metal freely.

Electrical relations of tin and iron.—Tin is feebly negative to iron at all temperatures between 62° and 203° Fahr. in distilled water, and positive to it at 212° Fahr. It is positive to iron at all temperatures between 62° and 212° Fahr. in a saturated solution of boracic acid ; also the same between those temperatures, in a strong solution of phosphoric acid in distilled water, or in one measure of oil of vitriol, mixed with either nine or ninety-six of distilled water ; or in a mixture of one measure of this acid, and 192 of distilled water, from 73° to 158° Fahr., and negative to iron above that to 212° Fahr.; it is positive to iron from 72° to 212° Fahr. in a mixture of equal measures of hydrochloric acid and water ; it is negative to iron from 70° to 77° Fahr., and positive above that to 212° Fahr., in a mixture of one measure of hydrochloric acid, and nine of distilled water ; it is negative to iron from 70° to 212° Fahr. in a mixture of one measure of hydrochloric acid, and ninety of distilled water, and positive to iron from 68° to 212° Fahr., in one measure of hydrofluoric acid, and nine of water; it is positive to iron in one measure of nitric acid, and nine of water from 70° to 111° Fahr., and negative from 111° to 212° Fahr. ; and it is positive to iron from 82° to 212° Fahr. in a mixture of one measure of nitric acid, and ninety-six of water.

Deposition of tin by simple immersion.—(See also p. 79.) —I found that crystals of silicon, did not deposit tin from a solution of stannous fluoride containing free hydrofluoric acid ; and that zinc immersed in a solution of stannic fluoride, evolved gas, and produced a flocculent precipitate.

A remarkable instance of deposition of tin, is mentioned by M. Henri Loewel. He added metallic tin to a solution of green crystallised chloride of chromium (which did not contain an excess of acid), in a closed glass vessel, and boiled the mixture ten or twelve minutes, and allowed it to cool. During the heating, the tin dissolved, and took chlorine from some of the chromium salt, forming proto-chloride of tin and protochloride of chromium. But during the cooling, a reverse action occurred, the protochloride of chromium removed the chlorine from the other protochloride, and the tin was deposited in the form of numerous small metallic plates ('The Chemist,' part viii. May, 1854, p. 476).

Magnesium deposits stannic acid, and spongy tin, from a solution of stannous chloride (Commaille, 'Chemical News,' vol. xiv. p. 188). A tin tree, is produced by immersing a rod of zinc, or a spiral of zinc wire, in ten or twenty ounces of water to which have been added three drachms of stannous chloride, and ten drops of nitric acid, and allowing the liquid and zinc to remain undisturbed.

To coat brass pins, and other small articles of copper or brass, with tin, they are placed in layers between sheets of grain tin, in a saturated solution of cream of tartar, and the liquid boiled. A little stannous chloride may also be added if necessary.

Clean articles of copper, bronze, or brass, in contact with cuttings of tin, in a boiling solution of peroxide of tin in caustic potash, become coated in a few minutes with a beautiful layer of metal. The solution may also be used for tinning iron, by the battery process, with large anodes of tin, and may be made to give a very fine deposit, but it precipitates its metal gradually, in the form of a white powder, by contact with the air (see p. 264).

C. Paul tins articles of zinc, iron, brass, copper, &c., in the following manner :—The zinc or iron articles, are immersed in a mixture of ten parts of water, and one of sulphuric or nitric acid, and a dilute solution of cupric sulphate is then slowly added, with stirring. After a thin layer of copper is deposited, the articles are removed, washed, wetted with a solution, composed of one part of crystals of stannous chloride, two of water, and two of hydrochloric acid, and then shaken with a mixture of finely powdered chalk, and sulphate of copper and ammonium, which is prepared by dissolving one part of cupric sulphate in sixteen of water, and adding aqueous ammonia, until a clear dark blue liquid is formed. The articles are now tinned by immersion in a solution, composed of one part of crystals of stannous chloride, and three of white argol, dissolved in water ('Journal of the Chemical Society,' vol. xi. p. 955).

To coat articles of iron or zinc with tin, dissolve one part of fused stannous chloride, and thirty of ammonium-alum, in 2,000 of water, heat the solution to boiling, and immerse the previously cleaned articles in it until they attain a fine white colour; add to the solution as it becomes weaker, small quantities of the stannous chloride. According to Roseleur, articles of zinc may also be tinned by simple immersion, in a solution composed of one part of fused stannous chloride, and five of pyrophosphate of sodium, dissolved in 300 of distilled water. Tin which has been dissolved from the surface of tinned iron, is sometimes reduced to metal, by immersion of pieces of zinc in the solution.

According to Becquerel, copper, and iron, do not coat

themselves with tin, in a dilute solution of the double chloride of tin and sodium, at 160 Fahr., but are readily tinned in that liquid by contact with zinc ('The Chemist,' vol. v. p. 408).

Depositing tin by contact with a second metal.—(See also p. 83.)—For coating articles of iron with tin, by means of contact with zinc, Roseleur recommends the two following liquids:—No. 1. Take equal weights of distilled water, stannous chloride, and cream of tartar, dissolve the tin salt in one-third of the cold water, warm the remainder of the water, dissolve the cream of tartar in it, and mix the solutions; the liquid is clear, and has an acid reaction. No. 2. Dissolve six parts of crystals, or four of fused stannous chloride, and sixty of pyrophosphate of potassium or sodium, in 3,000 of distilled water, and stir the mixture; the liquid is clear. Each solution is used hot, and kept in constant motion. The articles are immersed in contact with fragments of zinc, the total amount of surface of which is about $\frac{1}{30}$th that of the articles. The process of deposition occupies from one to three hours. Equal weights of pyrophosphate, and of fused stannous chloride, are added occasionally.

According to F. Weil, copper, and coppered metals, as well as iron and steel, may be tinned, by dissolving a salt of tin in a strong solution of potash or soda, and immersing the articles in the liquid in contact with zinc; the solution being at from 50° to 100° C.: the deposit, however, contains zinc. To obtain a pure deposit, of increasing thickness, place in the vessel containing the tin solution, a porous cell containing the alkaline liquid (without tin-salt) and the zinc. Put the article to be tinned, in the outer liquid, and connect it with the zinc by a wire. To revive the inner liquid, precipitate the dissolved zinc, by addition of sulphide of sodium ('Chemical News,' vol. xiii. p. 2).

Dr. Hillier uses for tinning metals, a solution composed of one part of stannous chloride and twenty of water, to which is next added a solution of two parts of caustic soda, in twenty of water; the mixture is heated. The articles are placed upon a perforated plate of block tin in the hot liquid, and agitated with a rod of zinc until they are sufficiently coated ('Chemical News,' vol. xx. p. 84).

Tinning iron wire: by M. Heeren.—The wire is first cleaned in a hydrochloric acid bath in which a piece of zinc is suspended. The cleaned wire is then brought into contact with a plate of zinc, in a bath in which two parts of tartaric acid are dissolved in 100 of water, with further addition of three parts of stannous chloride, and three of soda ; and after remaining about two hours in the liquid, the wire is brightened by drawing it through a hole in a steel plate ('Journal of the Chemical Society,' vol. xiii. p. 672).

F. Stolba uses, for tinning, a carefully made solution of protochloride of tin, containing from 5 to 10 per cent of that salt, to which mixture a small pinch of cream of tartar has been added. The article, previously well cleaned, is rubbed over with the solution, and then with powdered zinc ; it is then washed, and polished with soft whiting ('Chemical News,' vol. xxiii. p. 21).

According to Raoult, gold or copper in contact with tin, in a concentrated and boiling solution of stannous chloride, receives a deposit of tin. But gold in contact with iron, nickel, antimony, lead, copper, or silver, receives no such coating, either in the hot or cold solutions ('Chemical News,' vol. xxvi. p. 240 ; vol. xxvii. p. 59; 'Journal of the Chemical Society,' vol. xi. p. 464).

Electrolysis of salts of tin.—Fused stannous chloride, yields tin at the cathode, and stannic chloride escapes in vapour at the anode (Faraday). Fremy electrolysed fused fluoride of tin in a platinum vessel; it was easily decomposed ; but the deposited metal alloyed with, and perforated, the vessel in a few minutes ('The Chemist,' new series, vol. ii. p. 548). Anhydrous tetrachloride of tin did not conduct a current from 8,040 cells of W. de la Rue's chloride of silver battery ('Proceedings of the Royal Society,' vol. xxv. p. 325).

I electrolysed a saturated non-acid solution of stannous fluoride, by means of large platinum electrodes, and a current from ten large Smee's elements ; the conduction was sparing ; a little oxygen was evolved from the anode, and long feathery crystals of tin were slowly formed upon the cathode. No gas was evolved from the cathode, nor deposit formed upon the anode. In another experiment, with one Smee's cell, and a copper cathode, the deposit of tin was white, and beautiful crystals of the metal soon reached across the liquid and touched the anode.

By passing a current from six Grove's elements, by means of platinum electrodes, through a strong solution of stannic fluoride, containing little or no free hydrofluoric acid, a grey deposit of metallic tin soon occurred. There was free conduction, much gas from the anode, and heat evolved in the liquid. The anode was not corroded nor received any solid deposit.

To obtain crystals of tin by electrolysis.—The crystallisation of tin, is a phenomenon conspicuously striking, under some conditions, in a solution of stannous chloride. The crystals of tin formed upon the cathode, increase so rapidly in length, as to. grow across the solution, and touch the positive pole in a few minutes. And if the solution and current are strong, and the cathode small, quite a mass of crystals will soon fill the liquid, and converge towards the anode. If the anode be drawn farther away in the solution, the crystals follow it. The largest crystals are produced by slow action : to produce them, a platinum capsule is covered with an outer coating of wax, leaving the bottom uncovered, and then set upon a plate of amalgamated zinc in a porcelain vessel. The capsule is then filled completely with a dilute and not too acid solution of stannous chloride, whilst the outer vessel is filled with water (containing one-twentieth its bulk of hydrochloric acid), up to such a height, that the two liquids come into mutual contact. The electric current generated, reduces the salt of tin, and in a few days the crystals upon the interior of the capsule are well developed, and should be washed with water and dried quickly (F. Stolba, ' Chemical News,' vol. xxx. p. 177).

Deposition of tin by separate current process.—(See also p. 89.)—There are many solutions for electro-tinning by means of a separate current, but only a very few have been extensively used. Most of them alter in property by contact with the atmosphere, and deposit their metal as a white oxide. Roseleur uses a solution, composed of five parts of fused (or six of crystals) stannous chloride, and fifty of pyrophosphate of potassium or sodium, added to 5,000 of distilled water; the chloride is dissolved in a portion of the water, and added the last, and the liquid is stirred until it is clear. A very large surface of anode is employed, and a strong electric current. As less tin is dissolved than is deposited, it is necessary to add occasionally, equal weights of the pyrophosphate and fused chloride.

Fearn's patent process for tinning, has been worked in Birmingham. The liquids employed are prepared as follows:—

No. 1. A solution of stannous chloride (not containing much free acid) is first made, containing three ounces of metallic tin per gallon. Thirty pounds of caustic potash are also dissolved in twenty gallons of water, and thirty ot pyrophospnate of sodium in sixty gallons of water. Two hundred ounces by measure of the tin solution, are poured slowly (whilst stirring with a glass rod) into the twenty gallons of potash liquid ; the precipitate formed re-dissolves quickly; into this liquid is poured, first all the cyanide solution, and then all the pyrophosphate, and the mixture stirred.

No. 2. Fifty-six pounds of salammoniac are dissolved in sixty gallons of water, and twenty of pyrophosphate of sodium in forty gallons of water ; and into the latter is poured 100 ounces by volume of the chloride of tin solution, and the mixture stirred ; the precipitate soon re-dissolves. The salammoniac solution is then added to the mixture, and the whole stirred.

No. 3. One hundred and fifty pounds of salammoniac are dissolved in 100 gallons of water ; and 200 fluid ounces of the tin solution poured into it, and the mixture well stirred.

No. 4. Four hundred ounces of tartrate of potassium, are dissolved in fifty gallons of water, and 1,200 ounces of solid caustic potash in another fifty gallons ; 600 fluid ounces of the tin solution are then added slowly, with stirring, to the liquid tartrate ; and then the caustic potash solution poured into the mixture, with continual and thorough agitation, to re-dissolve all the precipitate.

The first solution is used at 70° Fahr. with a current from two Bunsen's cells. The second is worked at 100° to 110° Fahr. with a weaker current. The third is used at 70° Fahr. And the fourth solution may be used cold. The first and fourth solutions, yield thick deposits, without requiring alternate deposition and scratchbrushing. As during working, more tin is deposited than dissolved, the oxide or other compound of the metal, must be added occasionally, except in the case of the third solution, which acts upon the anode more freely than the others. Articles of cast iron require to be covered with a thin film of copper, previous to being tinned in these liquids. Articles of zinc are tinned in No. 1 solution. Further particulars respecting the means necessary for keeping each particular mixture in proper working order, are given in the specification of the patent. The process is worked by the ' Electro-stannous Company,' in Birmingham.

Mr. Joseph Steele, coats zinc, iron, steel, copper, and brass, with tin, in his patent solution, by the battery process, thus :—Dissolve sixty pounds of common soda, fifteen of pearlash, five of caustic potash, and two ounces of cyanide of potassium, in seventy-five gallons of water, and filter the resulting solution ; then add two ounces of acetate of zinc, and sixteen pounds of peroxide of tin ; stir the resulting mixture until all is dissolved ; it is then ready for use. Work it by means of a separate current, with an anode of tin, keeping the liquid at 75° Fahr.

De Lobstein's patent solution, for tinning by means of the battery, is composed of 500 gallons of water, eighty pounds of caustic soda, thirty-four ounces of cyanide of potassium, and twenty-two of salts of tin, *i.e.*, stannous chloride.

The acetate and oxalate of tin, also oxide of tin dissolved in a solution of cyanide of potassium, have been tried for tinning, but do not appear to yield satisfactory results.

Processes of electro-tinning are not extensively used, because there appears to be no great demand for electro-tinned articles.

Electro-deposition of alloys of copper and tin.—A colour similar to that of bronze, is imparted to articles of iron or steel, by agitating them for a long time, in a solution composed of four to five parts of sulphate of copper, and four to five of crystallised stannous chloride, dissolved in 100 of water.

F. Weil coats articles of iron, steel, and other metals, with true bronze (*i.e.*, an alloy of copper and tin), by adding to his tartrate of copper bath (see p. 204), some stannate of sodium, or a solution of chloride of tin previously treated with a sufficiency of soda ; and immersing the articles in the mixture in contact with zinc (see ' Chemical News,' vol. xiii. p. 2). Salzede patented (Sept. 30, 1847) a liquid for depositing bronze by the battery process ; it consists of carbonate of potassium, chloride of copper, chloride of tin, nitrate of ammonium, and cyanide of potassium, dissolved in water, and used at a temperature of 77° Fahr. A solution patented by Newton (July 29, 1850), for a similar purpose, consists of the tartrates of copper, tin, and potassium.

M. Weis-Kopp imparts a bronze appearance to electro-coppered articles of cast iron, by rubbing them with a mixture of four parts of salammoniac, one of oxalic, and one of acetic acid, dissolved in thirty of water ('Chemical News,' vol. xxi. p. 47).

29. **Cadmium.**—Elec.- chem. eqt. $= \dfrac{112}{2} = 56 \cdot 0$. The most usual salts of cadmium, are the oxide, nitrate, chloride, bromide, iodide, sulphide, and sulphate. The nitrate, chloride, iodide, bromide, and sulphate, are colourless, and may be easily prepared, by saturating the respective acids, with either cadmium or its oxide, and evaporating the solutions until they crystallise. Fluoride of cadmium may be made, by adding the carbonate to an excess of dilute hydrofluoric acid, and evaporating the mixture to dryness. I have found that crystals of silicon, heated with this salt, separate the metal.

From a solution of the chloride, magnesium deposits, with strong action, a mixture of cadmium and an oxychloride of that metal. (Commaille, 'Chemical News,' vol. xiv. p. 188.)

Deposition of cadmium by contact with another metal.— According to Raoult, gold, or copper, in contact with cadmium, in a concentrated and boiling solution, of cadmium sulphate, or chloride, decomposes these salts, and quickly deposits a white, brilliant, and firmly adherent, but thin film of cadmium, upon the gold or copper, even when the solution is not acidulated and no hydrogen is evolved. The experiment does not succeed with the nitrate. But gold, in contact with iron, nickel, antimony, lead, copper, or silver, in cold or boiling, acid or neutral, solutions of salts of cadmium, receives no such deposit ('Chemical News,' vol. xxvi. p. 240; vol. xxvii. p. 59; 'Journal of the Chemical Society,' vol. xi. p. 464).

Deposition of cadmium by means of a separate current.— According to Smee, it is difficult to obtain firm, coherent, deposits of this metal, from solutions of either its chloride or sulphate, but it may be easily deposited in a reguline flexible state, from a solution of the ammonio-sulphate, prepared by adding sufficient aqueous ammonia, to a solution of sulphate of cadmium, to re-dissolve the precipitate.

T

A patent was taken out (March 19, 1849), by Messrs. Russell and Woolrich, for the electro-deposition of this metal, and the following is their description of the process :— 'Take cadmium, and dissolve it in nitric acid diluted with five or six times its bulk of water, at a temperature of about 80° or 100° Fahr., adding the dilute acid by degrees until the metal is all dissolved ; to this solution of cadmium, one of carbonate of sodium (made by dissolving one pound of the ordinary crystals of washing soda in one gallon of water) is added until the cadmium is all precipitated ; the precipitate thus obtained, is washed four or five times with tepid water; next add as much of a solution of cyanide of potassium as will dissolve the precipitate ; after which one-tenth more of the solution of potassium salt is added to form free cyanide. The strength of this mixture may vary, but the patentees prefer a solution containing six troy ounces of metal to the gallon. The liquid is worked at about 100° Fahr. with a plate of cadmium as an anode.' Very little has yet been done in the practical electro-deposition of this metal.

M. A. Bertrand recommends, for depositing cadmium, a solution of its bromide, containing a little sulphuric acid ; also one of its sulphate ; and says the deposit obtained is white, adheres firmly, is very coherent, and is capable of receiving a fine polish ('Chemical News,' vol. xxxiv. p. 227).

30. **Zinc.** — Elec. chem. eqt. $= \dfrac{65}{2} = 32\cdot5$. The most common salts of zinc are the oxide, chloride, carbonate, and sulphate. The oxide may be formed, by precipitating a solution of the sulphate with aqueous ammonia, and washing the precipitate ; the carbonate, by precipitating such a solution by means of washing soda. The various soluble salts, such as the nitrate, chloride, bromide, sulphate, acetate, &c., may be easily formed, by digesting, until it ceases to dissolve, an excess of metallic zinc, its oxide, or carbonate, with the

corresponding acid, diluted with water, and evaporating the clear solution until it crystallises.

I have found by experiment that a solution of potassic cyanide will dissolve only about one half as much cyanide of zinc as of cyanide of copper. Zinc oxide dissolves somewhat freely in a boiling solution of cyanide of potassium. Cyanide of zinc dissolves freely in a solution of sesqui-carbonate of ammonium. Ferro-cyanide of zinc is but feebly soluble in a boiling solution, either of ferro-cyanide (yellow prussiate), or of ferrid-cyanide (red prussiate) of potassium, but dissolves freely in a boiling solution of potassic cyanide. Zinc deposits spread over black-leaded surfaces by the battery process, in the same manner as copper.

Deposition of zinc by simple immersion (see p. 79).—Zinc is too electro-positive a metal to be readily set free by the simple immersion process, except by means of metals more electro-positive than itself, such as magnesium. Silicon separates zinc from its fluoride : I heated together 1·5 grains of crystals of silicon, and 10·25 of perfectly dry fluoride of zinc, in a porcelain crucible to a red heat. Chemical action occurred throughout the mass, and vapour of zinc escaped ; some of the zinc remained in the solid state on cooling, and evolved bubbles of hydrogen on adding dilute hydrochloric acid. According to A. Cossa, aluminium separates metallic zinc from an alkaline solution of zinc (Watts, ' Dictionary of Chemistry,' vol. v. p. 54). From a solution of the sulphate, magnesium deposits, with energetic action, a mixture of zinc, the hydrated oxide, and subsulphate (Commaille, ' Chemical News,' vol. xiv. p. 188). From slightly acid solutions of salts of zinc, magnesium deposits the pure metal, and hydrogen gas (Roussin, 'Chemical News,' vol. xiv. p. 27).

V. Roque coats articles of wrought and cast iron with zinc, in the following manner :—Mix together 1,000 measures of water, 550 of hydrochloric acid, fifty of sulphuric acid, and twenty of glycerine. Clean the iron in this mixture, and place it in a solution of one part of carbonate of potassium and ten of water. Then immerse it during from three to twelve hours (according to the thickness of the coating required), in a mixture composed of 1,000 parts of water, ten of chloride of aluminium, eight of bitartrate of potassium, five of chloride of tin, four of chloride of zinc, and four of acid sulphate of aluminium ('Chemical News,' vol. xxi. p. 288).

Depositing zinc by contact with a second metal (see p. 83).— Raoult states that gold or copper in contact with zinc, in a concentrated and boiling solution, of chloride or sulphate of zinc (but not in the nitrate), acquires a metallic deposit. But gold in contact with iron, nickel, antimony, lead, copper, or silver, in cold or boiling, acid or neutral, solutions of salts of zinc, receives no such coating ('Chemical News,' vol. xxvi. p. 240 ; vol. xxvii. p. 59 ; 'Journal of the Chemical Society,' vol. xi. p. 464).

Articles of copper, or brass, cleaned with hydrochloric acid, and immersed in contact with zinc, in a boiling saturated solution of salammoniac, or chloride of zinc, acquire in a few minutes, a specular covering of metal; but in a solution of cream of tartar, no such deposit occurs (R. Böttger, 'Gmelin's Handbook of Chemistry,' vol. i. p. 501). Another process is as follows :—Powdered zinc is added to a concentrated solution of salammoniac, and the liquid heated to boiling. The articles of copper or brass to be coated, are placed in the hot liquid in contact with zinc, and become covered with a brilliantly-white layer of adherent metal (Dr. R. Böttger, 'Chemical News,' vol. xxii. p. 108). F. Weil coats copper, or coppered metals, with zinc, by immersing them in a concentrated solution of potash or soda (heated to 100° C.), in contact with metallic zinc. The deposit is fixed and brilliant ('Chemical News,' vol. xiii. p. 2).

Deposition of zinc by means of a separate current (see p. 89).—Zinc may be deposited from its sulphate, ammonio-sulphate, chloride, ammonio-chloride, acetate, tartrate, &c., by the separate current process. As with nearly all other metals, the nitrate forms a bad depositing solution. By proper management, good coherent metal may be obtained from the sulphate, acetate, and chloride. A solution of zinc in caustic potash is not a good conductor ; a zinc anode does not readily dissolve in it ; similarly with the potassio-tartrate, and potassio-cyanide (Smee). A solution of one part of the sulphate in five to ten of water, with a large zinc anode, may be made to yield a good deposit, by a current from two small Smee's cells feebly charged.

Many years ago, sheets and other articles of iron, were coated with zinc by electrolysis, in order to protect them from rusting ; but this process has been entirely superseded by the so-called 'galvanising,' which is not a galvanic process at all, but consists of dipping the previously cleaned iron into a bath of melted zinc ; the latter being covered with a layer of saline flux, in order to prevent oxidation, and also to dissolve any trace of oxide which may be upon the iron articles. Such a coating of zinc is a much more effectual preventive of rusting than an electro-deposited one, because the heat expels all moisture from the pores of the iron, and the layer of zinc is homogeneous and not granular or porous, whilst that formed by voltaic action is always more or less porous and very liable to contain traces of the depositing liquid ; the surface beneath the electro-deposit, not having been heated before receiving the coating, is also liable to contain moisture and acid, absorbed during the preparatory processes of cleaning, &c.

Alexander Watt patented, in the year 1855, a process by means of which 'tough reguline' zinc might be deposited. He first makes a mixture, composed of twenty gallons of distilled water, 200 ounces of cyanide of potassium, and eighty by measure of the strongest aqueous ammonia. He then fills several large porous cells, with a solution composed of sixteen ounces of cyanide of potassium to each gallon of water, and partly immerses them in the other liquid. In the porous cells, he places sheets of copper or iron to act as cathodes, and in the outer liquid, clean pieces of zinc to act as anodes, and connects the battery in the usual way, until about sixty ounces of zinc are dissolved, and then stops the current and removes the porous vessels. He next dissolves eighty ounces of carbonate of potassium in a part of the zinc solution, and returns it to the original portion, and stirs the mixture thoroughly. After the sediment formed has subsided, he decants the clear liquid for use. Articles of iron may be coated in this liquid. Anodes of zinc are employed, and a little cyanide of potassium, and liquid ammonia, are occasionally added, if necessary. The battery preferred, is composed of two Bunsen's cells.

MM. Person and Sire employ a mixture composed of one part of oxide of zinc, dissolved in 100 of water containing ten of alum, at a temperature of 15° C. They use a single battery cell, and an anode of the same amount of surface as that of the articles. 'The deposition proceeds as easily as that of copper, and takes place indifferently on any metal— on platinum as well as upon copper and iron' ('Chemical News,' vol. ii. p. 275).

Estimation of zinc by means of the battery.—J. M. Merrick electrolysed known weights of pure double sulphate of zinc and ammonium, in aqueous solution, in a covered platinum crucible, with a platinum wire for the anode, and the crucible for the cathode (using a current from two or three Grove's cells), until all the metal was deposited ; and then weighed the deposit. In two analyses, 16·16 and 16·31 per cent of zinc was obtained ; theory requiring 16·20 per cent. The deposits were washed with alcohol, and cautiously dried ('Chemical News,' vol. xxiv. pp. 100, 172).

Electro-deposition of alloys of copper and zinc.—There are various solutions for depositing brass. As early as the year 1841, M. de Ruolz deposited it by the battery process, from the mixed cyanides of zinc and copper, dissolved in a solution of cyanide of potassium. Whenever zinc is electrodeposited upon perfectly clean copper, the first film deposited, produces a yellow colour, by uniting with the copper. In accordance with this, copper articles may be superficially brassed, by immersing them either in a boiling solution of bi-tartrate of potassium, containing zinc amalgam, or in the same liquid, after some dilute hydrochloric acid has been added to it. Thicker coatings may be formed upon articles, by de-positing upon them, alternate thin films of zinc and copper, by the separate current method. Processes have also been patented, for coating iron and steel with brass, by depositing a layer of copper, and then one of zinc, upon them, and heat-ing them until the two metals more perfectly alloy with each other (see MM. Person and Sire's process, ' Chemical News,' vol. ii. p. 275).

A good solution for brassing by means of a separate current, with an anode of brass, may be made by dissolving nine or ten ounces of the strongest aqueous ammonia, six-teen to twenty of cyanide of potassium (with or without the addition of twenty of the strongest aqueous hydrocyanic acid, 'Scheele's strength,') in 160 (i.e. one gallon) of water, and saturating the hot liquid with brass by means of an electric current ; it must be used at 212° Fahr.

Brunel, Bisson, and Gaugain's formula, for an electro-brassing solution, consists of fifty parts of carbonate of potas-sium, two of chloride of copper, four of sulphate of zinc, and twenty-five of nitrate of ammonium, dissolved together in 200 parts of cold water, and used with a brass anode, and a strong battery. They also give a second formula, viz. :—
Take twelve and a half gallons of water, and dissolve in it, ten ounces of chloride of copper, twenty of sulphate of zinc, twenty-four of cyanide of potassium, and 160 of carbonate of potassium ; add the cyanide the last.

Salzede's patent, dated September 30, 1847 :—To form the solution, take 5,000 parts of water, dissolve twelve parts of cyanide of potassium in 120 parts of it, then add 610 of carbonate of potassium, forty-eight of sulphate of zinc, and twenty-five of chloride of copper, to the remainder of the

water, and heat the mixture from 144° to 172° Fahr.; and
when the salts are entirely dissolved, add 305 parts of nitrate
of ammonium, allow the liquid to remain undisturbed for
twenty hours, and then add the solution of cyanide of potassium; allow it to remain again till clear, and then draw off
the transparent liquid, which is ready for use; work the solution with a large brass anode and a strong battery. Another
liquid which he uses for brassing, consists of 5,000 parts of
water, 500 of carbonate of potassium, thirty-five of sulphate
of zinc, fifteen of chloride of copper, and fifty of cyanide
of potassium.

Russell and Woolrich's patent, dated March 19, 1849 :—
Take ten pounds of acetate of copper, one of acetate of zinc,
ten of acetate of potassium, and five gallons of hot water;
dissolve the salts in the water, add as much of a solution of
cyanide of potassium as will precipitate the mixture and just
re-dissolve the precipitate; and then add about one-tenth
more of the cyanide.   Use a brass anode, or else two anodes,
one of zinc and one of copper.

Joseph Steele's patent, dated August 9, 1850 :—Dissolve
two and a quarter pounds of American potash in six gallons
of hot water, and filter the solution; also dissolve two and a
half ounces of acetate of copper in half a pint of strong
liquid ammonia, and add it to the first liquid, with stirring;
then add four or five ounces of sulphate of zinc, and stir till
dissolved; and finally add two ounces of cyanide of potassium; filter the resulting solution, and use it at 100° Fahr.,
with a brass anode.   To obtain a dark-coloured brass, add
more acetate of copper; and to obtain it of a lighter colour,
add more sulphate of zinc.

Mr. Wood's solution is composed thus:—Dissolve one
pound (troy weight) of cyanide of potassium, two ounces of
cyanide of copper, and one of cyanide of zinc, in one gallon
of distilled water, and add two ounces of salammoniac.
For coating smooth articles, use the solution at 160° Fahr.

with a battery of from three to twelve Grove's cells. It is suitable for coating iron ('Scientific American').

Mr. Watt gives the following formula for a brassing liquid :—Acetate of copper five parts, cyanide of potassium eight, sulphate of zinc ten, liquid ammonia forty, and caustic potash seventy-two parts. Reduce the copper salt to powder, and dissolve it in eighty parts of water ; then add twenty of the aqueous ammonia. Dissolve the zinc salt in 160 parts of water at 180° Fahr., add the remaining twenty of ammonia to it, and stir the mixture strongly. Dissolve the potash in 160 parts of water, and the cyanide in 160 of hot water. Add the solution of copper to that of zinc, then add the caustic potash, and then the cyanide. Dilute the mixture to eight gallons by addition of water, and thoroughly stir the solution. Use a strong battery, add a little ammonia occasionally, and when it works slowly, add cyanide. Keep the brass anode clean.

According to Dr. Heeren, a brassing solution may be formed, by taking a mixture containing a great excess of zinc and very little copper, thus :—Dissolve one part of sulphate of copper, eight of sulphate of zinc, and eighteen of cyanide of potassium, in separate portions of warm water. Mix the copper and zinc solutions, then add the dissolved cyanide and a further quantity of 250 parts of distilled water, and stir the mixture. The bath is used at a boiling temperature, with a current from two Bunsen's cells. Very rapid deposits of brass have been thus obtained upon articles of copper, zinc, brass, and Britannia-metal ('The Chemist,' No. 16, January 1855, New Series, p. 342).

Morris and Johnson's patent, dated December 11, 1852. According to this, dissolve one pound of cyanide of potassium, one of commercial carbonate of ammonium, two ounces of cyanide of copper, and one of cyanide of zinc, in a gallon of water, and use the solution at 150° Fahr. with a large anode of brass and a powerful battery. Or a solution may be taken of one pound of cyanide of potassium, and one of carbonate of ammonium, dissolved in one gallon of water, and saturated with copper and zinc to the requisite degree by means of a strong current, a large brass anode and a small cathode, until the latter receives a good deposit of brass, the liquid being at a temperature of 150° Fahr. To increase the proportion of copper in the deposit, either add cyanide of potassium, or raise the temperature of the liquid ; and to increase that of zinc, either add carbonate of ammonium, or lower the temperature.

Roseleur gives the following recipes for making brassing solutions. No. 1.—Dissolve in 1,000 parts of water, twenty-five of sulphate of copper, and twenty-five to thirty of sulphate of zinc ; or twelve and a half of acetate of copper, and twelve and a half to fifteen of fused chloride of zinc. Precipitate the mixture, by means of 100 parts of carbonate of sodium dissolved in plenty of water, and stir the mixture. Wash the precipitate several times by adding water to it, stirring, allowing the precipitate to subside, and pouring the clear liquid away. Add to the washed precipitate, a solution composed of fifty parts of bisulphite of sodium, and 100 of carbonate of sodium, dissolved in 1,000 of water, and, whilst stirring with a wooden rod, add a strong solution of ordinary cyanide of potassium, until the precipitate is just all re-dissolved. Then add two and a half or three parts of free cyanide. No. 2.—To form a cold bath for brassing all metals, dissolve in 200 parts of water, fifteen of cupric sulphate, and fifteen of sulphate of zinc, and then add a solution of forty parts of carbonate of sodium dissolved in 100 of water, and stir the mixture. Allow the precipitate to subside, throw away the clear liquid, and wash the sediment by addition of water, followed by subsidence, and decantation. Add to the drained and wet precipitate, 900 parts of water, containing twenty of bisulphite and twenty of carbonate of sodium dissolved in it. Dissolve twenty parts of cyanide of potassium, and two-tenths of a part of arsenious acid (white arsenic), in 100 of water, and add it to the previous liquid; this decolourises the mixture, and completes the brassing solution. The arsenious acid causes the deposit to be bright, but if added in too large a quantity, whitens it; this latter effect, however, soon disappears by working the liquid. In using this bath, if the deposit is dull, add a little of the arsenic; if it looks earthy or ochreous, add cyanide; if too red, add zinc salt, and if necessary also cyanide; if too white, add copper and cyanide; if its action is very slow, add copper and zinc salts, and if necessary also cyanide. As the brass is deposited faster than it dissolves, salts of copper, and of zinc, with cyanide, must be added occasionally. When by addition of these substances, the specific gravity of the solution becomes greater than 1·091, the solution must be diluted with water to a sp. gr. not below 1·036. No. 3.—For coating steel, cast iron, wrought iron, and tin :—Dissolve two parts of bisulphite of sodium, five of cyanide of potassium (of 75 per cent.), and ten of carbonate of sodium, in eighty of distilled water; and add to the mixture, one part of fused chloride of zinc, and one and a quarter parts of acetate of copper, dissolved in twenty of water. No. 4.—For coating articles of zinc :—Dissolve twenty parts of bisulphite of sodium, and 100 of cyanide of potassium (of 75 per cent.), in 2,000 of water; and dissolve thirty-five parts of chloride of zinc, thirty-five of acetate of copper, and forty of aqueous ammonia, in 500 of water; mix the two solutions, and filter the mixture. Too strong a current in these solutions usually makes the deposit white, and too weak a one, or keeping the articles in motion, makes it red. The proportion of zinc in the baths may be increased by using a zinc anode, and of copper by employing a copper one. The brass anodes may be kept free from undissolved oxide of zinc, by adding the minimum quantity of ammonia; but that must not be added to cold solutions used for brassing iron. To preserve the colour of the deposit, rinse the coated articles in water, then in water made slightly alkaline by addition of caustic lime, and then thoroughly dry them in a stove.

For coating cast iron with brass, Walenn recommends a bath, composed of an aqueous solution of equal parts of ammonic-tartrate and potassic-cyanide. After addition of cyanide of copper, and cyanide of zinc, in certain proportions, the oxides of the metals are added to the solution. If upon trial, hydrogen is evolved at the cathode, a little ammoniuret of copper is added to the cold bath. The temperature at which this liquid is used, determines the colour of the brass, and may vary from 60° C. to nearly the boiling point ('Chemical News,' vol. xxi. p. 273). He also makes the following remarks on the electro-deposition of copper and brass. 'A solution containing one pound of cupric sulphate, and one of sulphuric acid, to a gallon of water, deposits the metal in a solid compact mass, with, a somewhat botryoidal surface. The addition of one ounce of zinc sulphate (as recommended by Napier) prevents this botryoidal form, and renders the deposit tough, compact, and even. From a solution containing a greater proportion of zinc sulphate, the copper is deposited in tufts of needles standing at right angles to the surface of the metal. Ordinary electro-brassing liquids show the same peculiarity in even a more marked degree, and this makes it impossible to produce a good deposit of more than ·01 to ·03 inch in thickness. This form of deposit is owing chiefly to a copious evolution of hydrogen taking place during its formation.' However the author has found, that by employing a solution containing both the oxides and the cyanides of the two metals, together with some neutral tartrate of ammonium, this evolution of hydrogen may usually be avoided, or, should it, nevertheless, take place to a slight extent, it may be entirely stopped by the addition of some cupric ammonide. Such a solution yields brass of a uniform character, and the deposit, which may be obtained of any desired thickness, is tough, and has a compact, even texture. As there is no evolution of hydrogen, no electric force is wasted, and perfect results may be obtained with a single Wollaston's or Smee's cell ; but in practice a little stronger current than this is employed, in order to hasten the process (' Philosophical Magazine,' 4th Series, vol. xli. p. 41 ; ' Chemical News,' vol. xxii. pp. 1, 181 ; ' Journal of the Chemical Society,' vol. x. p. 103).

Deposition of other alloys of zinc.—Watt states that he has ' succeeded in depositing an alloy of copper, zinc, and nickel, forming a very good quality of german-silver, by dissolving german-silver in nitric acid, precipitating with an alkali, and re-dissolving with cyanide of potassium ' (' Electro-Metallurgy,' by A. Watt ; Weale's ' Elementary Series,' p. 91).

Newton patented (July 29, 1850) a solution for depositing an alloy of copper, tin, and zinc ; it consisted of the double cyanide of copper and potassium, in combination with ' zincate ' and stannate of potassium, ' or the double tartrates of the metals and potassium, may be used.'

Messrs. Morris and Johnson (according to their patent), deposit german-silver by the following process :—Dissolve one pound of cyanide of potassium, and one of carbonate of ammonium, in a gallon of water, heat the solution to 150° Fahr., immerse a large anode of german-silver in the liquid, and a small cathode of any suitable metal, connect the two with a powerful battery, and pass the current of electricity until a considerable amount of metal is dissolved, and a bright cathode receives a deposit of good colour ; the solution is then ready for use. If the deposit becomes too red, add carbonate of ammonium ; if too white, add cyanide.

Separation of copper and zinc by electro-deposition.—According to M. Hautefeuille, the oxides of copper and zinc, dissolved in aqueous ammonia, may be separated by the following process :—Add an excess of acetic acid to the solution, immerse a clean sheet of lead, and boil the mixture two hours, or until it becomes quite colourless ; the whole of the copper is then precipitated in the metallic state (' The Chemist,' New Series, part xviii., March 1855, p. 334).

CLASS V.   EARTH AND ALKALINE EARTH METALS.

MAGNESIUM—CERIUM—LANTHANUM—DIDYMIUM— GALLIUM — ALUMINIUM — GLUCINIUM — CALCIUM — STRONTIUM — BARIUM.

31. **Magnesium.**—Elec.-chem. eqt. $=\dfrac{24\cdot3}{2}=12\cdot15$. The ordinary salts of this metal are the oxide (calcined magnesia), the nitrate, chloride, carbonate, and sulphate; the most frequent impurity in them is lime.   As the metallic magnesium of commerce is free from calcium, and very pure, extra pure salts of the metal may be made by means of it.   The nitrate, chloride, and sulphate, may be made by saturating the corresponding diluted acids, with the metal, the oxide, or carbonate, and evaporating the solutions.   In evaporating a solution of the chloride to dryness, towards the end of the operation, some of the salt is decomposed by the watery vapour, hydrochloric acid being formed, and oxide of magnesium left ; and to obviate this, a portion of salammoniac is added when the solution has become very concentrated.

Magnesium is highly electro-positive ; it precipitates as metal, from solutions of their salts, bismuth, platinum, gold, silver, mercury, copper, lead, thallium, tin, and cadmium (Roussin, 'Chemical News,' vol. xiv. p. 27).   According to Phipson, magnesium deposits nearly all metals from their neutral solutions (even iron and manganese from ferrous and manganous salts) in the metallic state.   It deposits platinum, gold, silver, bismuth, mercury, copper, nickel, cobalt, iron, lead, thallium, tin, cadmium, and zinc, but not aluminium (Watts, ' Dictionary of Chemistry,' vol. v. p. 795).

I melted to a perfect liquid, at nearly a white heat, a mixture of six grains of magnesic fluoride, and four of calcic fluoride; then added two grains of crystals of silicon, the crystals were not dissolved, and there appeared no signs of metallic magnesium having been separated.

Electrolysis of salts of magnesium.—When moistened sulphate of magnesium is formed into a cup upon a platinum plate, the cup filled with mercury and made the cathode, and the platinum the anode, with a current from a powerful battery, magnesium is deposited into the mercury.

Bunsen electrolysed fused chloride of magnesium, at a red heat, in a deep and covered porcelain crucible, divided by a vertical partition of porous porcelain, which extended half-way down the vessel. The electrodes were of carbon, and introduced through openings in the cover ; and the current was from ten cells of a zinc and carbon battery. As magnesium is a light metal, it would rise to the surface of the mixture and burn in the air ; and, in order to prevent this as much as possible, the cathode was notched (see fig. 28), so that the melted metal collected in the notches. According to Matthiessen, a fused mixture, in the proportions of four

FIG. 28.

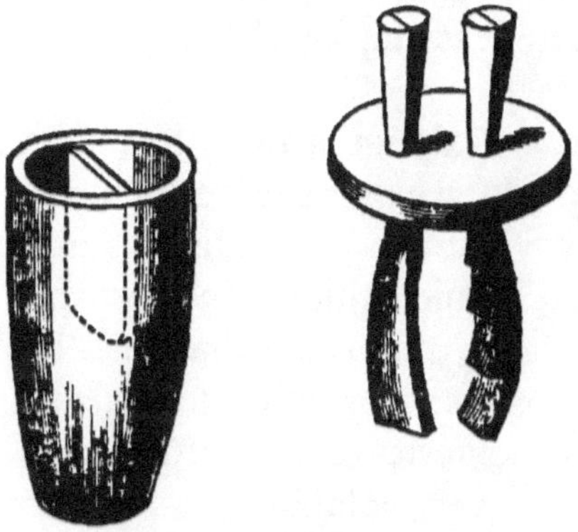

molecules of magnesic, and three of potassic chloride, forms a much better electrolyte, because the magnesium sinks in it. 'A solution of the chloride of sodium or ammonium, electrolysed with electrodes of magnesium wire, deposits a black powder of suboxide of magnesium upon the anodes (W. Beetz, 'Watts' Chemical Dictionary,' Supplement, p. 796)

According to M. A. Bertrand, a strong current deposits magnesium in a few minutes upon a sheet of copper, in an aqueous solution of the double chloride of magnesium and ammonium. The deposit is homogeneous, strongly adherent, and polishes readily ('Chemical News,' vol. xxxiv. p. 227).

32. Cerium, lanthanum, and didymium.—To deposit either of these metals, its chloride is mixed with salammoniac (both as dry as possible), and the mixture heated to redness in a platinum crucible to expel all the salammoniac. A porous clay vessel, of the best quality, is filled with the residue, then placed in a Hessian crucible, surrounded by a cylinder of sheet iron (with a long projecting strip for connexion), to serve as the anode, and the space between the two vessels filled with a previously melted mixture of an equal number of equivalents of the chlorides of potassium and sodium. A thick iron wire, enclosed in a clay pipe, has a coil of very fine iron wire at its extremity to serve as the cathode, and is immersed in the fused salt in the inner vessel. The fusion is effected by preference in a fire of glowing charcoal, to prevent as far as possible the presence of aqueous vapour, and a strong current is employed (Bunsen, 'Electrical News,' vol. i. p. 184).

33. Gallium.—Gallium is allied to aluminium. It melts easily (at $29 \cdot 5°$ C.), when held between one's fingers; but does not easily volatilise or oxidise, even when heated to bright redness. Liquid gallium is whiter than mercury; the solid metal is hard, and somewhat malleable, of specific gravity $4 \cdot 7$, not oxidised by cold nitric acid, but dissolves in cold dilute hydrochloric or hot nitric acid The oxide of this metal is not very soluble in aqueous ammonia, but a solution of caustic potash dissolves a large quantity of it. The metal has been obtained by the electrolysis of both these solutions, formed by adding an excess of those alkalies to the sulphate of gallium. It is deposited upon the platinum cathode as a dead, whitish-grey coating, formed of minute globules like mercury ('Chemical News,' vol. xxxiii. p. 193).

34. **Aluminium.**—Elec. chem. eqt. $= \dfrac{27\cdot5}{3} = 9\cdot16$. The
oxide (alumina), chloride, sulphate, alumand (a double sulphate of aluminium and an alkaline metal, usually ammonium, or potassium) are its most common salts ; the chloride
sulphate, and alum, are freely soluble in water.

Magnesium does not deposit aluminium as metal from
its solutions (Roussin, ' Chemical News,' vol. xiv. p. 27).

Electrolysis of salts of aluminium.—With regard to the
separation of aluminium by means of electrolysis, M. H.
Sainte-Claire Deville says :—' It appeared to me impossible
to obtain aluminium by the battery in aqueous liquids. I
should believe this to be an impossibility, if the brilliant experiments of M. Bunsen on the production of barium, did
not shake my conviction. Still I may say, that all processes
of this description which have recently been published for the
preparation of aluminium, have failed to give me good results.
It is of the double chloride of aluminium and sodium, of
which I have already spoken, that this decomposition is
effected. The bath is composed of two parts, by weight, of
chloride of aluminium, with the addition of one part of dry
and pulverised common salt. The whole is mixed in a
porcelain crucible, heated to about 392° Fahr. The combination is effected with disengagement of heat, and a liquid
is obtained which is very fluid at 392° Fahr., and fixes at
that temperature. It is introduced into a vessel of glazed
porcelain, which is to be kept at a temperature of about
392° Fahr. The cathode is a plate of platinum, on which
the aluminium (mixed with common salt) is deposited in
the form of a greyish crust. The anode is formed of a
cylinder of charcoal, placed in a perfectly dry, porous
vessel, containing melted chloride of aluminium and sodium.
(The densest charcoal rapidly disintegrates in the bath, and
becomes pulverulent, hence the necessity of the porous
vessel.) The chlorine is thus removed, with a little chloride
of aluminium proceeding from the decomposition of the

double salt. Thïs chloride would volatilise and be entirely lost, if some common salt were not in the porous vessel. The double chloride becomes fixed, and the vapours cease. A small number of voltaic elements (two are all that are absolutely necessary) will suffice for the decomposition of the double chloride, which presents but little resistance to the electricity. The platinum plate is removed when it is sufficiently charged with the metallic deposit. It is suffered to cool, the saline mass is rapidly broken off, and the plate replaced' ('The Chemist,' New Series, No. 13, October 1854, p. 12).

M. Duvivier states, that by passing an electric current from eighty Bunsen's cells, through a small piece of 'laminæ disthene' between two carbon points, the disthene melted entirely after two or three minutes; the elements which composed it, were partly disunited by the power of the electric current, and the aluminium freed from its oxygen. Several globules of the metal separated, one of which was as white and as hard as silver ('The Chemist,' New Series, No. xi., August 1854, p. 687).

Bunsen electrolysed fused chloride of aluminium and sodium, by a similar process to that already described in depositing magnesium (see p. 287). The salt fused at 356° C., and readily yielded the metal. The temperature of the liquid should be raised nearly to the melting-point of silver; the particles of liberated aluminium then fuse, unite together, and form globules, which, being of greater specific gravity than the salt, fall to the bottom of the crucible. I electrolysed a strong solution of pure fluoride of aluminium, containing free hydrofluoric acid, using the platinum containing vessel as the cathode, and a sheet of platinum as the anode; gas was evolved freely from the latter, and the liquid became heated.

M. Corbelli, of Florence, deposits aluminium, by electrolysing a mixture of rock-alum, or sulphate of aluminium, and the chlorides of calcium or of sodium, the anode being formed of iron wire, coated with an insulating material, and dipping into mercury placed at the bottom of the solution ; and the cathode of zinc immersed in the solution. Aluminium is then deposited upon the zinc, and the chlorine which is eliminated at the anode, unites with the mercury, and forms calomel (Watts, ' Dictionary of Chemistry,' vol. i. p. 152).

Thomas and Tilley took out a patent Dec. 26, 1854, for depositing aluminium from a solution, composed of freshly precipitated alumina, dissolved in boiling water containing cyanide of potassium; and another, Dec. 6, 1855, for depositing it from a solution of calcined alum in aqueous cyanide of potassium ; also from several other liquids ; and for depositing alloys of aluminium and silver; aluminium, silver, and copper; aluminium and tin ; aluminium, silver, and tin ; aluminium and copper ; aluminium and nickel ; aluminium and iron, &c.

J. B. Thompson states, that he has for more than two years, been depositing aluminium on iron, steel, and other metals, at a temperature of about 500° Fahr., and also depositing aluminium bronze of various tints, from the palest yellow to the richest gold colour (' Chemical News,' vol. xxiv. p. 194).

Jeancon patented a process, for depositing aluminium from an aqueous solution of a double salt of that metal and potassium, of sp. gr. 1·161, by means of a current from three Bunsen's cells, the solution being at 140° Fahr. (' Telegraphic Journal,' vol. i. p. 308). T. Bell also patented a process, for depositing aluminium upon other metals, from the double chloride of that metal and potassium (' Chemical News,' vol. v. p. 153). M. A. Bertrand states, that he has deposited aluminium upon a plate of copper, in a solution of the double chloride of aluminium and ammonium, by employing a strong current ; and that the deposit was capable of receiving a brilliant polish (' Chemical News, vol. xxxiv. p. 227).

According to C. Winckler, plating with aluminium cannot be effected by electro-deposition ('Chemical News,' vol. xxvi. p. 157; 'Journal of the Chemical Society,' vol. x. p. 1134). Sprague also states his inability to deposit that metal (Sprague, 'Electricity,' p. 309).

Aluminium used as an anode in electrolysing dilute sulphuric acid, stops the current; but not if it is employed as the cathode ('Telegraphic Journal,' vol. iii. p. 59; 'Chemical News,' vol. xxxi. p. 99).

35. **Glucinium.**—Elec.-chem. eqt. $= \dfrac{9\cdot3}{2} = 4\cdot65$. The commonest salts of this metal, are the carbonate and sulphate; the latter is freely soluble in water; the nitrate and chloride are also very soluble, and may be easily made, by adding sufficient of the carbonate to the respective acids, and evaporating the solutions.

Becquerel deposited the pure metal from a concentrated solution of its chloride (by means of a current from twenty cells), in the form of brilliant, steel-grey, crystalline laminæ (Gmelin's 'Handbook of Chemistry,' vol. iii. p. 293).

36. **Calcium.**—Elec.-chem. eqt. $= \dfrac{40}{2} = 20$. The common salts of calcium, are the oxide (caustic lime), the nitrate, fluoride, chloride, bromide, carbonate (chalk or whiting), sulphate, phosphate, &c. Ordinary caustic lime is often impure; a much purer kind may be obtained, by heating the pure carbonate to full redness; the latter may be obtained as follows: Add an excess of clear lime-water to a solution of nitrate of calcium; filter the liquid, and precipitate it with a mixture of ammonia and carbonate of ammonium, dissolved in water, and wash the precipitate thoroughly. Pure soluble salts of lime may be obtained, by neutralising the respective pure acids with the pure carbonate, and evaporating the clear solutions. The sulphate of calcium is only sparingly soluble in water, the fluoride and phosphate much less so; caustic lime also does not dissolve very freely.

Electrolysis of salts of calcium.—Sir H. Davy first deposited impure calcium by electrolysis into a cathode of mercury during the year 1808; and since that period, several other investigators have effected the same object, by means of much less powerful electric currents.

M. Frèmy fused pure fluoride of calcium in a platinum crucible, and electrolysed it; a brisk effervescence occurred in the mass, and a gas was evolved at the positive pole, which corroded glass; metallic calcium was deposited upon the cathode, and was at once converted into lime by the atmosphere. It was difficult to make the observations, and the crucible was soon alloyed, and leaked, and the melted fluoride escaped ('The Chemist,' New Series, vol. ii. p. 548).

Matthiessen melted and electrolysed a mixture, in the proportions of two molecules of chloride of calcium, and one of chloride of strontium, with a small quantity of sal ammoniac, in a porcelain crucible, with an anode of gas-carbon, and a cathode formed by winding a thin iron wire round a thicker one, and dipping it only just into the fused mixture. The calcium was deposited in beads upon the fine wire.

Bunsen electro-deposited metallic calcium, in a similar manner to that employed for manganese and chromium (see pp. 250, 253), except that the density of the current employed was greater. He acidulated a concentrated and boiling hot solution of the chloride, with hydrochloric acid, poured the boiling liquid into the porous cell, and employed as a cathode, an amalgamated platinum wire. The calcium was deposited as a grey layer upon the amalgamated surface. The process is difficult, because the calcium quickly oxidizes to a layer of lime, which covers the cathode, and stops the current. The deposit must be frequently removed, and the wire freshly amalgamated, each time before re-immersion; and even then but a small amount of the metal is obtained ('The Chemist,' New Series, vol. i. part ii., August 1854, p. 686).

Herschel observed, that in a solution of chloride of calcium undergoing electrolysis, the cathode evolved gas, and became covered with caustic hydrate of lime.

37. **Strontium.**—Elec.-chem. eqt. $= \dfrac{87 \cdot 5}{2} = 43 \cdot 75$. The ordinary salts of strontium, are the nitrate, chloride, carbonate, and sulphate ; the two former are freely soluble in water, and the two latter insoluble.

Silicon does not separate strontium from its fluoride. I heated to redness, a mixture of crystals of silicon and fluoride of strontium ; the crystals suffered no loss of weight, did not appear corroded, and no signs of free strontium were obtained.

Electrolysis of salts of strontium.—Sir H. Davy first deposited this metal by electrolysis, in the year 1808, by forming into a cup, a pasty mass of its carbonate with water, and placing it upon a platinum dish as an anode, the cup being filled with mercury to act as the cathode. On passing a current from a 500-cell battery through it, the strontium was deposited upon, and absorbed by, the mercury.

Bunsen obtained strontium, in a precisely similar way to that of obtaining calcium (p. 293), using a salt of strontium instead of one of that metal (Watts, ' Dictionary of Chemistry,' vol. ii. p. 437). Matthiessen obtained it from the fused chloride in the following manner :—A small porous cell was placed in a porcelain crucible, and both vessels nearly filled with anhydrous chloride of strontium, the level of that in the porous cell being the highest. The salt was melted so that a crust appeared on its surface. The cathode consisted of a thick iron wire, enclosed in the stem of a tobacco-pipe, so that only $\frac{1}{20}$th of an inch of it projected at the lower end, round which a very thin iron wire was coiled. The anode was a cylinder of sheet iron placed in the outer space. The cathode was immersed in the inner vessel, and the current passed ; the metal collected upon it beneath the crust (Watts, ' Dictionary of Chemistry,' vol. ii. p. 438).

Strontium is electro-negative to potassium and sodium in water, but positive to magnesium.

38. **Barium.**—Elect.-chem. eqt. $= \dfrac{137}{2} = 68\cdot5$.   Its ordinary salts are, the oxide, hydrated oxide (both termed caustic baryta), nitrate, chloride, carbonate, and sulphate. The oxide, nitrate, and chloride are soluble in water; the carbonate and sulphate are insoluble.

Crookes deposited barium, by simple immersion of an alloy of sodium and mercury in a saturated solution of barium chloride at 93° C.   The deposited metal dissolved in the mercury, and formed an amalgam ('Chemical News,' vol. vi. p. 194; Watts, 'Dictionary of Chemistry,' vol. v. p. 252).

Electrolysis of salts of barium.—Davy first deposited barium, by passing a powerful electric current through a concentrated solution of hydrate of baryta, into a cathode of mercury; the deposit formed an amalgam with that metal.

Bunsen obtained barium, by electrolysis of a boiling hot, concentrated, and acidulated solution of its chloride, in a similar manner to that of separating calcium (see p. 293); it was more easily obtained.   The deposit upon the mercurial surface formed an amalgam, which was silvery-white, and very crystalline ('The Chemist,' New Series, vol. i. p. 686; Watts, ' Dictionary of Chemistry,' vol. i. p. 500).

Matthiessen obtained barium from its chloride, in a similar manner to that in which he deposited strontium (see p. 294).

A solution of nitrate of barium, electrolysed by means of platinum electrodes, yields nitric acid at the anode, and baryta at the cathode (Sir H. Davy).

CLASS VI.   ALKALI-METALS.

LITHIUM—SODIUM—POTASSIUM—RUBIDIUM—CÆSIUM—
AMMONIUM.

39. **Lithium.**—Elec.-chem. eqt. $= 7$.   The most common salt of lithium, is the carbonate, and from this the other salts

may be easily prepared, by adding an excess of it to the par-. ticular acid, filtering the mixture, and crystallising the solution. All the salts of lithium (except the phosphate), formed by adding the carbonate to the common mineral acids, are freely soluble in water.

I added crystals of silicon to a fused mixture of the fluorides of lithium and sodium, but the silicon was not corroded or diminished in weight, and lithium was not deposited.

Electrolysis of salts of lithium.—Bunsen was the first person who electro-deposited this metal (Watts, ' Dictionary of Chemistry,' vol. iii. p. 727). By electrolysing fused chloride of lithium, with a current from four or six Bunsen's cells, an anode of gas-coke, and a cathode of iron wire, he deposited silver-white metal upon the wire (Watts' 'Dictionary of Chemistry,' vol. ii. p. 437, and vol. iii. p. 727). Schnitzler also electrolysed a mixture of the fused chlorides of lithium and ammonium, with a current from twelve Bunsen's cells, and a cathode of iron wire ('Journal of the Chemical Society,' vol. xii. p. 961).

FIG. 29.

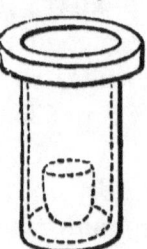

I fused some fluoride of lithium in an open platinum crucible, within a partially covered clay muffle (see fig. 29), and electrolysed it by means of a current from six Smee's elements, and two flat platinum wire helices as electrodes, during thirty minutes. The conduction was free, and much gas was evolved from the anode only, all the time. The anode was not corroded; a small amount of lithium was deposited upon the platinum cathode, and alloyed with it. By electrolysing a larger mass of the salt with a current from six Grove's cells and a thick platinum wire cathode, enclosed within, but insulated from, a platinum tube, to exclude the air from contact with the deposited lithium, the action was copious ; with a gold anode, the gold was corroded freely, and particles of it in large quantity, floated in the liquid and united the electrodes. The cathode swelled greatly, and its lower end, bent itself towards the anode, became quite grey in colour, and split in the direction of its length.

40. **Sodium.**—Elec.-chem. eqt. = 23. Nearly all the common salts of sodium are freely soluble in water, and may be readily obtained in a pure state, by neutralising the pure acids with pure carbonate of sodium, and evaporating the solutions. The fluoride is one of the least soluble.

Soon after Sir H. Davy first isolated sodium, Gay-Lussac and Thenard showed, that iron at a white heat, set free the metal from caustic soda. Other investigators soon found that carbon acted similarly. I have also observed that crystals of silicon, thrown into melted fluoride of sodium, evolved small bubbles of vapour, which exploded and burned with a yellow flame, on arriving at the surface of the liquid. In a second experiment, seven grains of the dry fluoride in powder, and one grain of the crystals, were mixed, and heated to redness ; the silicon lost ·15 of a grain in weight. .

Electrolysis of salts of sodium.—Sir H. Davy first electro-deposited sodium in the year 1807, by moistening its hydrate with water in a platinum capsule, which acted as the anode, dipping a platinum wire cathode in the salt, and using a current from a battery composed of 100 to 200 cells. He also deposited it more easily into mercury, in a similar way to that already described under magnesium (see p. 287), and thus obtained an amalgam of the two metals.

According to Faraday, fused borax is decomposed by electrolysis into oxygen and sodium, and the latter takes oxygen from the boracic acid and sets free boron (see p. 46). According to Burckhard, ' Carbonate of sodium in a fused state is a good conductor ; by electrolysis it is decomposed into carbonic acid and soda, but a small portion of carbon is also set free.' He also states, that fused pyrophosphate of sodium, electrolysed with platinum electrodes, yields phosphide of platinum, but the chief result is, that the salt splits up into oxygen, phosphorus, and soda ('Chemical News,' vol. xxi. p. 238). In the electrolysis of melted sulphate of sodium with platinum electrodes, sodium is deposited, and combines with the cathode (Brester, 'Chemical News,' vol. xviii. p. 145). A fused mixture of the chlorides of calcium and of sodium, yields a deposit of the latter metal, when electrolysed in a certain manner (Matthiessen, Watts, 'Dictionary of Chemistry,' vol. i. p. 715).

By electrolysing a solution of common salt, Higgins and Draper observed, that chlorine was set free at the anode, and hydrogen gas and soda at the cathode; but if the cathode consisted of mercury, sodium amalgam was produced.

Hisinger and Berzelius, electrolysed a solution of common salt, with silver electrodes; gas was evolved at the cathode, and after a time at the anode also; the anode became covered with argentic chloride, the liquid near it contained dissolved chlorine, and the solution near the cathode contained free soda; with lead electrodes. the negative wire evolved gas, and received a deposit of crystals of lead, and the anode became coated with plumbic chloride.

I electrolysed a saturated aqueous solution of sodic fluoride, by a current from six Grove's cells, with platinum electrodes; gas was evolved from the anode, and emitted an odour of ozone powerfully. From the electrolysis of sulphide of sodium, Buff concluded, that the sodium travelled towards the cathode, and all the sulphur to the anode, ('Chemical News,' vol. xv. p. 279). A solution of ordinary phosphate of sodium, is resolved by electrolysis, into soda at the cathode, and phosphoric acid at the anode.

41. **Potassium.**—Elec.-chem. eqt. = 39·1. As the salts of potassium, like those of sodium, are very numerous, and are rarely electrolysed for the purpose of depositing their metal, I need only remark, that most of them may be prepared in a pure state, by neutralising the respective acids by pure potassic carbonate; and that nearly the whole of them are freely soluble in water; and I must refer the reader to a work on chemistry for special chemical information respecting them.

Soon after Sir H. Davy first isolated potassium by electrolysis, Gay-Lussac and Thenard showed, that the metal was
also set free by the contact of iron with melted potash at a
white heat. Other investigators found subsequently, that
carbon acted similarly. According to H. St.-Claire Deville,
even silver deposits pctassium, when immersed in fused
potassic iodide ; it also renders an aqueous solution of iodide
of potassium alkaline, and forms argentic iodide, by a similar reaction ('The Chemist,' New Series, vol. iv. p. 329).

Electrolysis of salts of potassium.—Sir H. Davy first
deposited this metal in the year 1807, by the influence of an
electric current upon wet hydrate of potash, in a precisely
similar way to that which he employed for depositing sodium.
Since that time it has been found, that even a feeble electric
current from two or three cells of any ordinary battery,
passed through a solution of the common salts of potassium
(not, however, including the nitrate, or chlorate ?, bromate ?,
or iodate ?) into a cathode of mercury, converts that
metal more or less into an amalgam of potassium ; and if
the cathode is composed of a metal which does not readily
absorb the deposit, the latter at once decomposes the water,
setting free hydrogen, and forming potash. For instance,
according to Faire and Roche, in the electrolysis of solutions of alkaline carbonates or bicarbonates, the molecule
splits up in such a way, that an atom of potassium or sodium
is set free at the cathode and liberates hydrogen ('Chemical News,' vol. xxx. p. 63; 'Journal of the Chemical Society,'
vol. xii. p. 861). Buff also electrolysed potassic sulphides,
and concluded that the metal travelled towards the cathode,
and the sulphur towards the anode ('Chemical News,'
vol. xv. p. 279).

Faraday found, that by passing an electric current through
melted iodide of potassium, iodine was set free at the anode,
and potassium at the cathode. Jaquin observed, that fused
sulphide of potassium yields, by electrolysis, potassium at
the cathode. Mathiessen states, that a fused mixture of the

chlorides of calcium, potassium, and sodium, yields a deposit of potassium, when electrolysed in a certain manner (Watts, 'Dictionary of Chemistry,' vol. i. p. 715). According to Brester, in the electrolysis of melted caustic potash, an anode of either platinum, silver, or copper, dissolves in the fused alkali, and the respective metals are deposited upon the cathode. The electrolysis of melted chlorate of potassium, with a platinum anode, yields potassium, which unites with a cathode of copper or platinum. Chlorine and oxygen (having an odour of phosphorus), are set free at the anode, and form thick white vapours by contact with water ('Chemical News,' vol. xviii. p. 145).

The following experiments of mine, now first published, bear upon the simultaneous liberation of potassium and fluorine, by means of electro-deposition:—I fused 130 grains of pure fluoride of potassium in a platinum crucible, within a partially covered clay muffle (see fig. 29, p. 296) inserted in the hole in the top of one of my gas furnaces, and electrolysed it during two and a half hours, by means of a current from six Smee's cells, and two flat helices of platinum wire as electrodes. There was free conduction, and much gas (of an odour like that of hydrofluoric acid), evolved from the anode, but none from the cathode, and no signs of any deposit. The anode was not corroded, nor altered in weight. I also electrolysed some of the same salt in a state of fusion, by means of a current from six Grove's cells, with a thick platinum wire as the anode, and the platinum vessel as the cathode; great heat was evolved, and violent electrolytic action occurred; nearly white-hot metallic globules also accumulated, and exploded repeatedly. The end of the anode fused, and particles of platinum ramified from it in white-hot threads, and a short electric arc (about one-tenth of an inch in length) was produced.

I also perfected and used, a somewhat elaborate platinum apparatus, by means of which the gas from the anode, was prevented from coming in contact with the cathode, and might be collected ; the electrodes being enclosed within (but isolated from) two wide platinum tubes. One thousand grains of the perfectly pure salt was electrolysed in this apparatus, by means of a current from six Grove's cells. The anode, which was a solid rod of platinum, was rapidly corroded, and was thus cut off at the level of the liquid, and stopped the current ; the corroded surface was very bright, as if fused. Potassium was deposited upon the cathode. Much spongy platinum was diffused in the melted salt; and the apparatus was a little corroded at the surface of the liquid. No gas was evolved at the anode. The deposited potassium did not alloy with the stout rod of platinum used as the cathode. 55·35 grains of grey metallic platinum were found in the saline mass. A salt of platinum appeared to have been formed at the anode, then dissolved or diffused throughout the liquid, and decomposed by the heat, and thus the liberated fluorine did not escape at the anode, but was evolved in the mass of the liquid generally, and came into contact with the liberated potassium.

Having ascertained the electrical relations of palladium, gold, platinum, and iridium in the fused fluoride, palladium being the most positive, and iridium the most negative, I repeated the experiments with an anode of iridium, and a current from three Grove's cells. Copious clouds descended at once from the anode, and made the liquid opaque ; there was also a violent action at the anode, it became black, and a little gas was evolved from it, accompanied by an acid odour, like that of a mixture of sulphurous anhydride and hydrofluoric acid. Potassium was freely liberated at the cathode, and produced occasional explosions. With a current from six cells, the anode dissolved rapidly, and soon lost thirty-eight grains. I then put a pure gold anode, and employed two cells. Gas, of a feebly acid odour, was freely evolved at the anode ; and with a current from six cells, was very copious, and smelt much like sulphurous anhydride. The gold dissolved much less rapidly than the iridium. With a palladium anode, and a current from six cells, the anode rapidly dis-solved, potassium was deposited, and exploded frequently; and an odour like that of hydrofluoric acid, was strong, much gas being liberated. 33·3 grains of free metal was found in the saline mass. The platinum cathode was not corroded. The platinum anode was dissolved as if melted; the iridium one was black; the palladium one was oxidised of various colours. The platinum vessel was cut into, at the level of the surface of the liquid, evidently not by the fused fluoride of potassium, but by some substance, set free at the anode by electrolysis. In another instance, I electrolysed the pure fused fluoride with a large platinum anode, small platinum cathode, and a current from three Grove's cells, during half an hour. Much gas, having an odour of ozone and hydrofluoric acid, was evolved from the anode; and the latter dissolved rapidly, and lost thirty-seven and a half grains in weight. The gas reddened test-paper. The platinum containing vessel was corroded at the line of surface of the liquid, and lost about eleven grains. About fifty-one grains of free metallic platinum, in loose powder, was found in the saline residue. Each of these experiments shows that a very corrosive substance was liberated at the anode.

I electrolysed the fused salt with a gas-carbon anode, and a platinum wire flat helix as cathode, with a current from six Smee's cells. Free conduction occurred, and much gas was set free from the anode only. The part of the anode in the liquid was not visibly corroded.

I also electrolysed about eight ounces of pure double fluoride of hydrogen and potassium (KF, HF) in a fused state, during half an hour, at about 300° Fahr., with a current from ten Smee's cells, and electrodes of stout sheet platinum. There was copious conduction, and abundance of hydrogen evolved at the cathode, but no gas from the anode, which was rapidly corroded away, with a rough surface, and lost 9·37 grains. The salt became less fusible by loss of hydrofluoric acid, which escaped freely all the time. The saline residue contained a small amount of dissolved platinum salt, and nearly nine grains of free metallic platinum. In a second experiment, lasting half an hour, the salt was kept only just fused, and a small gold anode was employed. The conduction was free, and much gas was evolved from the cathode, and a film of bright yellow gold spread over the surface of the salt, and connected the electrodes, unless the liquid was continually stirred. The anode rapidly dissolved (more quickly than that of platinum), and the salt of gold at once decomposed, and set free finely divided gold as a dull red-brown powder at the anode. No gas appeared at the anode at any time ; that from the cathode, detonated on applying a light. There was loose red-brown powder of gold, weighing 1·4 grains, upon the cathode, but only a faint gilding, weighing ·05 grain. The anode was corroded, dull and rough, and it lost 6·80 grains. The saline residue contained no dissolved gold, but 5·85 grains of red-brown powder, containing 5·30 grains of gold. In a third similar experiment, by using a large sheet platinum anode, and a small platinum cathode, and a current from ten Smee's cells, during two hours, the phenomena were the same as in previous experiments. The anode lost twenty-eight grains ; much loose platinum collected on the cathode, which was neither corroded, nor alloyed. The saline residue contained a trace of dissolved platinum salt, and nearly all the corroded platinum in a metallic state. In a fourth experiment I continued the action during three and a half hours ; the results were as before. The loss of the anode was 35·73 grains. The saline residue contained a small quantity of dissolved double fluoride of platinum and potassium, which, after being well washed, was dried, and heated to redness ; it then shot about as if gas was evolved from it. In a fifth similar experiment, lasting four and a half hours, at the lowest possible fusion temperature, more of the brown platinum salt formed at the anode, and dissolved in the liquid. The anode lost 64·81 grains. In a last experiment I electro-

lysed a gently fused mixture of 900 grains of the pure double
salt, and one hundred grains of pure argentic fluoride, with
a large anode of platinum, and a large cathode of silver.
Conduction was complete with ten Smee's cells. No gas
was evolved at either electrode. The surface of the anode
disintegrated rapidly, and lost 49·84 grains in four and a half
hours' action. The separated platinum dissolved only to a
small extent in the liquid, and subsided, in admixture with
the silver, to the bottom of the vessel, as a fine black powder,
weighing 73·93 grains, which lost less than two per cent. when
heated to redness. Some grey silver powder was deposited
upon the cathode. In all these experiments with the acid
fluoride, films continually formed upon the surface of the
liquid ; they came from the cathode, and were more abundant, the deeper the cathode was immersed.

According to Faraday, an aqueous solution of nitrate of
potassium conducts electricity very well, yielding hydrogen
gas at the cathode (it also gives ammonia at the cathode
[Daniell]). He also observed that an aqueous solution of
cyanide of potassium, yielded by electrolysis, hydrogen and
potash at the cathode ; at the anode no oxygen was evolved,
but the adjacent liquid became brown; that solutions of
sulphocyanide, and ferrocyanide of potassium behaved similarly ; also that fused cyanide behaved like its aqueous solution. (Gmelin's ' Handbook of Chemistry,' vol. i. p. 458.)

I electrolysed a nearly saturated aqueous solution of
pure fluoride of potassium, by means of a current from six
Grove's cells, with large platinum electrodes ; conduction was
copious, and the liquid acquired a nearly boiling temperature.
Much gas (having an odour like that of a mixture of ozone
and chlorine) was evolved at the anode. A saturated solution
of the same salt, electrolysed by a current from ten large
Smee's cells, with large platinum electrodes, evolved gas at
each electrode ; that from the anode smelt powerfully of
ozone, and re-inflamed a red-hot splint. Several other experiments, with variations in the sizes of the electrodes,

were made, and with addition of hydrofluoric acid; but the results were similar.

I saturated some pure dilute hydrofluoric acid of 40 per cent. at 60° Fahr., with pure double fluoride of hydrogen and potassium, and electrolysed the solution by a current from ten Smee's cells, a gold anode, and a platinum cathode, during five and a half hours. Gas was evolved freely from both electrodes, and a strong odour of ozone was observed. The anode lost 1·73 grains; and the cathode acquired first a gilded appearance, and then a black coating, and the liquid became black with finely divided matter.

Bourgoin electrolysed a concentrated solution of neutral tartrate of potassium; gas was evolved from each electrode, and alkali set free at the cathode. The anode became coated with acid tartrate, and evolved nitrogen, oxygen, carbonic oxide, and carbonic anhydride. The effect of the current upon a solution of the neutral tartrate, mixed with free alkali, was very different; oxygen, acetic acid, carbonic anhydride, carbonic oxide, hydride of ethylene, and acetylene, were set free at the anode ('Chemical News,' vol. xvii. p. 33).

42. **Rubidium.**—Electro-chem. eqt. = 85, and **Cæsium**, electro-chem. eqt. = 133. These metals have also been electro-deposited into a cathode of mercury, in a similar manner to potassium, sodium, and other highly electropositive elements.

43. **Ammonium.**—H_4N. Electro-chem. eqt. = 18. Nearly all the salts of ammonium are freely soluble in water, and may be made, by adding aqueous ammonia, or a solution of the carbonate, to the respective acids, and evaporating the solutions.

Electrolysis of anhydrous ammonia.—Anhydrous ammonia, liquefied by pressure, has been electrolysed by Bleekrode, with a current from eighty Bunsen's cells; gas was evolved, and the liquid became of an intensely blue colour. He also operated with a current from 3,240 cells of the chloride of silver battery of Mr. W. De la Rue; the anode became

x

black, much gas was evolved, and the liquid became deep blue. On stopping the current, the liquid became colourless. By dissolving rubidium, potassium, sodium, or lithium in such a liquid, I also obtained deep blue solutions ('Proceedings of the Royal Society,' No. 141, 1873; also vol. xxv. p. 323); probably, therefore, metallic ammonium was set free and dissolved.

Electrolysis of salts of ammonium.—Sir H. Davy electrolysed a saturated solution of salammoniac, with an anode of platinum and a cathode of mercury; chlorine was evolved at the former, and the mercury became alloyed with ammonium, and swelled to a very large bulk.

Hisinger and Berzelius found, that on electrolysing a solution of salammoniac with silver electrodes, hydrogen was evolved at the cathode, and oxygen at the anode, and the latter electrode acquired a coating of argentic chloride. They also electrolysed a mixture of aqueous ammonia and sulphate of ammonium. Hydrogen was evolved at the cathode, and nitrogen, mixed with some oxygen, at the anode. A gold anode, dissolved in such a liquid, became covered with brown fulminate of gold, and the cathode was gilded (Gmelin's 'Handbook of Chemistry,' vol. i. p. 458).

According to Seebeck, a moistened cup of carbonate of ammonium, filled with mercury as a cathode, yields by electrolysis, the ammoniacal amalgam. Faraday electrolysed fused nitrate of ammonium; hydrogen gas, mixed with a little nitrogen, was deposited at the cathode. The aqueous solution of that salt, similarly treated, yielded the same mixture at the cathode, and oxygen at the anode. I electrolysed gently fused ammonium fluoride, by means of a current from six Grove's cells, a thick platinum wire anode, and a large platinum sheet cathode. The conduction was copious, and heat was evolved. Much gas was liberated at the anode, but no odour of ozone.

Dry nitrate of ammonium condenses gaseous ammonia, and becomes a liquid; the solution is a good electrolyte, and yields ammonia and hydrogen at the anode. Anodes of silver, copper, lead, zinc, and magnesium, dissolve in it, but one of mercury becomes coated with an insoluble compound. When the anode is corroded, no nitrogen is liberated from it (Divers, ' Chemical News,' vol. xxvii. p. 37; ' Proceedings of the Royal Society,' vol. xxi. p. 109).

According to A. Favre, under the influence of the current, ammonic-oxide is decomposed thus :—1st. $3(NH_4)_2 O = 3(NH_4)_2 + O_3$. The three equivalents of ammonium set at liberty, decompose the water like potassium or sodium, thus :—2nd. $3(NH_4)_2 + 3H_2O = 3(NH_4)_2O + 3H_2$. The oxygen of equation No. 1, reacting upon the ammonium, gives :—3rd. $O + NH_4 = N + 2H_2O$. The first equation represents the electrolysis proper (' Journal of the Chemical Society,' vol. ix. p. 985).

CLASS VII. DEPOSITION OF METALLOIDS.

TITANIUM — SILICON — BORON — CARBON — PHOSPHORUS — SELENIUM—SULPHUR—IODINE—BROMINE—CHLORINE — FLUORINE—OXYGEN—NITROGEN.

Although this book is one on the electro-deposition of metals, it can hardly be considered complete, unless something is said about the deposition of the metalloids or non-metallic elementary substances, because the two classes of bodies are so closely related to each other in the electrolysis of their compounds.

44. Deposition of titanium.—Elec.-chem. eqt. $= \dfrac{50}{4}$ $= 12\cdot5$. This element does not appear to have been yet electro-deposited. Titanic acid is the most usual compound of it obtainable in a state of purity; it dissolved very slowly in pure dilute hydrofluoric acid, and by evaporating the solution to dryness, a white and somewhat deliquescent salt remained. I found that a heap of crystals of

nitro-cyanide of titanium, conducted freely, an electric current
from sixty Smee's elements, and that a single crystal pressed
between the terminal wires, became red-hot, and then exhibited a splendid white light.

45. **Silicon.**—Elec.-chem. eqt.$= \dfrac{28}{4} = 7\cdot0$. In some ex-

periments with silicon (which I had fused into lumps) I found
that the pieces conducted sparingly, a current from twelve
Smee's elements ; and that in dilute sulphuric acid, the
silicon was strongly, but only temporarily, electro-positive to
platinum. Silicon is generally strongly electro-positive to
other substances in fused fluorides, and like carbon, is much
more electro-positive at very high temperatures ; at such
temperatures, it sets free sodium from its fluoride (see p. 297).
Some of the electrical relations of silicon have already been
given (see pp. 66, 67). Phipson states that magnesium
separates silicon from silica at a high temperature ('Proceedings of Royal Society,' vol. xiii. p. 217; 'Chemical News,'
vol. ix. p. 219).

According to Becquerel, crystals of silicon may be deposited upon a platinum cathode, in a saturated solution of
gelatinous silica in hydrochloric acid ('Chemical News,' vol.
xii. p. 4). According to Golding Bird, silicon may also be
deposited from a solution of its fluoride in alcohol ('Philosophical Transactions of the Royal Society,' 1837, p. 37 ;
Golding Bird's 'Natural Philosophy,' 5th edition, p. 408).
I electrolysed fused pure silico-fluoide of potassium in a
platinum vessel, with platinum electrodes; silicon was deposited upon the cathode, and formed a fusible alloy with it.

46. **Boron.**—Elec.-chem. eqt. $= \dfrac{21}{3} = 10\cdot33$. Phipson

states, that magnesium in contact with boracic acid in a fused
state, liberates boron ('Proceedings of the Royal Society,'
1864, vol. xiii. p. 217; 'Chemical News,' vol. ix. p. 219).
. 'Boron was first electro-chemically isolated by Davy.
He states that when boracic acid is exposed between two

surfaces of platinum, receiving at the same time all the action of a current from 300 cells, an olive-brown matter is formed upon the negative surface, gradually increasing in thickness, and finally becoming black. The isolated body is boron.' ('Chemical News,' vol. xii. p. 3.) Fused borax yields oxygen gas at the anode, and boron at the cathode. The boron is separated by indirect action; the current resolves the soda into oxygen and sodium, and the latter separates boron from the boracic acid (Faraday, Gmelin's 'Handbook of Chemistry,' vol. i. p. 460). According to Burckhard, pure boracic acid in a state of fusion is a non-conductor, but fused borax conducts, suffers electrolysis, and a series of compounds are formed or volatilised; but the chief result is, that the salt is resolved into oxygen at the anode, and soda and boron at the cathode ('Chemical News,' vol. xx. p. 238). I have found that by the electrolysis of pure borofluoride of potassium in a fused state, with platinum electrodes, the cathode became rough and brittle, by being converted into a compound with electro-deposited boron.

47. **Carbon.**—Elec.-chem. eqt. $= \dfrac{12}{4} = 3\text{·}0$. According to Phipson, magnesium in contact with carbonate of sodium at a high temperature, sets free carbon abundantly ('Proceedings of the Royal Society,' 1864, vol. xiii. p. 217; 'Chemical News,' vol. ix. p. 219). Deville states, that metallic aluminium deposits carbon, from carbonate of potassium in a state of fusion ('Chemist,' New Series, vol. iv. p. 481).

I have electro-deposited perfectly pure carbon, from the pure carbonates of potassium and sodium in a state of fusion, adding in some cases a little silicate of potassium, or fluoride of silicon and potassium ; the deposit was black, non-crystalline, insoluble in all acids, burned with a glow, and left no residue. According to P. Burckhard, fused carbonate of sodium is a good conductor, and yields by electrolysis, carbonic anhydride, soda, and a small portion of carbon ('Chemical News,' vol. xxi. p. 238). Researches on the electro-deposition of carbon are interesting, with reference to the possibility of the artificial formation of diamonds.

I have found that carbonic anhydride, liquefied by pressure, does not conduct a current from forty Smee's elements; and is a powerful insulator of electricity ('Transactions of the Royal Society,' 1861, p. 85). According to Bleekrode and W. De la Rue, neither liquefied cyanogen (C_2N_2), liquid carbonic anhydride (CO_2), carbonic bisulphide (CS_2), nor benzine (C_6H_6), were decomposed by a current from 5,640 cells of a chloride of silver battery ('Proceedings of the Royal Society,' vol. xxv. pp. 324-326).

48. **Phosphorus.**—Elec.-chem. eqt.$= \dfrac{31}{3} = 10\cdot33$.   The electric current resolves a solution of ordinary phosphate of sodium, into soda at the cathode, and phosphoric acid at the anode.   The electrolysis of concentrated phosphoric acid, produces a metallic phosphide upon a cathode of copper or platinum (H. Davy).   Acid phosphate of sodium in a state of fusion, yields hydrogen at the cathode (Faraday, Gmelin's 'Handbook of Chemistry,' vol. i. p. 460).   According to Burckhard, fused pyrophosphate of sodium, yields by electrolysis, phosphorus, oxygen, and soda; and, with a platinum anode, phosphide of platinum is formed ('Chemical News,' vol. xxi. p. 238).

49. **Selenium.**—Elec.-chem. eqt. $= \dfrac{79\cdot5}{2} = 39\cdot75.$
During the electrolysis of an aqueous solution of selenate of nickel, containing selenate of sodium, and free selenic acid, I have repeatedly observed an abundant deposit of bright red selenium upon a platinum cathode; the deposition of selenium ceased on neutralising the acid with ammonia.

50. **Sulphur.**—Elec.-chem. eqt. $= \dfrac{32}{2} = 16.$ A yellow solution of sulphide of potassium, yields a quantity of sulphur at the anode, and hydrogen gas at the cathode; fused sulphide of silver is also slightly decomposed by electro-

lysis, into sulphur at the anode, and silver at the cathode
(Faraday, 'Gmelin's Handbook of Chemistry,' vol. i. p. 456)·
An aqueous solution of sulphurous anhydride, yields sulphur
and hydrogen at the cathode (De la Rive).

51. **Iodine.**—Elec.-chem. eqt. = 127. Liquefied anhydrous hydriodic acid (HI), does not conduct the current
from eighty Bunsen's cells (Bleekrode, 'Proceedings of the
Royal Society,' vol. xxv. p. 323). An aqueous solution of
iodic acid, yields by electrolysis, oxygen at the anode, and
iodine alone at the cathode. Concentrated hydriodic acid,
yields iodine alone at the anode; but the dilute acid gives
iodine and oxygen. According to Faraday, fused iodide of
potassium, or iodide of lead, yields iodine at the anode.

52. **Bromine.**—Elec.-chem. eqt. = 80. Anhydrous hydrobromic acid (HBr), in the liquid state, does not conduct the
current from eighty Bunsen's cells (Bleekrode, ' Proceedings
of the Royal Society,' vol. xxv. p. 323). Aqueous hydrobromic acid, deposits by electrolysis, bromine at the anode,
and hydrogen at the cathode.

53. **Chlorine.**—Elec.-chem. eqt. = 35·5. Liquefied anhydrous hydrochloric acid (HCl), does not conduct the
current, even from as many as 5,640 cells of a chloride of
silver battery (Bleekrode and W. De la Rue, ' Proceedings
of the Royal Society,' vol. xxv. p. 325). Concentrated hydrochloric acid gives hydrogen at the cathode, and chlorine
alone at the anode; and only after considerable dilution
with water, does oxygen begin to be deposited along with
the chlorine. An aqueous solution of salammoniac yields
chlorine at the anode, and hydrogen and ammonia at the
cathode; and one of common salt, gives chlorine at the
anode, and hydrogen and soda at the cathode. According
to Faraday, fused chloride of lead, and chloride of silver,
yield chlorine at the anode, and the metal at the cathode.

54. **Fluorine.**—Elec.-chem. eqt. = 19·0. I have repeatedly observed, that aqueous solutions of metallic fluorides,
yield oxygen at the anode, because water is decomposed

more readily than fluorides ; but with certain fluorides in a
state of fusion, a highly corrosive substance is evidently
liberated.   (See the electrolysis of various fused fluorides,
already described, especially that of fluoride of potassium,
pp. 121, 300-303.) By the electrolysis of pure anhydrous hydrofluoric acid (see p. 96), I found that the acid conducted
readily a current from ten Smee's cells, but evolved no
fluorine from an anode either of palladium (see p. 115),
platinum (see p. 120), or gold (see p. 126), and anodes of
the densest varieties of carbon were instantly disintegrated.
And by electrolysis of the pure aqueous acid, with electrodes of platinum, ozone and ordinary oxygen were alone
evolved at the anode ('Phil. Trans. Roy. Society,' 1869,
p. 173).

55. **Oxygen.**—Elec.-chem. eqt. $=\dfrac{16}{2}=8$. Water, to which
almost any acid or salt has been added (in not too great
quantity), in order to make it conduct, yields by electrolysis,
oxygen at an anode of platinum.

56. **Nitrogen.**—Elec.-chem. eqt. $=\dfrac{14}{2}$ 4·66. A concentrated aqueous solution of ammonia, with electrodes of
iron, yields hydrogen at the cathode, and pure nitrogen at
the anode (Hisinger and Berzelius).

SPECIAL TECHNICAL SECTION.

SECTION B.

HAVING described all the methods in practical use for depositing metals, and as briefly as I have been able, the circumstances under which almost every known metal has been deposited, so that any one wishing to apply to practical uses, the deposition of metals not yet so employed, may be able to make a successful commencement, I will now give a number of special technical points of information, necessary for the successful prosecution of the art, which could not be so conveniently described in the preceding sections of the book.

FIG. 30.

General workshop arrangements.—Before commencing, on a practical scale, the art of electro-metallurgy, it will be necessary to provide a depositing-room, vats for solutions, scouring and cleaning apparatus, batteries, a magneto-electric machine, or other source of electric power ; the various chemicals necessary for making and reviving depositing liquids acids for cleaning and 'stripping' ; materials for making moulds, and preparing their surfaces, &c., &c.

The establishment should consist of several rooms, and an open yard ; i.e., a room for depositing copper, another for silver, and a smaller and more private one leading out of it, for gilding. The rooms should be upon the ground-floor, on account of the weight of the vats containing the solutions, and should be provided with a cemented floor, and a drain running into a small cemented well, to recover valuable liquids which may be accidentally spilled. They should be well lighted and ventilated, because of the noxious vapours sometimes evolved ; and should contain conveniences for the placing of the vats, washing-troughs, and scratch-brush lathes ; and be provided with a plentiful supply of water. An outhouse for containing a large iron-boiler ; also a covered shed in a yard (for the processes of dipping), will be necessary. The yard is required for precipitating solutions, from which the poisonous vapour of prussic acid is evolved. Instead of an outhouse, a separate, but adjoining room, may be used, in which to erect the iron boiler, for containing caustic potash solution, for cleaning greasy and other articles. If voltaic batteries are much employed, they are best placed outside the plating-room, because the vapour arising from them is unhealthy, and also tarnishes the articles. If a magneto-electric machine is used, it is also best to have it, and the engine which drives it, at a distance from the cleaning liquids, or in an adjoining dry apartment.

Accessible from each of the rooms, should be erected a low furnace, having a long horizontal flue covered with plates of iron, upon which are placed several large iron trays filled with hot sawdust, in which the wet articles are to be dried. Each depositing-room should be supplied with a water-tap, and several large wooden tubs or troughs filled with water, for washing the articles. The 'pickling' and 'stripping' liquids are best kept in large stoneware pans, under the open roof in the yard. In the gilding-room, will be placed iron vessels for containing the gilding liquids; these vessels are usually of enamelled iron, either wrought or cast, and should be supported on iron frames, with large Bunsen burners beneath, for the purpose of heating the liquids ; flues should also be provided to convey the products of combustion from the burners into the open air. Accessible also to each of the rooms, should be placed several scratch-brush lathes, for scouring and brightening the articles. Round the walls of the coppering and silvering rooms, should be fixed well-insulated stout copper wires, to convey the electric currents from the batteries or magnetic machines to the vats. For the gilding-room these will not be required, because gilding is usually effected, by means of a small voltaic battery, or thermo-electric pile, placed close at hand.

Vats for solutions.—The construction of the vats for containing silver solutions, has been already described (see p. 169) ; those for containing sulphate of copper solution, are usually made of wood, lined with a thick sheet of gutta-percha, so that the liquid shall not come into contact with the wood. According to Berthoud, a good mixture for lining vats to contain sulphate of copper solution, is composed of six parts of Burgundy-pitch, and one of gutta-percha, cut into very small pieces ; the whole being thoroughly mixed by melting and kneading. The silver-plating vats are sometimes placed in the middle of the room, but more frequently against a wall where the sunlight does not fall directly upon the solutions.

Cleaning articles for receiving a deposit.—All articles which are to receive a deposit, require to be made scrupulously clean, especially if it is wished to make the coating adhere firmly to the receiving surface. It is the practice before plating an article, to make its surface not only perfectly clean, but also smooth, by means of the revolving 'scratch-brush,' and by other methods. Articles of copper are usually, not 'scratch-brushed,' but dipped.

The processes of cleansing, are both of a mechanical and chemical nature. The mechanical means, are the usual ones of filing, scrubbing, and scouring, with various gritty materials. Emery-cloth is employed when the articles are dry, and fine silver-sand and a hand-brush, or piece of canvas, when they are wet. In addition to this, an instrument called a 'scratch-brush' is continually used, and cannot be dispensed with.

A 'scratch-brush' is merely a bundle of fine and hard brass wires, about six or eight inches long, bound round very tightly with other wire, except at the ends (see fig. 31).

FIG. 31.

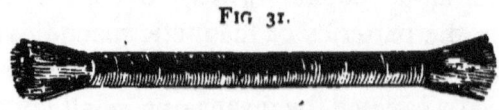

These wires are of various degrees of fineness, and are also annealed to different degrees, to suit the various kinds of work. Four of such brushes are usually fixed in grooves upon the outside of the chuck of a lathe, so that the wires are parallel with the axis of the chuck (see fig. 32). Another form of scratch-brush, in which the wires are radial instead of parallel is shown in fig. 33.

FIG. 32.

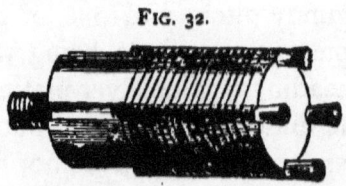

To use these brushes, a lathe is required. A 'scratch-brush lathe,' suitable for cleaning small articles, is represented in the annexed figures 33 and 34.

Above the revolving brush, is placed a cistern containing stale beer, a little of which is allowed to dribble upon the articles during the process of brushing, and the brushes are surrounded by a screen, to prevent splashing.

The chemical methods of cleaning, consist in immersing the articles for a greater or less period of time, in various acids or alkalies, according to the nature of the metals. Alkalies are usually employed hot, and are generally used for removing greasy, tarry, or resinous matters; and acids

FIG. 33.

are generally used cold, after the greasy matters have been removed. The alkalies are kept in iron vessels, and the acids in stoneware pans, &c.

The alkali commonly employed is caustic potash, because it is the strongest. A solution of it is prepared, by adding freshly-made cream of lime, to a boiling solution, composed of about half a pound or a pound of pearlash, to each gallon of water, contained in an iron boiler, until a small quantity of the clear liquid gives no effervescence, on adding to it a few drops of dilute hydrochloric acid. The precipitate formed in the mixture, is carbonate of lime, and may be thrown away. As this liquid rapidly absorbs carbonic acid from the air, it should be kept covered as much as possible; and a small quantity of the cream of lime, should be added to it occasionally, to renew its full degree of causticity. The

FIG. 34.

articles to be cleaned, are immersed for a short time in the boiling hot liquid; copper only requires to be immersed a few seconds. Copper articles, joined by solder containing tin, must not remain long in the liquid, or the tin will dissolve, and be deposited upon the adjoining parts of the copper, and blacken them.

Several kinds of acid liquids are employed, viz., dilute

sulphuric, strong nitric, and various mixtures of them. Nitric acid for dipping, contains about 10 per cent. of sulphuric acid, and has a sp. gr. of about 1·52.

The special methods of cleaning, depend both upon the nature of the impurities upon the surface, and of the metal beneath. All greasy articles, of whatever metal they are composed, are always dipped into the potash solution, and then usually thoroughly swilled in water. Articles composed of lead, tin, Britannia metal, or pewter, are dipped in the caustic potash, and, with or without swilling, transferred at once to the depositing solution. Those of zinc are sometimes treated similarly ; and at other times, are, after swilling, dipped in dilute sulphuric acid, washed again, and transferred to the plating liquid. For cleaning iron articles, a cold mixture of about twenty measures of water, and one of sulphuric acid, is frequently used ; but a better liquid is composed of one gallon of water, and one pound of sulphuric acid, with one or two ounces of zinc dissolved in it ; to this is added half a pound of nitric acid. This mixture leaves the iron quite bright, whereas dilute sulphuric acid alone, leaves it black, or of a different appearance at the edges. For glassy patches upon cast iron, (which usually consist of silicate of iron), hydrofluoric acid is used ; it is kept in a bottle of gutta-percha closed by a bung of indiarubber ; it must not be allowed to come into contact with glass vessels, nor must the mouth of the bottle be left open. The fumes from it are extremely dangerous to inhale. Articles of iron which have been cleaned in acids, and the adhering acid washed away with water, may be protected from rusting, by continued immersion in lime-water, a solution of washing-soda, or in water containing any caustic alkali, until required.

Articles of pure silver, are best dipped in a heated state, in dilute boiling sulphuric acid, after having been immersed in the alkali and swilled ; or they may be dipped cold, in strong and pure nitric acid, and then in distilled water. New

anodes of rolled silver are often greasy, and have a film of
oxide of iron upon them ; they should be scoured with caustic
alkali, or be heated to redness, before placing them in the
plating solution

For articles of copper, brass, or German silver, a series of
liquids is used ;—first, strong nitric acid; second, 'dipping
liquid' (consisting of sixty-four parts of water, sixty-four of
sulphuric acid, thirty-two of nitric acid, and one of hydrochloric acid) ; and, third, 'spent' liquid, i.e., either nitric
acid or dipping liquid, which has become weak. Such
articles are often partly cleaned, by heating them to dull
redness, and then plunging them into dilute sulphuric acid.
(Those having solder upon them are not heated thus ;
articles of cast bronze are also not heated in this way, because they would be liable to crack.) They are then soaked
in old aqua-fortis, until, after rinsing, they look uniformly
metallic ; they may then be dipped in strong aqua-fortis for
a few seconds, and swilled. The straw-coloured aqua-fortis
acts the best : the white acts too feebly, and the red too
strongly. It is necessary to have a large bulk of the acid, in
order that it may not become too warm by the action.

To dip gilding metal bright.—Immerse it in weak aqua-
fortis until there is a black scale formed; then dip it in 'strong
pickle' for a few minutes. (N.B. 'strong pickle' is exhausted
aqua-fortis ; 'weak pickle' is the same diluted with the wash-
ings.) Then dip it quickly into strong aqua-fortis, and then
into several waters in succession. There are various mixtures, which may be employed for imparting a bright lustre
by dipping ; the following is one of them : one measure of
nearly exhausted aqua-fortis, two of water, and six of hydrochloric acid ; the articles of copper, brass, or German silver
should be immersed in it a few minutes, or until, after washing off the black mud which entirely covers them, they look
bright ; they are then cleaned and dipped again. To obtain
a dead lustre, the articles of copper or its alloys, are dipped
into a cold mixture of one volume of oil of vitriol mixed

with two volumes of yellow aqua-fortis, and a little common salt then added ; the articles must remain some time in the bath, and then be quickly dipped into the liquid for producing a bright lustre, and immediately rinsed. Articles composed of German silver, are more difficult to impart a proper appearance to, by the process of dipping, than those of copper or brass.

Old aqua-fortis is revived to a certain extent, by addition of oil of vitriol and common salt; the sulphuric acid decomposes the nitrate of copper in it, and also the common salt, and sets free nitric and hydrochloric acids ; and crystals of sulphate of copper form at the bottom of the liquid. All the nitric acid may be utilised in this manner.

FIG. 35.

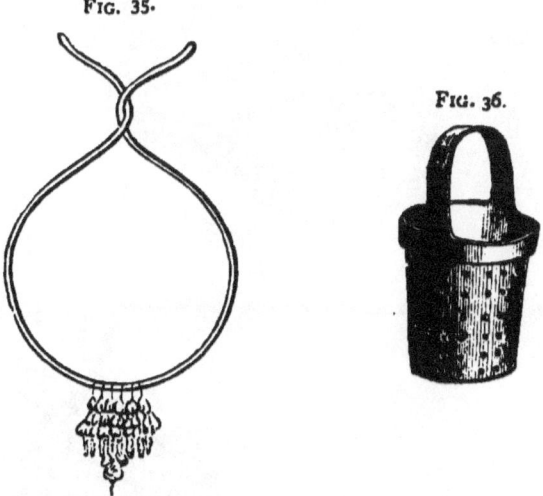

FIG. 36.

Small articles are either strung upon wires (see fig. 35) of the same or similar metal, or they are put into a stoneware basket (see fig. 36), and then dipped. Hooks and strong rods of copper, brass, &c., of the annexed forms (see figs. 37, 38), are necessary for suspending articles upon, for the purpose of dipping them into the various liquids. It is best that these hooks should be of the same material as the articles, because they are then less liable to cause a

stain. Very small articles are placed in a basket or perforated bowl, of stoneware or gutta-percha (see fig. 39), or a
tray of platinum wire gauze (see fig. 40), to be dipped.
Cleaned ones of brass, are immersed in a solution of argol,
to keep them from oxidising. There should also be a series
of vessels, containing water, for effectually swilling the articles. The pans for containing pickling, dipping, stripping,

FIG. 37.

FIG. 38.

FIG. 39.

FIG. 40.

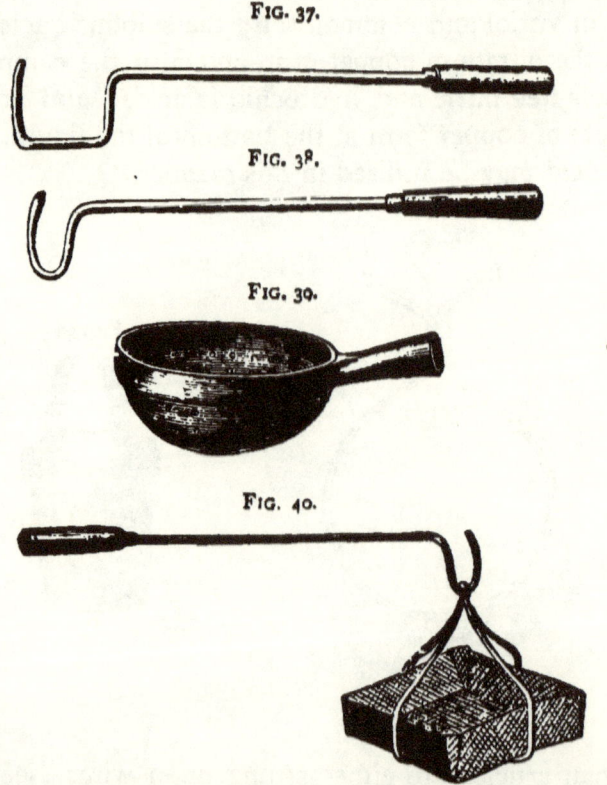

and quicking liquids, should be of the very best quality of
salt-glazed stoneware.

Sometimes, in order to assist in cleaning the articles,
they are suspended for a short time in a suitable acid or
cyanide liquid, in contact with the positive pole of a battery;
this dissolves the surface, and loosens the impurities, unless

they are very foul. In every case, they should be well rinsed with water, to remove the adhering acid, &c., before dipping them into the ' quicking' solution, or immersing them in the depositing vat. All objects which are to have a definite amount of metal deposited upon them, are weighed, and their weight noted, after they have been cleaned.

' *Stopping-off* ' *to prevent deposition.*—Many articles which are to receive deposits, require to have portions of their surface ' stopped-off,' to prevent the deposit spreading over those parts ; for instance, in taking a copy of one side of a bronze medallion, the opposite side must be coated with some kind of varnish, wax, or fat, to prevent deposition ; or in gilding the inside of a cream-jug which has been silvered on the outside, varnish must be applied all round the outer side of the edge, for the same reason. For gilding and other hot solutions, copal varnish is generally used ; but for cold liquids and common work, an ordinary varnish, such as engravers use for a similar purpose, will do very well. In the absence of other substances, a solution of sealing-wax dissolved in naphtha may be employed. (See also pp. 182, 226.)

' *Quicking* ' *the surfaces of articles.*—' Quicking ' means coating the surfaces with a film of mercury, for the purpose of causing the deposited silver, &c., to adhere firmly ; the mercury acts, by offering a perfectly clean surface to receive the deposit, and, by dissolving to a minute extent, both the surface of the article, and that of the deposit, enables them to mutually interpenetrate, and alloy with each other.

Solutions of nitrate or of cyanide of mercury, are used for preparing the surfaces of copper, brass, and German silver, for receiving adhesive deposits of silver. The nitrate solution is prepared, by adding one ounce of mercury to sufficient nitric acid, diluted with three times its bulk of distilled water, to dissolve it ; no more mercury must be added than the liquid will take up ; when completely dissolved, add about one gallon of water (see also pp. 143, 166). To prepare

the cyanide solution, dissolve one ounce of mercury as just
stated, dilute it with water, and add a solution of cyanide of
potassium to it, exactly as long as a precipitate is produced ;
filter it, add a small quantity of water to the precipitate in
the filter, and, when thoroughly drained, take out the precipitate, and add to it, with stirring, a strong solution of
cyanide of potassium, until it is all dissolved, then add a
little more cyanide solution, and finally dilute it with water,
until the whole measures one gallon. Another solution is
composed of one part of pernitrate of mercury, and two
parts of nitric or sulphuric acid, dissolved in 1,000 parts of
distilled water; or, take nitric acid of specific gravity 1·383,
add to it half its weight of mercury, and heat the liquid
nearly to 100° C. until yellow fumes are no longer evolved ;
the solution should not be crystallized : dissolve one part by
weight of this liquid in 1,000 parts of water, with which two
parts of sulphuric acid have been previously mixed. Or,
dissolve two ounces of mercury in two ounces of cold nitric
acid, and then add three gallons of water; this forms a good
solution.

    Almost any salt of mercury ('red precipitate' for instance)
may be dissolved in a solution of cyanide of potassium, to
form a ' quicking liquid.' Such a liquid is frequently made,
by adding the cyanide to the nitrate, and not troubling to
wash the precipitate. The objection to a solution of nitrate
of mercury alone is, that as the quicking liquid cannot be
readily washed completely away from hollow articles, the
traces remaining in crevices, cause the silver to strip from
those parts. Oxide of mercury, dissolved in a solution of
cyanide of potassium, is often used as a ' quicking solution,'
but it is not as good for copper articles, as pernitrate of
mercury containing a little hydrochloric acid. The solution,
when prepared, is kept in a large stoneware vessel, with a
pan of ' dipping liquid,' and two others containing water, near
it ; and each placed near the scratch-brush lathe and depositing vats, in the silvering-room.

'Quicking solution' should only contain sufficient dissolved mercury, to make a copper surface immersed in it a few seconds, become white ; if the copper becomes black, the silver deposited upon it will not adhere ; it also shows that the solution of mercury is either exhausted, or not in a proper condition. Too much ' quick ' causes the silver to ' strip ; ' and usually too little can hardly be put on, but the amount varies in different cases.

Articles which are to receive a thick coating of gold or silver, require a stronger mercurial solution, than those which are to receive a thin deposit, and they should be perfectly white and bright like silver, on coming out of the mercurial bath ; if the ' quicking ' has succeeded, they will have an uniform appearance. The solution will last a long time ; when it gets nearly exhausted, it is liable to turn the articles which are dipped into it, of a dark colour; it is then better to prepare a fresh liquid, than to revive the old one.

All articles, while still wet from the cleaning and quicking processes, should be quickly immersed into the depositing vat. The practical minutiæ of preparing the surfaces of different metals, for receiving adhesive deposits, vary in almost every manufactory, and much information yet remains to be developed upon this point ; for want of this knowledge, the most skilful operators sometimes fail in producing perfect adhesion, especially upon zinc, cast iron, steel, and Britannia metal.

Wireing articles.—The articles have wires of copper attached to them, to suspend them by when in the vat. The wires differ in size ; with small objects, such as spoons, knives, forks, snuffers, teapots, jugs, &c., size No. 20 or 22 of the Birmingham wire guage, and about eighteen or twenty inches long, are used ; very large ones, such as fire-irons, fenders, hat-stands, and pieces of ornamental iron-work, are suspended by strong copper or brass hooks. In some cases, where a powerful and certain connection is required, the wires are soldered to the articles.

Voltaic batteries.—There are but few kinds of voltaic batteries usually employed in electro-metallurgy, and those which are used, are not often employed for operations of the greatest magnitude; in such cases, magneto-electric machines are rapidly superseding voltaic batteries, because they furnish electricity at much less expense, and their action is more uniform. I shall, therefore, only briefly describe such as have been commonly employed.

Those most used in electro-deposition are, the old Wollaston battery of zinc and copper plates in dilute sulphuric acid, Smee's, Daniell's, Bunsen's, and Grove's.

Wollaston's battery.—The one which has been most employed for electro-deposition upon the large scale, is represented in fig. 41. In consists of a large stoneware jar, nearly filled with a mixture of about ten parts of water and one of oil of vitriol Across the top of the jar, is a moveable bar of well varnished wood, with a longitudinal and vertical groove in it, within which a thick plate of zinc may be raised and lowered, by means of a weight with a cord passing over a pulley: the great use of this is, to regulate the quantity of the current. To the edges or sides of the bar are fixed two sheets of copper, connected together by a copper band at their corners, and so attached, that they may be occasionally removed and cleansed; they extend nearly to the bottom of the liquid. Vertical rods of varnished wood, are fixed upon the under surface of the cross-bar, to prevent the zinc touching the copper. The copper plates should not be allowed to remain many hours in the liquid, when the battery is not in action, because they then corrode, and form a small amount of cupric sulphate, which dissolves in the liquid, and this acts upon the zinc plates, and causes them to waste rapidly, because the zinc precipitates the copper upon itself, and thus a local battery is formed. If the copper plates remain long in the

Fig. 41.

air in a wet acid state, they become covered with a badly conducting blackish film of oxide, and should be scrubbed with sand and a hard brush, and washed before being again used.

Smee's battery.—This one has been extensively used for small operations, and is very convenient. It consists of amalgamated zinc and platinized silver, immersed in dilute sulphuric acid ; and is usually of the form shown by fig. 42. Two plates of zinc z z are held together (with a bar of varnished wood between them), by means of a clamp binding-screw (see fig. 50, p. 335), and the sheet of platinized silver s, is fixed in a groove in the under side of the wooden bar, and attached to a pillar binding-screw (see fig. 46, p. 335). The silver and zinc are prevented from mutual contact, by means of pieces of cork placed between them at their lower ends. It is important in this battery (and to a less extent in that of Wollaston) that the sulphuric acid employed should be free from nitric acid; also that the negative plate should not come into contact with mercury. Platinized silver (i.e. silver coated with black platinum in a state of very fine division) is much more effective than silver alone, because with the latter metal, the bubbles of hydrogen evolved, adhere to its surface, and diminish the action, whilst with platinized silver they escape to the surface of the liquid rapidly. Platinized silver is also more electro-negative than silver alone, and still more so than copper, and therefore produces a stronger current. The mode of platinizing has already been described (see p. 118).

FIG. 42.

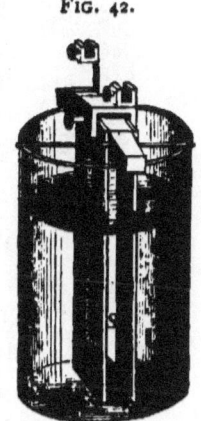

Daniell's battery.—This one has also been largely used in electro-deposition, but its use for that purpose has diminished. It consists essentially of amalgamated zinc in dilute sulphuric acid, and copper in a nearly saturated solution of cupric sulphate, the two liquids being prevented from mixing (but allowed to touch each other) by means of a porous partition.   One of its forms is that shown in fig. 43, in which

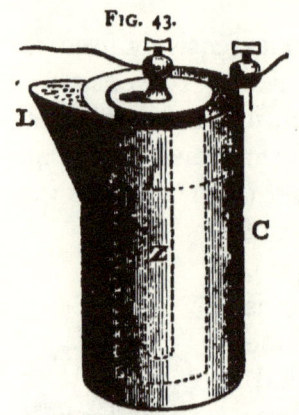

FIG. 43.

c, is a copper vessel forming the negative metal, and containing the cupric solution, and z, a bar of cast zinc, supported in the acid and water within the porous cell, by the wooden lid of that vessel.   The copper cell has a large lip L, which is kept full of crystals of blue vitriol, to supply the loss of copper deposited upon the vessel ; it may also be used for the purpose of pouring out the solution.

The great advantage of this battery, is the uniformity of its action, and it is therefore called the ' constant ' battery. It is sometimes constructed with the acid and water outside, and the copper plate and solution inside ; in that case a cylinder of rolled plate zinc is employed; it is also occasionally made of a rectangular form, with the porous cell of a flat shape.

FIG. 44.

Bunsen's battery.—This kind is often employed for gilding, two or three large cells being commonly used.   It consists of amalgamated plate zinc in dilute sulphuric acid, and gas-carbon or Bunsen's coke, in strong nitric acid; the latter liquid being in a porous cell.   The gas-carbon is usually in the form of thick rectangular bars, and in such cases the nitric acid is in a cylindrical porous cell; but sometimes it is in the form of plates, and flat porous cells are then necessary.   As the carbon is a porous substance, the acid rises in it by capillary action, and corrodes the metallic connections; the most effectual way of obviating this, is by using very long pieces of the substance, a considerable portion of each piece being out of the liquid, and putting a coating of varnish or paraffin upon them a little way down. Sometimes, in order to form a more secure connection, the upper end of the bar is coated with copper by electro-deposition; or else it is encased with metal by dipping it into melted lead. Fig. 44 shews a bar of carbon with its bind- Fig. 45. ·ing screw attached.

Grove's battery is precisely similar to Bunsen's in its essential parts, except, that it has platinum instead of carbon. The nitric acid and sheet of platinum, are contained in narrow flat porous cells of the form shewn in fig. 45. It is one of the strongest of batteries, but emits noxious acid fumes after having been some time in action, and its power soon declines.

Relative strength of batteries.—The electro-motive force, or power of overcoming resistance (see p. 70), varies in different batteries, and is, according to Latimer Clarke, as follows :—

Grove's	.	.	. 100	Smee's (when in action) about 25	
Bunsen's	.	.	. 98	Wollaston's (copper and zinc	
Daniell's	.	.	. 56	in dilute acid) . .	46
Smee's (when not in action)	57				

(See ' Electrical Measurement,' p. 108, by L. Clarke).

From this table it will be observed, that the strength of a Smee's cell decreases during its working; this occurs very quickly after the current commences, because the internal resistance is increased by hydrogen gas adhering to the negative plate; after that has occurred, the current remains tolerably constant; a similar phenomenon happens with the Wollaston's element, but not with the Daniell's, because in the latter, the negative surface is kept free from that gas.

Relative advantages of different batteries.—Wollaston's is the most suitable one in cases where the resistance is not great, and where a large quantity of electricity, and long-con-

tinued action (as in depositing copper and silver) are required, because its electro-motive force is small; its action
(after once it has commenced) is tolerably uniform, and large
plates, and considerable bulks of exciting liquid, may be conveniently employed. Smee's is suitable for similar cases, but
where only a small quantity of electricity is required, because
large plates of platinized silver are expensive. Daniell's is
the best in cases where the resistance is greater, and a very
uniform current is necessary. Grove's and Bunsen's are the
most suitable where the resistance is still greater, and an occasional current of considerable electro-motive force, but not
of long continuance, is necessary, as in gilding, and preparing for gilding (*i.e.* brassing or coppering) small articles of
iron, steel, &c. in cyanide solutions.

Exciting liquids for batteries.—In all these batteries, the
zinc element is immersed in dilute sulphuric acid. The
strength employed of this mixture, varies from one measure
of acid and fifty of water, to one of acid and five of water;
the usual strength with batteries such as Grove's and
Bunsen's (which are soon exhausted), is one to five, but with
Daniell's, Smee's, or Wollaston's, one to ten or twenty is a very
good proportion. The price of concentrated sulphuric acid
(oil of vitriol) is about three-halfpence per pound. It is important that this liquid be free from nitric acid (which it sometimes contains), because that acid wastes the zinc, and in
Smee's battery also corrodes the silver. To test for nitric
acid, add to the suspected liquid, a small quantity of a solution of indigo in pure sulphuric acid, and boil the mixture;
if the colour of the indigo does not disappear, nitric acid is
not present. If the silver plates in a Smee's battery, become
covered with a dirty whitish film, a trace of nitric acid is
probably present. The nitric acid used in Grove's battery,
should be free from hydrochloric; otherwise, when it gets
warm by the action of the battery, it will corrode and
dissolve a little of the platinum plates. To ascertain if
hydrochloric acid is present, dilute some of it with dis-

tilled water, and add two drops of a solution of nitrate of silver : if a white cloud, or milkiness appears, that acid is present. Common oil of vitriol nearly always contains sulphate of lead dissolved in it, and when one measure of the acid is added to five or ten measures of water, the mixture becomes cloudy, and a greyish white powder (consisting of the sulphate) settles to the bottom of the vessel ; this powder should not be allowed to get into the battery cells, otherwise it will settle upon the zinc plates, and cause them to waste. In mixing oil of vitriol and water, it is highly important that the acid should be gradually added to the water *and not the reverse,* and also that the mixture be stirred during the addition; and it is especially necessary, that the water and acid be cold, because great heat is evolved by mixing them ; if water be added to oil of vitriol, an explosion may be produced by the heat; and more especially is it dangerous to add *hot* water to oil of vitriol. Brown oil of vitriol is that which has been made from iron-pyrites obtained from the coal measures, and its colour is due to particles of carbon ; it is sometimes also impure; but even the purest sulphuric acid is occasionally brown, from particles of organic dust getting into it. Strong sulphuric acid has a specific gravity of 1·845 ; if its gravity is less than this, it contains water. If the acid used in a battery is not sufficiently diluted, crystals of sulphate of zinc are apt to form upon the bottom ends of the zinc plates after a time, through want of water to dissolve them, and this impedes the current; a mixture of ten parts by measure of water to one of acid, is sufficiently dilute to prevent this; such a mixture has a specific gravity of about 1·10.

The only other liquid used in the batteries I have described, is a solution of sulphate of copper ; this salt is usually sufficiently pure, if a proper price (about sixpence per pound) is paid for it. Any green colour in it, is indicative of the presence of sulphate of iron, with which the cheaper varieties are contaminated.

Amalgamation of zinc.—Zinc rods and plates are always amalgamated, because it makes them more electro-positive (see p. 63), and because it also largely protects them from corrosion when the battery is not in action. The explanation of this is not very clear, but it probably is, that the mercury, by dissolving the surface of the zinc, and traces of foreign metals in it, renders the whole of that surface of uniform composition, and therefore no one part of it is relatively electro-positive or negative to another, and no local current can be generated. It is however dependent also upon the presence of a film of hydrogen upon the surface of the metal, for if a trace of nitric acid, or other liquid capable of oxidizing or removing such a film, is present, the mercury does not protect the zinc.

Zinc rods or plates may be well amalgamated, by immersing them in dilute sulphuric acid until gas is freely evolved, then pouring mercury upon them, and rubbing them until they are bright all over. If the plates are new, they are probably greasy from the process of rolling, and should first be dipped in the caustic potash solution and swilled, before putting them into the acid, or they should be scraped. After having been amalgamated, they should be placed on their ends to drain off the superfluous mercury, and then the residuary mercury wiped off them. Ruhmkorff amalgamates zinc plates, by dipping them into a solution made as follows:—Dissolve one part of mercury in five parts of aqua regia (i.e. one part of nitric and three of hydrochloric acids), and then add five parts of hydrochloric acid. Another plan, is to put some mercury into a coarse flannel bag, dip the bag occasionally into dilute hydrochloric acid, and rub it upon the zinc plate or rod.

Roseleur uses an amalgamating salt, prepared by boiling an aqueous solution of mercuric nitrate, with an excess of a powder, composed of equal parts of mercuric chloride and mercuric sulphate, cooling the mixture, and using the liquid only. The liquid is added to the mixture of sulphuric acid and water, in those batteries only where two liquids are employed.

The mercury used for amalgamating should be pure ; if it contains tin, lead, bismuth, or copper, &c., these metals will adhere to the zinc, and cause great waste, by what is termed 'local action,' which means, that the zinc and the particles of foreign metal, being in contact in an acid liquid, constitute a multitude of little voltaic couples, which generate electric currents (by corrosion of the zinc and waste of the acid), when the principal current is not circulating. For a similar reason, the zinc also should be free from metals less positive than itself. New zincs require frequent amalgamation, because the mercury soaks into them, but as they get old and thin by use, this mercury is left upon their surface, and therefore they rarely need to be amalgamated. When zinc plates become so .thin as to fall to pieces on handling, new ones should be substituted, and the old ones may be melted, and cast into rods for Daniell's batteries ; or be broken up, put in an iron retort, and the mercury distilled from them at a strong red heat, through a wide and wet tube of leather, into a vessel of water.

Selection of zinc for batteries.—The best kind of zinc for batteries, and the kinds chiefly in use by electro-platers, are the Belgian and Silesian. The thickness of the plates should vary with the size of the battery; the smallest should not be much less than one-eighth of an inch thick, on account of its brittleness when amalgamated ; large ones are generally about three-sixteenths or one-quarter of an inch in thickness. Zinc bolts for Daniell's batteries are sometimes made, by melting together a number of old worn-out pieces of battery plates, and casting in a suitable mould. The wholesale price of unrolled (cake) zinc, is usually from twenty to thirty shillings per hundredweight. As all zinc contains traces of less positive metals; when the former dissolves away, the latter come to the surface, and form an amalgam, and diminish the protective power of the mercury; such a coating should occasionally either be scraped off, or removed by means of a very hard brush, and pure mercury applied. Cast zinc is not so good for electrical purposes as rolled zinc ; it is also less easy to amalgamate. Plate zinc may be cut by means of a saw with fine teeth, or by drawing a line across it repeatedly (using great pressure), with the end of a triangular file which has been ground to a sloping point. It may also be bent into cylinders whilst it is hot.

Battery cells.—These are either of stoneware, glass, gutta-percha, or ebonite. For large cells, stoneware is nearly always employed : for small ones, glass is very good, and so is gutta-percha, but the preference is generally given to ebonite, especially for Grove's battery, because it is not brittle like glass, and does not become softened like gutta-percha by the heat generated in the battery.

Porous cells for batteries.—These vary very greatly in quality: some are so slightly porous, that they very seriously hinder the passage of the electricity ; most excellent ones are manufactured by Messrs Wedgwood & Co. Formerly, porous cells of wood were employed, but now, only those of earthenware are used ; they should always be kept in clean water when not in use, to remove nitric acid and salts of the battery liquids from them, to prevent their cracking, and to preserve them always fit for immediate use. The degree of porosity of two cells may be compared by drying them, and then simultaneously filling them with water, and observing the appearance of their outer surfaces after one or two minutes.

Binding-screws. — These are employed for connecting and holding together the plates, connecting wires, &c. of a battery. That shown by fig. 46 is for attaching to the wooden cross-bar and platinized silver plate of a Smee's cell ; 47 and 48 are for attaching to zinc or copper plates ; 49 is for joining zinc and platinum, or zinc and copper plates together ; 50 is for attaching to the top of a thick bar of carbon, or for holding together the zinc plates of a Smee's battery ;

and 51 is a screw I have devised and employed for joining together the ends of copper wires.

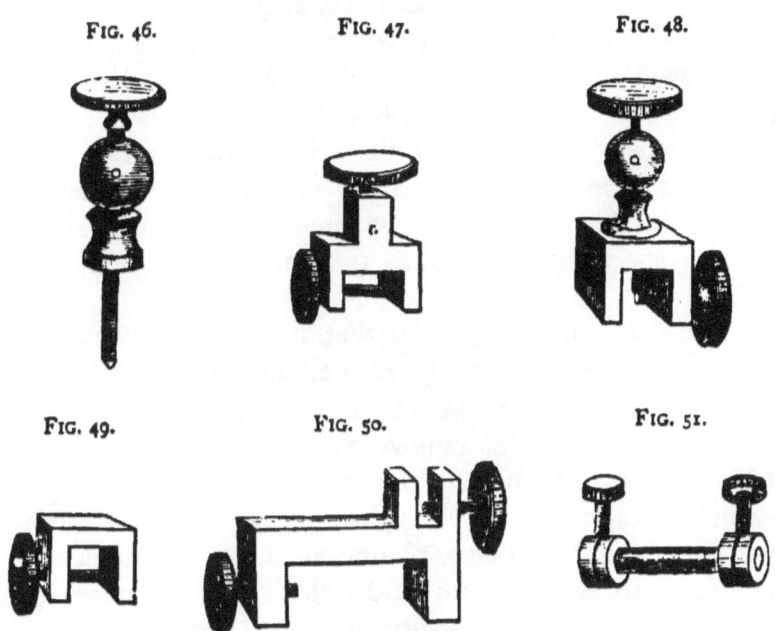

FIG. 46.   FIG. 47.   FIG. 48.

FIG. 49.   FIG. 50.   FIG. 51.

Management of batteries.—If the acid liquid in contact with the zinc is very strong, the zinc plates require frequent watching, to see that there is no local action, and when gas is seen or heard rising from them, or when any dull patches appear upon them, where the acid has acted too strongly, they should be amalgamated ; if this is neglected, great holes will be quickly corroded in them. They should be taken out of the cells every evening, if the acid liquor is at all strong, unless deposition is required to continue all night.

After a Wollaston's, Smee's, or Daniell's battery, has been at work a few days, a small amount of sulphuric acid should be added, and the liquid stirred, and this should be done as often as the current becomes feeble, until at length the liquid acquires an oily consistence, and becomes nearly saturated with zinc salt, which crystallizes upon the cells and plates above the surface of the liquid; it is then time to remove the liquid, and charge the battery afresh.   If crystals of sulphate of zinc are required for depositing or other purposes, the exhausted solution may be set aside, and allowed to evaporate.   Sometimes in a Smee's or Wollaston's battery, a deposit of zinc forms upon some of the negative plates ; when this happens, it is a sign that the acid is exhausted in those cells ; either more acid, or a fresh mixture, should then be put in ; the deposit may also be removed, by immersing the negative plates in a separate portion of dilute sulphuric acid.

Great care must be taken, that no mercury comes into contact with the plates of copper or platinized silver, the latter especially, as it makes them brittle, and greatly diminishes the electric power.   To remove mercury from copper plates, the latter should be heated to redness, but with silver plates, a much less heat should be applied for a longer time, and then the plates should be re-platinized.   Copper plates should be frequently scoured with sand with a hard brush ; and the silver plates should be re-platinized when they become light in colour, which will happen after about six months' careful working.

In managing a Grove's or Bunsen's battery, it is highly important, not to allow any of the nitric acid to get into contact with the zinc, because it produces strong local action, and waste of that metal.   As the nitric acid cannot be prevented from passing through the porous divisions, such batteries cannot be kept in continual energetic action more than a day, in consequence of this circumstance.   The porous cells of such batteries should be soaked in water, the water being changed twice or three times (so as to extract all the nitric acid from them), before they are used a second time ; therefore, for continual use of such a battery, two or three sets of such cells are necessary, some being in soak whilst others are in use. In charging any two-liquid batteries, it is best to have the liquids level, and if they be either Grove's, Bunsen's, or

Daniell's, the liquid in the zinc division should be rather the higher.

It is best to employ separate batteries for each different depositing liquid. Each battery should be tested before it is used: this may be done in a rough, though usually sufficiently accurate way for the purpose, if the current is a strong one, by connecting one end of the battery to a file, and drawing the point of the wire from the other end of the battery along its surface; by the degree of brilliancy of the sparks produced, the strength of the current can be estimated. Before testing or using a battery, it is necessary to examine, and see that all the points of contact of the wires, screws, &c. are clean, and that the screws hold the wires firmly; it is also advisable to see that all the cells are connected in the right order, for if only one cell is connected the opposite way, it will not only be rendered ineffective, but will also neutralize the action of one of the others; and its negative plate will be liable to dissolve by the influence of the current from the remaining cells. Voltaic batteries should be kept in a place of moderate and uniform temperature; not where the liquids are liable to freeze, or rapidly evaporate.

Regulation of electric power.—This is always a matter of considerable importance, especially when depositing from solutions, which will not bear a great range of electric force, without spoiling the quality of the deposited metal. It may be effected in a variety of ways, viz., by making alterations either in the battery, in the depositing vessel, or in the wires connecting them. The electro-motive force (commonly called ' the intensity ') of the current may be increased, by adding to the *number* of cells in the battery; or by using cells of greater intrinsic pushing power, for instance Grove's instead of Smee's, &c. (see p. 327). As the electro-motive force is diminished by resistance, a diminution of resistance in any part of the circuit will increase it; this may be effected to a certain extent by making the depositing liquid hot, using larger

z

electrodes, or placing them nearer together. The quantity of the current may be increased by all these means, and also by immersing the battery plates, or only one of them, deeper in the liquid. The usual method, however, for regulating the electro-motive force of the current, is to alter the number of cells in the battery; and for regulating the quantity, to alter the depth of immersion of one of the battery plates (see p. 326); but sometimes the latter cannot be conveniently effected, and in that case, the anode is either increased, or diminished in size. As that also is usually inconvenient, a large piece of copper or brass is sometimes suspended to act as a cathode along with the article to be coated, and thus relieve it of part of the current. Galvanometers or voltameters (see p. 73) are very rarely employed to measure the electric currents employed in practical electro-deposition, chiefly, because the want of such instruments is not felt, and partly, because the processes are too coarse for the use of delicate apparatus.

Compound voltaic batteries are usually so constructed, that they may be used to supply either a current of less quantity and greater electro-motive force, or the reverse. By connecting a series, say of twelve cells, all in one row, with the metals alternating throughout, we obtain from the end wires, a current of a quantity of one, and an electro-motive force of twelve. By connecting them as a double row or series of six, the two end zincs being connected to one terminal wire, and the two end coppers to the other, we get a current of a quantity of two, and electro-motive force of six. By connecting them in a similar way in a treble row as a series of four, we obtain a current, the quantity of which is equal to three, and the electro-motive force equal to four. By arranging them in a quadruple row, and a series of three, we get a quantity of four, and electro-motive force of three. By placing them as a sextuple row, and as a series of two, we get a quantity of six, and electro-motive force of two. And finally by placing them in single row, connecting all the zincs together by one wire, and all the silvers by another, they all act as one pair of twelve times the surface of a single cell, and we obtain a quantity of twelve, and electro-motive force of one. To make such arrangements successfully, it is, however necessary, that all the plates be provided with suitable screws, also that all the cells be of a similar kind, and equal in electro-motive force, otherwise the currents from the stronger ones will be liable to pass partly through the weaker ones instead of through the plating solution, and also perhaps damage the battery, by causing some of the negative plates to be corroded ; it is therefore only occasionally that batteries are so arranged, i.e., not in single alternate series. The power of the current from magneto-electric machines, is usually regulated by interposing a piece of thin iron wire in the circuit.

Selection of depositing processes.—Different articles are electro-coated by different methods ; some are coated, as already stated, by simple immersion, others by simple contact with zinc, and others by means of a separate current ; but an electro-plater usually employs only the latter method. For very small articles of which there are a great number, such as buttons, hooks and eyes, pins, &c., and which require only a very thin deposit, the simple immersion or wash process answers very well, being both easy of execution, and cheap. But for all ordinary deposits, plating, &c., the separate current method is by far the best, because coatings of greater, and of sufficient thickness, of all ordinary metals, may be obtained by it, and the solutions do not usually (as in the other processes) require renewal.

'*Pyro plating.*'—A process termed 'pyro-plating' has during the last few years been introduced, and is stated to be specially suited for causing a coating of gold, silver, platinum, copper, nickel, brass, bronze, or aluminium-bronze, to adhere to metals, in cases where the metals to be plated, will not readily receive a film of mercury by the ordinary 'quicking' process, as with iron, steel, nickel, and aluminium.

The article of iron, steel, &c., is first made perfectly clean, by immersion in a boiling solution of caustic alkali, then brushed with emery, also with a steel brush in a stream of solution of washing soda; then suspended in a similar solution; next made the cathode in a hot solution of caustic alkali, with a strong current to evolve from it plenty of hydrogen, until its surface looks 'silvery;' and then transferred to a special solution of silver, and plated. A previously weighed metal plate, of equal amount of surface, is immersed as a cathode by its side, and weighed from hour to hour, until sufficient silver has been deposited. The original article is then removed from the vat, and (after washing?) heated in a furnace to 'drive' the coating of silver (or other metal as the case may be), into its surface; and if the article requires tempering, it is quenched in water. Pyrogilding is performed in a similar way to pyro-silvering, except that the whole of the metal is not put on at once, but in three successive layers, and heated in the furnace after each coating. The first, before being heated, looks perfect, but by the heating, the gold nearly all disappears, being driven into the under metal. The second, only partly disappears by the influence of the heat; and the third entirely remains. Pyro-gilding is specially recommended for coating articles of iron and steel ('Chemical News,' vol. xxvi. pp. 26 and 173).

Selection of depositing liquids.—The most important points to be observed, in selecting a liquid for the separate current process, are: first, that it should yield its metal freely, and in a reguline state; second, it should not decompose, or deposit its dissolved metal, by contact with the atmosphere, or by exposure to light; third, it should not act chemically to any great extent, upon the base metals, or upon those to be coated; fourth, it should dissolve the anode sufficiently freely; fifth, it should possess good electrical conducting power ; sixth, it should not evolve gas at the surface of the articles. The three first conditions are, I consider, indispensable, and if it fail in either, it is worthless or nearly so, for the purposes of electro-deposition.

Testing a depositing liquid.—From what has just been said, the mode of testing is obvious. To test it, pass a current from two or three Smee's cells through it, by clean and weighed anodes and cathodes, the latter being composed of the particular metal which it is intended to coat. Observe the quality of the deposit, the speed of deposition, and whether much gas is evolved from the electrodes. Set a portion of the clear liquid aside, in a colourless glass vessel, exposed to light and air, and observe if it acquires a film, deposits a sediment, changes in colour, evolves gas, or shows any other signs of decomposition. Immerse in a separate portion, for about a quarter of an hour (that will be abundance of time), a bright and perfectly clean piece of metal, of the kind to be deposited, and observe if it becomes coated with the dissolved metal, or changes in appearance in any way. A liquid which requires a strong current to make it yield its metal freely, or which liberates gas at the cathode, but has no other defects, does no harm except being wasteful of the electric power. One which evolves gas at the anode, becomes gradually deprived of its dissolved metal.

Practical management of depositing solutions (see also p. 90 *et seq.*).—Having obtained a good depositing liquid, we must manage to keep it so ; because a large vat of silver solution, or a vessel of gilding liquid, is valuable. The operator should as far as possible, avoid doing anything to such liquids which will alter their chemical composition ; many valuable ones have been injured and spoiled, by persons (unused to making careful experiments) adding substances to them, with the hope of improving them. The tales told by electro-platers of their experiences with depositors, in making and mending electro-plating solutions, should act as warnings to those about to commence in the art upon a large scale. One discovered that the operator, by using cyanide of potassium, regardless of its strength, to make cyanide of silver, re-dissolved about sixty ounces of silver, and threw it away in the wash-waters. Another had a similar mishap with eleven ounces of gold. A third had 450 ounces of silver converted into waste residue. A fourth had two large vats of silver solution rendered incurable, by addition of too much 'brightening' liquid; and many electro-platers have had similar mishaps. Others have found their anodes dissolve with extraordinary rapidity, through the use of too much free cyanide, or by allowing them to remain in contact with the iron vat, and have been surprised to find, that a solution, which when made, contained only an ounce of silver per gallon, held in solution more than four times as much. Others, by keeping a record of the silver dissolved and deposited, as well as of that found in the liquid by analysis, have missed a considerable quantity, and ultimately found that it had soaked into the sides of the thick wooden vats. The composition of a depositing solution should not be altered, except so far as it can be done with perfect safety, as by diluting it to a certain extent with water, or adding materials to exactly re-place those abstracted from it. The electric power should always be adapted to the liquid, and not the latter to the electric power. The electrodes should as a rule, be kept nearly equal in amount of surface, the anode being in some cases the largest; and the quality of the deposit should not usually be regulated by altering their proportionate extent of surface, but by altering the battery or other source of the current.

As a general rule, in order to prevent depositing liquids gradually becoming contaminated with foreign metals, any metal which will be corroded by a particular depositing liquid (and which will therefore coat itself by simple immersion in that liquid), should previously receive a coating of suitable metal in a preparing solution; for instance, iron articles which are to receive a thick coating of copper, are first coated with a thin film of that metal in a cyanide liquid.

Anodes of any metal may be formed of scraps, but that is not advisable, if better ones can be obtained.

Proper position of articles and dissolving plates in the vats.—Both the articles and the plates should be wholly submerged in the liquid; the former being a little the deepest. Both should be vertical, or nearly so; the plates may however overhang a little with advantage: it makes them dissolve more evenly. The horizontal position, with the dissolving metal above, although the most scientifically correct arrangement, does not succeed in practical working, because the metal used for dissolving is never quite pure (with nickel and copper especially), and the impurities from it, fall upon the surface of the receiving article beneath, and make it rough; in addition to this, the position of the article prevents its being easily removed or examined. If the object to be coated, has a very irregular outline, either the dissolving plate should be bent somewhat to its form, so that the two may be nearly equidistant at all parts; or the article should be often shifted in its position, so as to produce a nearly uniform thickness of coating all over. The nearer the receiving surface is to the dissolving plate, the more rapid is the deposition, and a large body of liquid, deposits more rapidly and more evenly than a small one. The greatest thickness of coating always takes place upon the most prominent places, i.e., upon those parts nearest the dissolving metal. If it is desired to prevent vertical lines in a thick deposit, the object must be kept in motion;—the means of doing this has been already described (see pp. 171–174).

Motion of the articles is very advantageous: it permits much more rapid deposition; it keeps the solution much more uniform in composition, prevents the lower portions of the objects being coated so much faster than their upper ones, and also prevents the upper parts of the anodes being dissolved so much more rapidly than their lower ones. In addition to this, by keeping the solution mixed, it greatly diminishes the electric conduction resistance, which would be produced by polarisation, due to layers of liquid of opposite electrical nature, collecting in contact with the electrodes (see p. 54).

As most of the deposit takes place upon the parts of the article nearest the dissolving plate, if other parts require also a thick deposit, the article must be so placed, or an anode must be employed of such a shape, as to effect that object.

Regulation of the deposit.—Regulation of the quality of the deposited metal is always an important matter, and with all metals, except a very limited number, it is one of the most difficult objects to effect. As a general rule, the greater the electro-motive force, and the smaller the quantity of the current, the harder and brighter is the deposited metal; but this of course only holds good in the case of a liquid which is capable of yielding such metal. The chief points are, first to obtain a good liquid, at the proper temperature, and second to adjust the density of the current (see p. 38) until the required kind of deposit is obtained. Some liquids are so constituted, (especially those of the more easily oxidizable base metals, such as manganese,) that if the current is only of sufficient density to deposit it in a bright reguline state, it is not sufficiently dense to prevent the metal at once taking up oxygen and forming a sub-oxide. If, in a good depositing solution of a non-readily oxidizable metal, such as copper, we are producing by means of a current of considerable electro-motive force, a black powder deposit, upon a very small article, a much larger article would receive by the same current a reguline deposit, and upon a very much larger one the deposit would be hard and crystalline. So much, however, depends in every case upon the special characteristics of the particular liquid, that these can only be considered as general instructions for the guidance of the electro-depositor. This part of the subject has also been already treated of in previous parts of this book (see pp. 35–39, 54–55, and 90–93).

The action of a current of great electro-motive power, but small in quantity, appears in some cases (for instance with copper), to confer upon the deposited particles, a kind of polarity, a power of grouping themselves into separate warty nodules or groups of crystals, each of which, as it becomes larger, appears to powerfully repel all particles in its neighbourhood, and thus causes the metal to spread rapidly; when this action is continued to a considerable thickness of deposit, especially in cold weather, the metal is exceedingly hard, and easily broken into a number of distinct grains or nodules, which are in the form of lumps with rounded edges. With a current from 100 pairs of Smee's battery, acting for a long period of time in cold weather, and the quantity of the current kept down to the lowest possible degree, I have seen a tough deposit of zinc spread over several square inches of clean gutta-percha; and in depositing copper by a current of rather high intensity, and small quantity, upon black-leaded gutta-percha medallions, I have repeatedly observed, that where there was a sunken boundary line near the edge, the deposit remained quite thin, as if powerfully repelled, whilst on each side of the line it was very thick, and on the outside edge accumulated in large masses, hard and distinctly separate, and containing as much metal as the whole of the medallion besides. The effect of lines is often seen in electro-copies of set-up type, and the deposits are very fragile at those parts.

With regard to the regulation of the quantity of the deposited metal, that part of the subject has been treated of in the theoretical division (see pp. 39–44, 74–75). We know that when all the arrangements, are properly made and carried out, the quantity of metal dissolved and deposited in the vat, is in direct proportion to the quantity of zinc dissolved, and acid consumed, in each alternation of the battery. With a perfect depositing liquid, good battery arrangements, and pure materials, for every equivalent of zinc, dissolved in each alternation of the battery, an equivalent of metal is dissolved on one side, and an equivalent deposited on the other, in the depositing vessel. For instance, for every equivalent $\left(\frac{65}{2} = 32\cdot5 \text{ parts}\right)$ of zinc so dissolved, and $\frac{98}{2} = 49$ parts, or one equivalent, of oil of vitriol consumed in the battery, an equivalent $\left(\frac{63\cdot5}{2} = 31\cdot75 \text{ parts}\right)$ of copper is deposited in the sulphate of copper solution, or an equivalent (108 parts) of silver in the cyanide of silver plating liquid, and a similar quantity of copper or silver dissolved at the anode. But in practical working, the materials are rarely if ever pure, or the arrangements perfect; the zinc nearly always contains a small proportion of other substances, the mercury contains tin or lead, and the sulphuric acid contains a little nitric acid, or plumbic sulphate. The acid liquid of the battery is often too strong; much of it is also thrown away before it is completely exhausted. The zinc plates are not kept well amalgamated, or the silver well platinized, or the plates are suffered to remain too long in the liquid when not in use. The metal of the anode is also frequently impure ; occasionally some of the deposit is allowed to re-dissolve, from the battery power becoming low, and from not stirring the solution; in some solutions, a part of the electric current is expended in evolving gas at the cathode; and finally, the repeated operation of 'scratching,' removes some of the deposit. Allowing for all these, and other unavoidable sources of loss, in practical working, about one pound only of copper, can be deposited in the ordinary sulphate solution, by the consumption of from one and a quarter to one and a half pounds of zinc, and an equivalent quantity of acid, in each alternation of the battery.

With regard to regulation of the speed of deposition, (see p. 337); with every liquid there is a limit of rate of deposition per given amount of surface, beyond which it is impossible to obtain good metal (see p. 38), and that limit differs with every different liquid, and probably with each liquid at every different temperature, besides being dependent upon the kind of receiving surface. It is well known to electro-depositors, that it is usually much more difficult to produce a reguline deposit upon rough surfaces, than upon smooth ones; upon cast-iron than upon most other metals, and that to obtain it at all upon that metal, the rate of deposit must be less, than upon a smooth surface of pure copper or silver.

Magneto-electric machines.—The fundamental principle of all magneto-electric machines, has been already stated and illustrated (see p. 57). As this is not a treatise upon dynamic electricity, but only upon the applications of it to metallurgical operations, and as the space at my command is only limited, and a clear and full description of magneto-electric machines would occupy too much space, I am only enabled to insert a very brief statement respecting these electro-motors.

Figures 52 and 53 represent Wilde's magneto-electric machine. It consists essentially of two electro-magnets, a small and a large one, with insulated copper wire coiled transversely upon them; and with armatures of soft iron (see fig. 53) (also with insulated copper wire coiled lengthwise upon them), revolving between their poles. The residual magnetism of the small (or upper) electro-magnet, excites a feeble current in the coil of its revolving armature. This current circulates through the wires of both the magnets, and increases the magnetism; and the increased magnetism of the small one, reacts upon the armature, and increases the

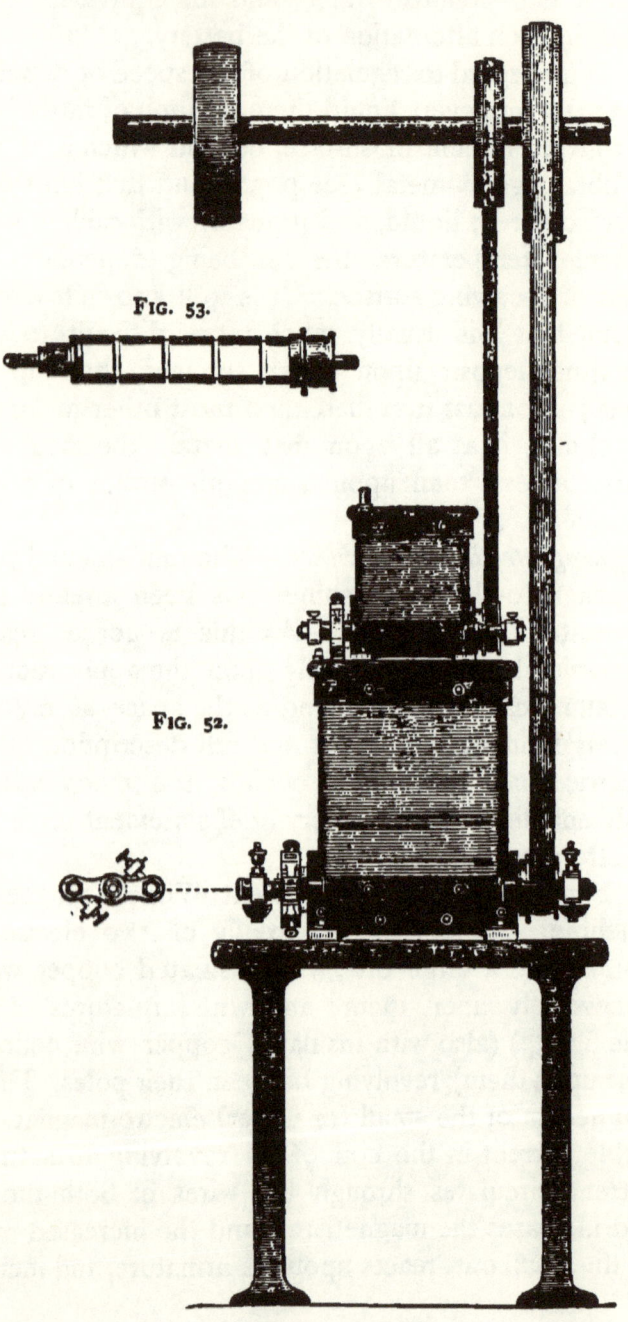

Fig. 53.

Fig. 52.

current, and so on, until both the magnets are saturated with magnetism at the expense of mechanical power. The current from the revolving armature of the large one alone, is used for electro-deposition, or other purposes. 'The armatures of both machines are driven at a speed of about 2,000 revolutions per minute, and at this rate, the current from the large one, deposits twenty-eight ounces of silver an hour, with an expenditure of two horse power.'

These machines are in extensive use at the works of Messrs. Elkington and Co., in Birmingham, for the purposes of depositing copper statues, and for general plating with silver ; also at the copper works of the same company at Pembrey, near Swansea ; for purifying by electrolysis upon the large scale, crude slabs of unrefined copper from the ordinary smelting process. A single 'multiple armature' machine of Wilde's (see 'Philosophical Magazine,' June 1873), at those works, deposits four and a half hundredweights of copper in twenty-four hours. These machines have also been successfully applied to the economic production of coppered iron rollers for calico-printing. To keep the armature cool, the ends of the large electro-magnet are made hollow, and a current of cold water caused to flow through the cavities.

Gramme's magneto-electric machine is shown in fig. 54. It consists essentially of a ring of soft iron, covered with a large number of coils of insulated copper wire, the respective ends of which are connected with the separate sections of two commutators fixed upon the axis of the machine. The ring with its coils and commutators, fixed upon the axis, revolves between the poles of an electro-magnet.

By this machine—'To deposit 600 grammes of silver requires one horse power, and a speed of 300 turns per minute ; the tension of the current being equal to that of two Bunsen's cells, and its quantity equal to thirty-two such cells of ordinary size. At a speed of 275 revolutions per minute, it has deposited 525 grammes of silver per hour; at 300 turns, 605 grammes ; and at 325 turns, 675 grammes. The weight of the copper wire on the fixed electro-magnets was

FIG. 54.

135, and on the moveable ones forty kilogrammes' ('Telegraphic Journal,' vol. i. p. 54). 'The present form of the machine as used for electro-deposition is composed as follows :—

Total weight	117·5	kilogrammes
Copper coils	47·0	,,
Total height	·6	metre
Total width	·55	,,
Deposits silver per hour . .	600·	grammes
Required power to work it . .	50·	kilogrammetres.'

('Telegraphic Journal,' vol. iii. p. 198). This machine is in use at Messrs. Christople's large electro-plating works in Paris.

The most recent form of magneto-electric machine, is that of Messrs Siemens and Alteneck, a description of which, with engravings of it, may be found in the 'Electrical News,' vol. i. p. 226.

The chief obstacle hitherto met with in the use of these machines has been, that after a few hours' action, the different parts are liable to become considerably heated, partly by the incessant molecular changes attending the variations of magnetism, and partly by the conduction resistance in the coils of wire. This has been largely overcome in Mr. Wilde's machine by the employment of several small machines instead of one large one, and by allowing a stream of cold water to run through the hollow ends of the magnet. In Gramme's machine, provided it is not worked too fast, the heat is reduced to a moderate amount; and in a large machine of Siemens and Alteneck's in the Vienna Exhibition, I also observed but little rise of temperature, after it had been in action a considerable time. Another objection to some of these machines, is the complexity of the commutator. The electric current from all these magnetic machines, is regulated for electro-metallurgical purposes, by interposing a piece of thin iron-wire in the circuit.

Thermo-electric piles.—The two most efficient kinds of this instrument, appear to be those of Nöe of Vienna, and Clamond of Paris. The former is the more quickly excited, and gives a powerful current; and the latter is the most strongly constructed.

Nöe's pile (see fig. 55) consists of small cylinders, about one and a quarter inches long, and three eighths of an inch diameter, of an alloy of about thirty-six and a half parts of zinc, and sixty-two and a half of antimony as the positive, and stout German-silver wire as the negative element. Twelve of these pairs have an electro-motive force of one Daniell's cell, and twenty of them that of one Bunsen. The resistance of twenty of them is about equal to one ohm (see pp. 70–73). With a great external resistance, twenty of them are equal to one Bunsen's, and with a small external resistance, twenty *quadrupled* ones are somewhat stronger than one of Bunsen's elements (Watts' 'Chemical Dictionary,' supplement, p. 458. Wiedemann's 'Galvanismus und Elektromagnetismus,' 1872, vol. i. p. 824. 'Journal of the Chemical Society,' vol. ix. p. 989, vol. xi. p. 465).

The construction of a few elements, is shown in the annexed figure. The junctions of the elements are heated by small gas-flames, and the alternate junctions are cooled

FIG. 55.

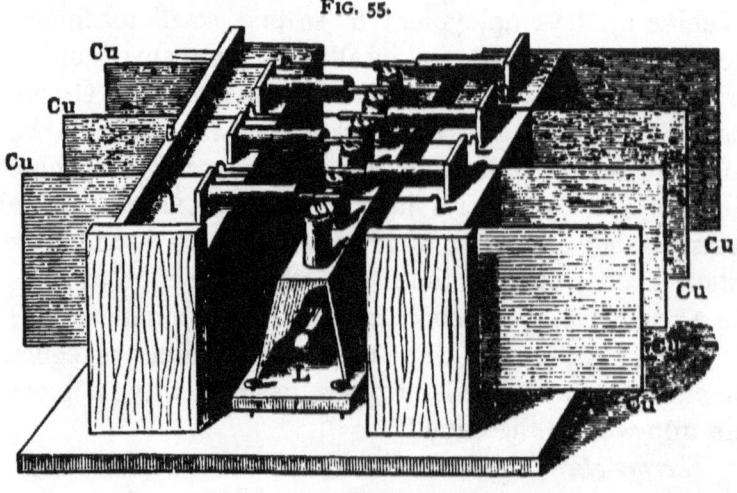

by the heat being conducted away by large blackened sheets of thin copper. To protect the German-silver wire from oxidation, it is enclosed in a tube of that alloy where the flame impinges against it ; and to prevent the ends of the positive cylinders being melted, they are faced with iron and a thin sheet of mica. The German-silver wire may be heated to low redness. The usual form of the apparatus is in ninety-six elements, which may be either used as ninety-six by one, forty-eight by two, or twenty-four by four ; an l instantly changed from one to the other of these arrangements, by means of a most ingenious and effective current transposer, which does not require cleaning. The current attains its maximum strength in about one minute ; that from the single series decomposes water rapidly ; and that from the quadruple series excites a large electro-magnet powerfully. I have used this apparatus with great satisfaction for many brief experiments. The instrument is made by W. J. Hauck, Kettenbrückengasse 20, Vienna ; also by P. Dörfell, Berlin. It is, I am informed, in use for electroplating in Dittmar's electrotype and lamp manufactory, Vienna.

FIG. 56.

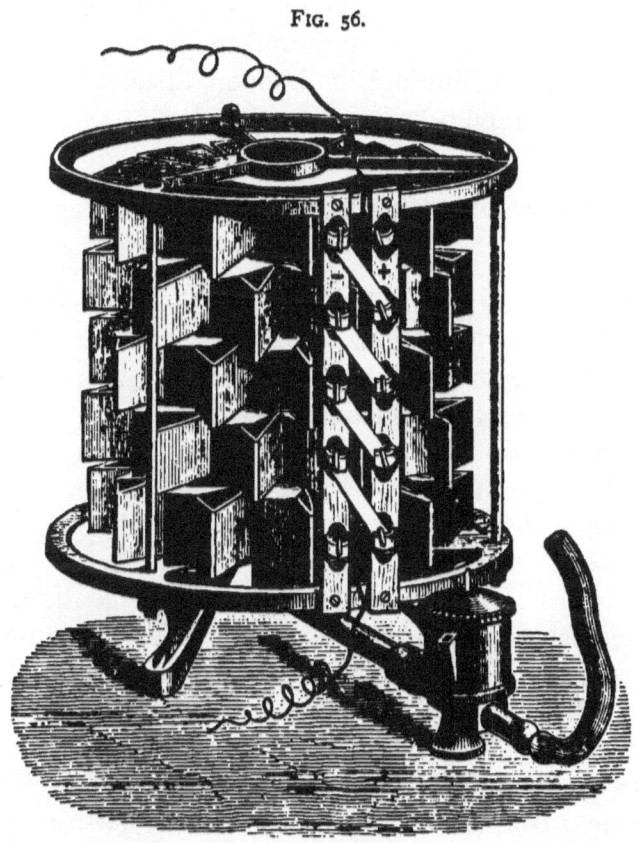

Fig. 56 represents a small Clamond's pile, connected for intensity (see also 'Telegraphic Journal,' vol. i. p. 12).

The elements are tinned sheet-iron as negative, and an alloy of two parts of zinc and one part of antimony as positive. A pile which consumes 150 litres of gas per hour, is capable of depositing one kilogramme of copper, at a cost of two francs fifty centimes ('Telegraphic Journal,' vol. iii. pp. 157 and 319). According to the inventors, 'a machine of 100 bars, with a consumption of 8 to 9 cubic feet of gas, deposits about an ounce of silver per hour. The same apparatus coupled for quantity, will deposit about one ounce of copper in the same time,' '100 bars, coupled for quantity, have an electro-motive force of about five volts, and an internal resistance of one ohm' (see pp. 70–75). Clamond's pile is being used for electro-plating and depositing, in various establishments in Birmingham, London, Sheffield, and other places. Its durability is being improved.

SPECIAL INFORMATION RESPECTING SUBSTANCES USED IN THE ART.

As there are various substances used in the different processes of electro-deposition, it will be useful to the practical operator in the art, to be acquainted with some special technical points of information respecting them, which may affect the success of his operations, and which have not already been given in the body of the book.

Water.—Distilled water is the most suitable for making solutions. It should give no cloud, on adding to separate portions of it, a few drops of solutions of argentic nitrate, chloride of barium, or oxalate of ammonium ; nor become brown on addition of sulphuretted hydrogen water. If distilled water cannot be conveniently obtained, filtered rain-water may usually be employed in its stead.

Nitric acid.—Called also aqua fortis. The pure acid for dissolving silver, &c., should be colourless, have a specific gravity of not less than 1·52 ; and separate portions of it, diluted with pure distilled water, should give no cloud with a single drop of solution of nitrate of silver, or of chloride of barium. It should be kept in a stoppered bottle, in a dark, cool, and dry place. If a drop of this or any other acid falls upon one's clothes, *diluted* aqueous ammonia should at once be freely applied.

All the pure strong acids should be kept in stoppered bottles, in a dry place. Carboys of common acids, and dipping liquids, should have stoneware stoppers, and be kept in an outhouse.

Hydrofluoric acid.—Called also fluoric acid. This liquid is always very impure. It should be kept in a bottle of gutta-percha, provided with a stopper of india-rubber, in a dry and cool place, and not in close proximity to glass vessels, because the vapour corrodes them. It is highly dangerous to breathe the fumes of this acid; and if a drop of it falls upon the skin it should be *thoroughly* washed off *at once*, otherwise after a few hours great pain will be suffered.

Hydrochloric acid.—Called also ' muriatic acid,' ' spirits of salt,' and ' smoking salts.' The pure acid should be colourless, of not less specific gravity than 1·20. It should be kept in a cool place.

Aqua regia.—Called also nitro-hydrochloric acid. This is a mixture of one volume of nitric, and from two to three of hydrochloric acid. It should not be prepared until required to be used, because it decomposes spontaneously.

Blacklead.—Called also plumbago and graphite. This substance always contains a little earthy matter, silica, oxide of iron, &c. The most suitable kind is usually very black, but without much lustre, until after rubbing. It should adhere to the articles, and not become detached when they are immersed in the solutions. The best can only be selected by means of actual trial, and should be gilded or silvered (see p. 217).

Sulphuretted hydrogen.—Called also hydric-sulphide. sulphide of hydrogen, &c. This substance is a gas, and may easily be prepared by putting some fragments of prepared sulphide [1] of iron ('sulphuret of iron '), into a flask with some water, and then adding sulphuric acid. The gas should be washed by passing it through a small quantity of water. Sulphuretted hydrogen water is prepared by passing the washed gas in bubbles through distilled water until the water is saturated. The water dissolves only about three times its bulk of the gas ; or one part by weight of the gas dissolves in about 250 parts of water. The solution soon decomposes.

Sulphurous anhydride.—Called also 'sulphurous acid.' This is best prepared, by heating in a glass flask, strong oil of vitriol, containing fragments of copper wire. The flask should be protected from direct contact with the flame by a sheet of iron wire gauze.

Sulphuric acid.—Called also 'oil of vitriol.' The pure acid should have a specific gravity of not less than 1·85, and be nearly or quite colourless. The least trace of dust or organic matter, imparts a darkness of appearance to it. It should be kept in a dry place. When diluting it, the water should not be poured into the acid, because that is dangerous, but the acid into the water, and that slowly (see also p. 330).

Bisulphide of carbon.—Called also 'sulphuret of carbon,' and 'carbon disulphide.' This is a very volatile and inflammable liquid, and a flame should not therefore be brought near its vapour. It should be kept in a well-stoppered or corked bottle, in a cool place.

Phosphorus.—This substance should be kept in a wide-mouthed stoppered bottle, filled with water, to keep the air from contact with it. The bottle should also be covered with black varnish, and kept in a dark place, because the light changes the phosphorus and makes it insoluble. Phosphorus should never be exposed to the air for more than a few seconds, or it may inflame ; and it should always be cut whilst under the surface of water.

[1] Containing one equivalent of sulphur to one of iron.

Phosphorus solution.—Called also 'Greek , fire.' This highly inflammable and dangerous mixture, composed of phosphorus dissolved in bisulphide of carbon, has been already described (see p. 218). It should only be prepared in small quantity; and the bottle containing it should be kept in a cool place, partly immersed in sand, in a stoneware vessel covered with a metal lid. It is extremely liable to spontaneous combustion, especially if any be spilt.

Arsenious acid.—Commonly called 'white arsenic.' Only a small quantity of this is required. The bottle containing it should be kept in a dry place, out of the reach of careless persons, and should be distinctly labelled ' poison.'

Antimony.—In purchasing this metal, what is known as the ' best star antimony' should be selected. It may be known by its whiter appearance, and by having crystalline markings, looking like fern-leaves, upon its surface.

Bismuth.—This metal varies a little in quality, and is liable to contain traces of arsenic, and sometimes also of copper. The purer kinds are very much higher in price than the common variety.

Chloride of platinum.—Called also 'platinic chloride,' ' muriate of platinum,' &c. As the substance sold in shops is liable to contain a variable proportion of platinum, it is best for the operator to prepare the salt himself, according to the directions already given (see p. 118).

Chloride of gold.—Called also 'muriate of gold,' and ' auric chloride.' It is better to prepare this than to purchase it, because the commercial article is liable to contain a variable proportion of gold ; it should contain 65·2 per cent. of that metal (see p. 122).

Silver.—This metal may be tested for copper, by dissolving it in warm dilute nitric acid, precipitating all the silver by means of a slight excess of hydrochloric acid, and then adding a drop of solution of ferrocyanide of potassium, or by adding ammonia to the solution of argentic nitrate, until all the precipitate first formed is re-dissolved. Now look down through a considerable depth of the clear liquid; if a blueness is visible, copper is present.

Nitrate of silver.—Called also 'argentic nitrate,' 'lunar caustic,' &c. It should be in colourless crystals, free from odour of nitric acid, entirely soluble in distilled water, and should contain 63½ per cent. of silver. To ascertain the latter point, simply melt it at a full red heat, with a little borax in an earthen crucible, and weigh the metal ; or precipitate its solution by a slight excess of dilute hydrochloric acid, wash, dry, and weigh the precipitate; 143½ parts of it equal 108 of silver.

Chloride of silver.—Called also 'argentic chloride,' 'horn-silver,' and 'muriate of silver.' This substance should contain 75¼ per cent. of silver. To ascertain its percentage, melt it, at a full red heat, with an excess of perfectly dry (i.e. anhydrous) carbonate of sodium, in an earthen crucible, and weigh the button of silver. The chloride is decomposed by light, and should be kept in an opaque bottle in a dark place.

Mercury.—Called also 'quicksilver.' Pure mercury is perfectly bright, and leaves no tail of drossy appearance, on pouring it all slowly out of a vessel : it also volatilises entirely by heat. It should be kept in strong bottles, and not allowed to come into contact with any metals, except iron, platinum, or aluminium.

Amalgam of gold.—To prepare it, heat pure mercury to about 200° C., and add to it the gold in foil or ribbon ; the gold is readily absorbed and forms the amalgam.

Sulphate of copper.—Called also cupric sulphate, 'blue vitriol,' 'blue-stone,' 'Roman-vitriol,' &c. The pure salt should be in large crystals of a deep blue colour, without any admixture of green ; the latter indicates the presence of iron. To test for iron, dissolve a little of the salt in distilled water, add aqueous ammonia with stirring, until all the blue precipitate is re-dissolved. After standing some time, pour away the clear blue liquid, add distilled water freely to the residue, and allow it to stand for some time again ; a residue of red brown powder, indicates the presence of iron.

Nickel.—This metal is always contaminated with silicon and carbon, which remain as a black powder on dissolving the metal in acids. The dried black powder, when fused with saltpetre, produces a mixture of silicate and carbonate of the alkali. The metal also frequently contains copper; to test for this, dissolve the metal in aqua regia, evaporate the solution to a small bulk, dilute with water and add sulphuretted hydrogen water; if a blackish cloud is not produced, copper is not present.

Sulphate of iron.—Called also 'green copperas, green vitriol,' &c. It should be in the state of clear green crystals, perfectly free from adhering water or acid, and with no brown or red powder about them. It must be kept dry, and in well-closed bottles.

Carbonate of lead.—Called also 'white-lead.' It is a heavy white powder, and should be entirely soluble in warm dilute nitric acid ; any white insoluble matter is probably sulphate of barium.

Tin.—This metal is often adulterated with lead; to detect which, cut the tin up as small as possible, digest it with warm dilute nitric acid ; evaporate the clear liquid part to a small bulk, dilute with water, and add sulphuretted hydrogen water ; a black colour or precipitate indicates the probable presence of lead or copper.

Chloride of tin.—Called also 'muriate of tin,' 'tin salt,' 'butter of tin,' 'stannous chloride,' &c. This should be freshly prepared, in nearly dry crystals ; and it should dissolve entirely in water, without making the water appear milky. The more milky the appearance, the longer has the salt been exposed to the atmosphere.

Caustic lime.—Called also 'lime,' and 'stone-lime.' The best quality is perfectly white, and after having been slaked, may be rubbed to a soft creamy mixture with water; gritty particles consist of silica. Lime should be kept in well-closed jars of stoneware ; if the damp gets in, the lime is apt to swell and burst the vessels. Avoid strong building limes. These, especially the hydraulic cements, always contain clay or iron in considerable quantity.

Carbonate of sodium.—Commonly called 'soda,' and 'washing soda.' This is usually sold in the form of clear colourless crystals, which lose their water, and their transparency, by exposure to dry air, and fall to a white powder. Two hundred and eighty-six parts by weight of the clear crystals, require fifty-six parts of pure anhydrous caustic lime, to convert them wholly into caustic soda.

Caustic potash.—Called also 'potash,' and *lapis infernalis.* This substance is sold in several forms, of different degrees of purity. It should be kept as much as possible from contact with the air, because it rapidly absorbs moisture and carbonic acid. A solution of it may be made, by converting fifty-six parts by weight of pure and dry caustic lime into a cream, by slaking it with water, and then stirring it with more water ; adding the creamy mixture to 138 parts of anhydrous pearlash dissolved in hot water, and boiling the mixture ; the lime subsides to the bottom in the form of a carbonate. A purified variety of caustic potash, is sold in the form of rods about six inches in length. Great care must be taken not to handle it, as it is very caustic, and makes most dangerous sores.

Carbonate of potassium.—Called also 'pearlash,' 'salts of tartar,' &c. It is a white salt. Strongly alkaline and deliquescent, and should be kept in well-closed bottles or jars.

Gaseous ammonia.—This substance may be easily prepared, by separately powdering, and then intimately mixing, equal weights of dry caustic lime and sal ammoniac, and heating the mixture in a glass flask.

Aqueous ammonia.—Called also 'volatile alkali,' 'spirit of hartshorn,' &c. This liquid is very volatile, and should be kept in well-stoppered bottles, in a very cool place. Its specific gravity should not be greater than ·880. It is dangerous to break the bottles.

Carbonate of ammonium.—Called also 'smelling· salts,' 'sal volatile.' The unchanged substance is in the form of *transparent* colourless pieces. By exposure to air it loses ammonia, and becomes opaque white. It should therefore be kept in well-closed bottles.

Hydrocyanic acid.—Called also 'prussic acid.' This is a colourless liquid, consisting of water, more or less impregnated with the gas. Water will dissolve a very large amount of the gas. The strongest usually sold, is known as 'Scheele's,' and contains about 5 per cent. of the actual substance ; the ordinary medicinal acid contains only 2 per cent. It is extremely poisonous, and dangerous to smell or inhale the vapour arising from it. It is decomposed by light, and should therefore be kept in an opaque bottle, in a dark and cool place.

Cyanide of potassium.—Called also 'prussiate of potash.' This substance also is a deadly poison, and almost as dangerous when absorbed by the skin, as when swallowed. It is strongly alkaline, and abstracts moisture rapidly, and should therefore be kept in well-covered jars or bottles.

Making cyanide of potassium.—As cyanide of potassium is extensively used in electro-gilding and electro-deposition generally, and especially in making electro silvering baths, it is desirable for the practical depositor to understand how it is made, and to possess information respecting ·its impurities, and the method of testing its quality. It is nearly always made by the following process :—Take ferrocyanide of potassium (yellow prussiate of potash), well crystallised, and free from sulphates ; reduce it to a fine powder, and gently heat it to 110° or 120° C. in an iron pan, with constant stirring, until quite dry. Heat to redness a nearly covered iron crucible provided with a lip, put some of the dry powder into it, and when that is melted add some more, and so on, until the crucible is three-fourths filled, keeping the crucible covered as much as possible by means of an iron lid ; gas will be evolved freely from the melting salt. Keep the salt melted about fifteen minutes, or until the end of an iron rod dipped into it shows a white sample. By allowing it to stand undisturbed a few minutes at the latter part of the operation and occasionally tapping the sides of the crucible, the iron, &c. which has separated from the ferrocyanide, will settle at the bottom as a fine black powder ; the colourless cyanide of potassium may then be poured off into a cold iron pan, or upon a thick and cold iron plate ; it should be broken up whilst still hot, and preserved in a well-stopped jar. The black sediment (which contains much cyanide of potassium) should be scraped out of the vessel while still soft, and preserved, as water will at any time dissolve the cyanide that is in it.

If the process has been well conducted, the product will be of a clear white colour, or at most but very slightly grey. The colour, however, is not a matter of importance. To prevent oxidation of the cyanide, and consequent formation of cyanate of potassium, some operators recommend the addition of a few fragments of charcoal, and a little powder of the same to the salt, before it is entirely melted. The white portion of the product, made according to these instructions, contains about 96 per cent. of actual cyanide, and the cyanide dissolvable from the black portion, by means of cold water, is nearly as pure. To obtain a cyanide of about 70 or 75 per cent., eight parts of the dried ferrocyanide, mixed with three of highly-dried carbonate of potassium, must be subjected to similar treatment ; it, however, requires a less high temperature for the fusion. By this plan, a larger total amount of cyanide of potassium will be obtained, than by the fusion of the ferrocyanide only, because in melting the latter alone, one third of the cyanogen escapes as gas ; but in fusing it with the carbonate, this portion of the cyanogen unites with the potassium, and carbonic acid gas escapes in its stead. Cyanide of potassium, from which the ferruginous matter has not been completely freed, is known as 'black cyanide.' Fifty-five parts of crystallised prussiate, become forty-eight by drying; and nineteen of the carbonate, become eighteen ; and the sixty-six parts of the dry mixture yield about thirty-eight of clean cyanide, besides about six parts contained in the black sediment.

By experiments with the commercial white cyanide, I have found, that 200 grains of it would dissolve in 230 grains of distilled water at 60° Fahr., and that it was more soluble in water containing hydrocyanic acid. The plan of purifying cyanide of potassium from foreign salts, by means of solution in alcohol, does not appear to effect the object perfectly. Dr. Schwarz recommends the purification of it from carbonate and cyanate of potassium, by digesting it in bisulphide of carbon, and recovering the solvent by distillation (see 'Chemical News,' vol. viii. p. 51); but this appears to be an unlikely process.

Testing cyanide of potassium.—According to Glassford and Napier, the quantity of pure cyanide in any given sample of cyanide of potassium, may be correctly ascertained thus— Make two solutions, one of the cyanide, and one of nitrate of silver, each containing known weights of the salts, say one ounce of the cyanide dissolved in distilled water in a graduated glass vessel, so as to form six ounces by measure of solution ; and 175 grains of the crystallised nitrate, dissolved in about two or three ounces of distilled water ; add the cyanide solution carefully and slowly to the nitrate of silver liquid, with continual stirring, until the precipitate first formed is exactly all re-dissolved. The amount of the solution required to effect this, with the above quantity of nitrate of silver, will have contained 130 grains of pure cyanide, and from the quantity used, we may easily calculate the amount of pure cyanide in the whole ounce. It is said by the authors, that 'when nitrate of silver is added to a solution of cyanide of potassium, so long as the precipitate formed is all re-dissolved, we obtain the *whole* of the cyanide of potassium in combination with the silver : none of the other salts in solution take any part in the action, even though they be present in a large proportion. This enables us to test the exact quantity of cyanide of potassium in any sample.'

I have employed this process on many occasions, and have found from 28 to 96 per cent. of actual cyanide in different samples. In what is termed ' black cyanide ' I have found from 17·65 to 23·40 per cent. of black insoluble matter, and of soluble salts, not cyanide, from 5·21 to 5·43 per cent., and in a grey specimen 1·35 per cent. of black solid matter, and 18·75 per cent. of soluble salts not cyanide. This black substance burned in a flame like iron filings, evolved an inflammable gas by addition of dilute sulphuric acid ; and after digestion in dilute hydrochloric acid, much black com- bustible powder was left : it doubtless consists of iron and carbon. The other impurities consist of carbonate, sulphide, chloride, cyanate, ferrocyanide of potassium, and silica. The chloride of potassium is derived from the original salts, and the sulphide from sulphate of potassium contained in them ; the silica occurs when the cyanide is made in an earthen crucible ; and even when the process is well conducted, and pure materials used, the product sometimes contains 20 per cent. of cyanate of potash, produced partly by the contact of the air with the melted mixture. The presence of even a small quantity of sulphates in the materials, is said to impart to the cyanate, a blue, green, or pink colour; probably in consequence of the production of an alkaline sulphide. The price of cyanide varies from about 2s. 6d. to 5s. per pound.

Ferrocyanide of potassium.—Called also ' yellow prussiate of potash.' This salt is in the form of large clear yellow crystals, and is used for making the simple cyanide of potas- sium.

Acetate of copper.—Called also ' crystallised verdigris.' It is in the form of dark green crystals, soluble in water. Common verdigris is in lumps or powder of a bluish colour, and contains a larger proportion of copper, but is insoluble in water : it dissolves in diluted acetic acid, and then forms the same liquid as the solution of crystallised verdigris.

Acetate of lead.—Known also as ' sugar of lead.' It is a colourless crystalline salt, with an appearance like that of loaf-sugar. It should be entirely soluble in distilled water ; if it is not so, add a small quantity of acetic acid (wood-vinegar).

Test-papers.—The most useful variety of these, is neutral-tint litmus ; the red and blue kinds may also be employed.

Thermometers and Hydrometers.—The operator will also require a couple of thermometers, and several hydrometers ; the latter should be suitable for testing the specific gravity, both of aqueous ammonia, and of strong sulphuric acid.

Syphons.—The most convenient are pieces of tubing of glass, gutta-percha, or lead, bent to the proper forms ; or a piece of india-rubber tubing. To cause them to act, they should be filled with the liquid to be decanted, the ends closed by the fingers, and then inverted, with the shortest leg plunged into the liquid.

Filters.—Small ones for filtering dilute acids or alkalies, and liquids generally, are made by doubling a circular sheet of filtering-paper (i.e. unsized or blotting-paper) twice at right angles, opening one of the outer folds, and placing the filter in a glass funnel. Large ones are usually formed by tying or nailing the edges of a piece of washed or unglazed calico to those of a square frame of wood, or of a wooden hoop. A filter for strong acids or alkalies, is made by placing a loose plug of asbestos in the neck of a glass funnel, or by filling the neck of the vessel with broken glass, and covering the latter with a layer of asbestos.

As various poisonous substances are employed in the art, it would be well for the operator to know their best antidotes.   If either nitric, hydrochloric, or sulphuric acid have been swallowed, the best remedies are, either to administer abundance of tepid water to act as an emetic, or to cause the patient to swallow milk, the whites of eggs, some calcined magnesia, or a mixture of chalk and water.   If those acids in a concentrated state, have been spilled upon the skin, the parts should be washed with plenty of cold water ; and, if necessary, a mixture of whiting and olive-oil then applied. A useful mixture for such cases is formed by slaking about an ounce of caustic lime with a quarter of an ounce of water, then adding it to a quart of water and shaking the mixture repeatedly; decanting the clear liquid, and beating it up with olive-oil to form a thin pomatum.   Acids spilled upon the clothes, should at once be treated with plenty of a quite dilute solution of ammonia or its carbonate, and then well washed with water.

In cases where hydrocyanic acid, cyanide of potassium, or the ordinary silvering or gilding solutions have been swallowed, almost instant death usually follows ; if it does not, *very cold* water should be allowed to run upon the head and spine of the sufferer, and the patient be made to swallow a dilute solution of either acetate, citrate, or tartrate of iron. If the poisoning arises from inhaling the vapour of hydrocyanic acid, cold water should be applied as above, and the patient be caused to inhale atmospheric air containing a *little* chlorine gas.   It is a dangerous practice to dip the naked hands or arms into cyanide solutions, (as workmen sometimes do, in order to recover articles which have fallen into them,) because those liquids are absorbed by the skin, and produce poisonous effects; they also cause very painful sores, which should be well washed with water, and the mixture of lime-water and olive-oil applied.

If alkalies, such as potash or soda, have been swallowed, a dilute solution of vinegar, some lemonade, or extremely dilute sulphuric acid, should be given; and, after about ten minutes, a few spoonfuls of olive oil.

If metallic salts have been taken, the patient should be made to vomit by means of tepid water, and then to swallow some milk, whites of eggs, precipitated sulphur, or some sulphuretted hydrogen water.

To remove stains of sulphate of copper, or of salts of mercury, silver, or gold, from the hands, etc., wash them first with a dilute solution, either of ammonia, iodide, bromide, or cyanide of potassium, and then with plenty of water; if the stains are old ones, they should first be rubbed with the strongest acetic acid, and then treated as above.

Grease, oil, pitch, or tar, may usually be removed from the hands, clothes, etc., by rubbing with a rag saturated with benzine, spirits of turpentine, or bisulphide of carbon.

APPENDIX.

LIST OF BOOKS ON ELECTRO-DEPOSITION.

1841. *Galvanoplastik Art*, by Dr. M. H. Jacobi, translated by W. Sturgeon.

1843. *Elements of Electro-Metallurgy*, 2nd Edition. A. Smee, F.R.S.

1844. *Manual of Electro-Metallurgy*, 2nd Edition. G. Shaw, F.G.S.

1853. *Electro-type Manipulation*, Part I. 16th Edition, Part II. 29th Edition. C. V. Walker.

1854. *Galvanoplastie. Encyclopédie Roret.* 2 vols. Paris.

1855. *Theory and Practice of Electro-Deposition.* G. Gore.

1856. *Repertorium der Galvanoplastik und Galvanostegie.* A. Martin. 2 vols. Vienna : Carl Gerold's Sohn.

1862. *Les Dépôts Métalliques.* Henri Bouilhet. Paris : Bonaventura et Ducessois.

1862. *De l'Orfèvrerie Electro-chimique.* V. Meunier. Paris : Savy.

1862. *Die Galvanische Vergoldung und Versilberung.* W. E. Rab. Leipzig : Abel.

1863. *History of Electro-Metallurgy.* H. Dircks.

1866. *Eléments d'Electro-chimie.* M. Becquerel. Paris.

1867. *Die Galvanoplastik.* G. L. v. Kress. Frankfort am Maine : Boselli.

1868. *Katechismus der Galvanoplastik.* T. Martius-Matzdorff. Leipzig.

1870. *Die Galvanoplastik.* A Hering. Leipzig : Waldow. .

1870. *Electro-chimie.* N. A. Renard. Nancy : Lordoillet et Fils.

1870. *Nouveau Manuel Complet de Dorure.* Paris : Matthey et Maigne.

1870. *The Electrotypers' Manual.* Buffalo : W. S. Spiers.

1872. *Galvano-plastic Manipulation.* A. Roseleur.

1873. *Kunst des Vergoldens.* C H. Schmidt. Weimar : Voigt.

1873. *Hydroplastie, Electro-chimie, Galvanoplastie.* A. de Plazanet. Paris : Lacroix.

1874. *Electro-Metallurgy*, 5th Edition. A. Watt.

B B

1876. *Manual of Electro-Metallurgy*, 5th Edition. J. Napier, F.R.S.E.
1876. *Handbuch der Galvanoplastik.* Von G. Kaselowsky. Stuttgard : Riegerische Verlagsbuchhandlung.
1878. *Journal of Electrotypy,* published in Chicago.
1879. *Katechismus der Galvanoplastik.* Von G. Seelhorst. Leipzig : J. J. Weber.
1880. *The Electro-Metallurgist,* a periodical. London : Brock & Co.
1880. *Stereo-typing and Electro-typing.* London : F. T. F. Wilson.
1880. *Electro-plating.* J. W. Urquhart.
1881. *Electro-typing.* J. W. Urquhart.
1881. *La Galvanoplastie, Electro-chimique sur Métaux.* G. Marius. Orleans : Puget et Cⁱᵉ.
1881. *Manuel de Galvanoplastie.* Madrid ; L. Monet.
1881. *Das Galvaniseren von Metallen.* W. Pfanhauser. Vienna : Lehmann und Wentzel.
1881. *Das Verzinnen, Verzinken, Vernickeln.* F. Hartmann. Vienna : Hartleben.
1881. *Manuel de Galvanoplastie.* Ferrini. Milano : Hoepli.
1882. *Electro-Metallurgy.* By C. Alker. Brussels : C. Marquardt.
1883. *Die Electrolyse, Galvanoplastik und Reinmetallgewinnung.* Von Edward Japing. Vienna : Hartleben.
1883. *Die Galvanoplastik.* Von Julius Weiss. Vienna : Hartleben.
1883. *Katechismus der Electrotechnik.* Von Th. Schwarze. Leipzig : J. J. Weber.
1883. *Die Elektrolyse.* Von Dr. Hans Jahn. Vienna : A. Hölder.

In addition to the above there are :—*Art of Electro-typing,* by Sturgeon ; *Instructions for the Multiplication of Works of Art by Voltaic Electricity,* by Spencer ; *Manuel Complet de Galvanoplastie,* by M. L. de Valicourt, 2 vols. ; *Traité de Galvano-plastie,* by J. L. ; *Manuel de Dorure et d'Argenture par la Méthode Electro-chimique et par Simple Immersion,* by MM. Selmi and Valicourt. Chapters on electro-deposition, in *Gmelin's Handbook of Chemistry,* vol. 1. *Birmingham and Midland Hardware District* (Hardwicke), 1865, pp. 477, 510. Sprague's *Electricity* (Spon and Co.), 1875, p. 267. *British Manufacturing Industries* (Stanford), 1876, p. 137. *Applications of the Physical Forces,* by A. Guillemin, translated by Mrs. Lockyer, 1877, p. 701. Also a very large number of original articles on different parts of the subject, scattered through the pages of various scientific periodicals, to which references have already been made in the body of this book.

LIST OF PATENTS RELATING TO ELECTRO-METALLURGY.

1836. June 24. G. R. Elkington. Gilding copper, brass, and other metals.

1837. February 17. H. Elkington. Coating metals with gold and platinum.

„ December 4. H. Elkington. Gilding and silvering certain metals.

1838. July 24. G. R. Elkington and O. W. Barratt. Coating copper and brass with zinc.

1840. March 3. J. Shore. Coating metals with copper and nickel.

„ March 25. G. R. and H. Elkington. Electro-silvering and gilding in cyanide solutions.

„ August 15. No. 8604. V. A. Fontainemoreau. Coating metals and alloys with silver, gold, platinum, &c.

„ October 7. T. Spencer and J. Wilson. Voltaic etching.

„ December 17. W. T. Mabley. Producing printing surfaces.

1841. January 14. A. Jones. Making copper vessels. Rendering surfaces conductible.

„ February 8. No. 8842. W. H. F. Talbot. Electrotype and photography.

„ March 8. T. Spencer. Making picture-frames. Depositing gold, silver, platinum, and tin.

„ March 29. A. Parkes. Production of works of art.

„ September 8. O. W. Barratt. Deposition of copper (from mineral waters), silver, gold, platinum, palladium, and zinc.

„ December 9. W. H. F. Talbot. Gilding, silvering, ornamenting, &c. Use of alkaline hyposulphites.

1842. January 15. E. Palmer. Producing printing and embossing surfaces (glyphography).

„ June 1. H. B. Leeson. Electro-depositing processes. 'Gelatine moulds;' 'positive wires;' 'guiding wires;' keeping articles in motion; cleaning articles, 'quicking' their surfaces. Claims 430 different salts.

„ June 4. E. Tuck. Deposition of silver.

„ August 1. J. S. Woolrich. Plating by means of magneto-electricity. Use of alkaline sulphites.

„ W. H. F. Talbot. Electro-gilding and silvering.

1843. April 11. J. Napier. Depositing copper upon fibrous materials.

,, May 4. No. 9720. E. Morewood and G. Rogers. Depositing tin upon iron and other metals.

,, May 25. M. Poole. Plating by means of thermo-electricity. Gold, silver, and copper solutions.

,, June 15. O. W. Barratt. Depositing gold, silver, platinum, palladium, lead, &c.

,, November 21. No. 9,957. A. F. J. Claudet. Producing printing surfaces from daguerreotypes.

,, December 8. J. Schottlaender. Electro-depositing upon felted fabrics.

1844. February 21. No. 10,063. A. Parkes. Deposition of metals and alloys.

,, July 31. No. 10,282. P. A. Fontainemoreau. Electro-brassing.

,, October 22. J. Napier. Depositing metals from fused minerals.

,, October 29. A. Parkes. Depositing gold and silver from their melted salts.

1845. October 9. A. Parkes. Embellishing metals.

1846. January 29. No. 11,065. G. Howell. Coating metals with platinum.

,, December 12. L. H. Piaget and P. H. Du Bois. Depositing gold, silver, and copper.

1847. March 23. M. Lyons and W. Milward. Bright silver deposition.

,, August 3. T. Fletcher. Depositing silver and copper upon the backs of glass mirrors.

,, September 9. J. C. Robertson. Separating sulphur, phosphorus, &c. from melted minerals.

,, September 30. C. De la Salzede. Deposition of brass and bronze upon iron, &c.

,, November 4. C. M. T. Du Motay. Inlaying metals.

1848. January 13. S. E. Morse. Production of printing surfaces.

1849. March 14. P. A. Fontainemoreau. Deposition of platinum, gold, silver, copper, brass, tin, and lead.

,, March 19. T. H. Russell and J. S. Woolrich. Deposition of cadmium and of alloys.

,, March 26. A. Parkes. Depositing printing-rollers, copper, silver, bismuth, tin, and lead.

1850. March 23. A. G. Roseleur. Deposition of tin.

,, August 9. J. Steele. Gilding, silvering, bronzing and brassing

1851. February 17. C. Cowper. Elastic moulds.

1851. May 3. No. 13,620. W. Cooke. Making soda and its carbonate.

,, August 23. No. 13,726. J. Palmer. Gelatine moulds for electrotype.

,, September 25. C. Watt. Depositing alkali metals. Separating and purifying metals.

1852. April 20. J. Ridgway. Coating glass and china.

,, August 26. A. Crosse. Separating copper from its ores.

,, October 1. W. Potts. Making sepulchral monuments.

,, October 2. J. J. Rousseau. Making door-plates.

,, October 12. F. Michel. Stereotyping in copper.

,, October 21. J. Bernard. Depositing printing surfaces for ornamenting leather.

,, November 13. W. Petrie. Refining metals.

,, November 29. J. D. Schneiter. Producing maps.

,, November 30. W. Jeffs. Making letters, figures, &c.

,, December 11. T. Morris and W. Johnson. Depositing brass and other alloys.

,, December 11. C. Griffin. Obtaining copper from mineral waters.

,, December 28. C. J. Junot. Depositing silicium, titanium, tungsten, chromium, and molybdenum.

,, December 29. J. Power. Silvering glass, &c.

1853. January 1. J. J. W. Watson and W. Prosser. Depositing carbon into iron to form steel.

,, May 11. W. Bradbury and F. M. Evans. Preparing printing surfaces.

,, July 29. W. E. Newton. Depositing metals, bronze, brass, and an alloy of manganese and zinc.

,, August 5. No. 1,836. W. Newton. Coating iron with metals and alloys (brass).

,, October 7. W. Ellis. Ornamenting china and porcelain.

,, October 8. W. Potts. Ornamenting mantelpieces.

,, November 7. H. Pershouse and T. Morris. Depositing metals and alloys.

1854. January 11. A. R. Brooman. Extracting gold from its ores.

,, January 19. G. H. Burrill. Extracting metals from minerals, slag, and jewellers' refuse.

,, January 30. W. Phillips. Making coffins.

,, February 1. R. and J. Jobson. Making moulds for casting metals.

,, February 28. T. Denny. Improvements in engraving.

1854. March 20. J. Perkins. Making printers' type.
,, April 13. G. Devincenzi. Making printing surfaces.
,, April 20. J. Reed. Extracting metal from amalgams.
,, April 27. C. C. Person. Coating with zinc.
,, July 4. J. H. Johnson. Coating iron with lead or copper.
,, July 15. M. F. Wagstaffe and J. W. Perkins. Extracting metals from their ores.
,, July 18. P. A. Fontainemoreau. Etching zinc plates for printing.
,, July 29. A. E. L. Bellford. Electro-engraving.
,, November 4. P. Pretsch. Making copper plates for printing.
,, December 21. J. H. Johnson. Making statuettes.
,, December 26. F. S. Thomas and W. E. Tilley. Coating metals with tin, nickel, or aluminium.
1855. January 3. J. H. Johnson. Coating iron with copper.
,, February 3. F. S. Thomas and W. E. Tilley. Deposition of silver, copper, nickel, and tin.
,, February 13. R. Cornfield. Coating iron with zinc.
,, March 17. T. Petitjean and L. Pétre. Making daguerreotype plates, &c.
,, April 2. G. W. Friend. Improvements in umbrellas.
,, April 11. L. and A. Oudry. Preserving wood, metal, and other substances.
,, June 5. F. Puls. Coating iron with zinc.
,, July 10. C. J. C. Elkington. Depositing alloys of nickel and silver, &c.
,, July 21. P. A. Fontainemoreau. Depositing copper upon carbon.
,, September 4. J. G. Taylor. Deposition of aluminium.
,, October 4. No. 2,215. H. Cornforth. Electro-coating hooks and eyes.
,, October 12. F. Puls. Deposition of zinc upon iron.
,, October 25. J. A. Richards. Making embossing surfaces for ornamenting leather.
,, November 14. A. V. Newton. Making surfaces for printing.
,, December 3. A. Watt. Coating iron and steel with zinc.
,, December 6. F. S. Thomas and W. E. Tilley. Depositing aluminium and its alloys.
1856. January 1. J. Calvert. Extracting metals from minerals.
,, January 12. C. Oudry. Preserving metals and other solids.
,, February 14. E. Morewood and G. Rogers. Deposition of zinc.

1856. March 10. L. Chablin and A. Hennique. Ornamenting china, &c.

,, March 25. G. and H. Cottam. Ornamenting chairs and bed-steads.

,, April 3. J. H. Glassford. Preparing surfaces for printing.

,, April 15. No. 899. E. R. Southby. Coating iron with copper.

,, June 4. R. A. Brooman. Electro-plating upon glass.

,, October 25. G. Ernst and W. Lorberg. Electro-etching.

,, December 3. No. 2,871. J. K. Cheatham. Electro-deposition with photography.

,, December 17. C. Cowper. Deposition of silver and copper upon base metals.

1857. January 1. E. T. Noualhier and J. B. Prévost. Coating glass, corpses, &c. with gold, silver, copper, platinum, or iron.

,, January 19. J. H. Johnson. Improvements in galvano-plastic processes.

,, January 22. J. Rubery. Electro-brassing the ribs of umbrellas.

,, January 27. No. 240. G. T. Bousfield. Coating metals with tin.

,, March 11. J. D. Cooper. Making printing surfaces.

,, March 13. E. J. N. Juvin. Making surfaces for printing.

,, March 31. S. Goode. Depositing alloys.

,, April 25. J. Burrow. Coating wrought iron with copper lead, tin, or zinc.

,, April 27. C. Cowper. Depositing gold and silver.

,, May 7. D. Morrison. Making printing rollers.

,, June 1. W. H. Walenn. Depositing gold, silver, copper, bronze, and brass.

,, June 24. No. 1,766. A. Parkes. Coating metals with metals.

,, July 1. W. E. Newton. Producing printing surfaces.

,, July 30. S. Coulson. Deposition of aluminium.

,, July 30. W. McKinley and R. Walker. Making moulds of soles of boots and shoes.

,, September 21. G. Schaub. Making printing cylinders.

,, October 16. J. Chadwick. Making printing rollers.

,, December 19. T. Newey, J. Corbett, and W. II. Parkes. Tinning steel-pens.

1858. January 19. No. 93. Otto von Corvin. Inlaying and ornamenting metals.

,, February 19. No. 317. J. M. Syers. Extracting metals from their ores.

,, February 22. No. 341. G. Schaub. Making printing types.

1858.   February 23. No. 353. E. C. Shepherd. Depositing metals
            and alloys.

,,   March 12. No. 507. L. F. Corbelli. Deposition of alu-
            minium.

,,   March 22. No. 594. G. Davies. Metallisation of objects
            for electrotype.

,,   March 29. No. 667. E. A. Jacquin. Coating printing sur-
            faces with iron.

,,   April 10. No. 785. A. C. Thibault. Making moulds for
            printing paper-hangin ɼs.

,,   April 16. No. 831. J. H. Johnson. Making printing surfaces.

,,   June 3. No. 1,255. Baron Justus Liebig. Protecting backs
            ؍of mirrors.

,,   June 8. No. 1,289. R. A. Brooman. Manufacture of copper
            pipes.

,,   June 22. No. 1,406. G. Schaub. Making door-plates, sign-
            boards, letters, &c.

,,   September 27. No. 2,161. W. Lander. Engraving and
            printing.

,,   October 23. No. 2,371. J. C. Martin. Manufacture of metal
            moulds, &c.

,,   October 28. No. 2,409. W. Munro. Making capsules, &c.

,,   December 16. No. 2,890. R. A. Brooman. Plating and
            gilding forks and spoons.

1859.   January 12. No. f03. C. Beslay. Depositing tin, zinc, or lead.

,,   February 5. No. 333. R. Tinkler. Improvements in churns.

,,   February 17. No. 444. B. Saillard. Making printing
            plates.

,,   April 26. No. 1,044. W. Mackenzie. Making printing sur-
            faces.

,,   April 30. No. 1,083. J. Toussaint. Moulds and moulding
            for deposition.

,,   August 29. No. 1,964. G. Edwards. Coating buttons.

,,   September 14. No. 2,095. C. Beslay. Making printing sur-
            faces.

,,   December 6. No. 2,764. F. Potts. Making tubes.

1860.   January 25. No. 187. T. Rampacher and C. F. Schmidt.
            Coating wire gauze.

,,   January 26. No. 204. W. E. Newton. Depositing crystal
            gold.

,,   February 21. No. 469. L. Sautter. Coating mica with
            metal for reflectors.

1860. March 10. No. 653. T. Morris. Improvements in vats for depositing.

,, March 22. No. 748. G. T. Peppe. Coating lead with tin.

,, April 9. No. 893. L. Eidlitz. Producing printing surfaces.

,, May 16. No. 1,209. C. M. Guillemin. Coating telegraph cables with copper.

,, June 6. No. 1,385. E. T. Hughes. Coating type and stereotype.

,, June 22. No. 1,523. N. Grattan. Gilding steel and other metals.

,, July 25. No. 1,800. M. A. F. Mennons. Etching surfaces for printing.

1861. January 7. No. 44. W. Bagley and W. Mincher. Coating metals.

,, January 19. No. 145. B. Piffard. Preparing non-conducting surfaces for deposition.

,, March 13. No. 619. J. Cimeg. Silvering glass, &c.

,, May 13. No. 1,214. T. Bell. Coating metals with aluminium.

,, May 17. No. 1,259. S. Tearne. Producing designs on metal articles.

,, June 8. No. 1,469. W. Clark. Rendering casks, &c., watertight.

,, July 16. No. 1,792. C. D. Abel. Depositing nickel, and making alloys.

,, August 3. No. 1,936. J. Lewis. Making surfaces for printing.

,, August 14. No. 2,023. R. A. Brooman. Coating wire with gold, silver, copper, &c.

,, August 15. No. 2,040. J. Faucherre. Making gold dials.

,, No. 2,314. J. Cimeg. Depositing silver and other metals on textile fabrics.

,, October 9. No. 2,521. H. B. Coathupe and F. H. Waltham. Making embossed surfaces.

,, No. 2,675. A. Dalrymple. Depositing metals.

,, October 28. No. 2,699. W. Clark. Producing printing surfaces.

,, November 5. No. 2,784. G. T. Rousfield. Depositing metals from concentrated solutions of cyanides.

1861. November 23. No. 2,944. J. Weens. Making metal tubes, and coating metals.

,, December 7. No. 3,074. T. Fearn and T. Cox. Coating the metal parts of umbrellas, &c.

,, December 9. No. 3,081. M. A. F. Mennons. Producing designs for printing and embossing, &c.

,, December 17. J. B. Bunney and T. Wright. Ornamenting bedsteads and other articles.

1862. February 22. No. 469. H. Chavasse, T. Morris, and G. B. Haines. Ornamenting bedsteads and other articles.

,, April 19. No. 149. A. Parkes. Coating surface condensers with silver.

,, May 20. No. 1,528. W. Petrie. Improvements in vessels for boiling acids, &c.

,, May 21. No. 1,538. W. E. Newton. Making metallised fabrics or surfaces.

,, June 28. No. 1,896. C. Beslay. Coating metals.

,, July 17. No. 2,044. J. Dickson. Making soda.

,, July 24. No. 2,101. J. Dickson. Extracting copper from ores and solutions.

,, August 12. No. 2,253. J. Dickson. Extracting zinc from ores and solutions.

,, August 12. No. 2,254. J. Dickson. Extracting lead from ores and solutions.

,, August 13. No. 2,265. J. Dickson. Making chlorine.

,, August 13. No. 2,266. J. Dickson. Deposition of sodium, with electrodes of carbon.

,, August 18. No. 2,314. J. Cimeg. Depositing silver and other metals.

,, August 30. No. 2,410. J. H. Johnson (from C. F. L. Oudry). Coating surfaces with copper.

,, September 9. No. 2,479. J. Maurice. Coating the bottoms of ships with copper.

,, October 3. No. 2,675. A. Dalrymple. Depositing metals.

,, November 4. No. 2,988. A. Wall. Purifying lead.

1863. January 20. No. 171. H. A. Bonneville. Ornamenting electro-deposited articles.

,, January 21. No. 180. F. A. Busch. Making vessels for containing liquids.

,, February 25. No. 529. W. E. Newton. Making stereotype plates.

,, March 26. No. 795. G. Davies. Engraving metals.

1863. April 21. No. 986. H. Rafter. Obtaining printing surfaces.

,, April 27. No. 1,048. J. J. Robert. Coating spoons and forks with silver.

,, June 24. No. 1,595. T. Skinner. Ornamenting plated articles.

,, August 22. No. 2,088. S. Moore. Improved apparatus for electro-plating.

1864. February 29. No. 497. F. Weil. Coating metals in alkaline solutions.

,, August 15. No. 2,029. S. Moore. Electro-gilding in cyanide solutions.

,, December 14. No. 3,095. J. B. Thompson. Coating iron with palladium, platinum, gold, and silver.

,, December 21. No. 3,164. H. A. de Brion. Varnish for electro-plated articles.

1865. March 10. No. 677. T. Reissig. Electrolysis with photography.

,, May 27. No. 1,457. R. A. Brooman. Producing copies of writings, &c.

,, June 5. No. 1,541. W. E. Newton. ' Photo-electro-typing process.'

,, July 6. No. 1,791. J. W. Swan. Producing printing surfaces.

,, August 15. No. 2,110. M. Henry. Electro-type with photography.

,, October 2. No. 2,521. T. Allan. Preparing iron for electroplating.

,, October 7. No. 2,592. J. B. Thompson. Depositing iron, and coating iron with platinum, gold, silver, and copper.

,, October 26. No. 2,762. H. Wilde. Apparatus for electrocoating.

,, November 16. No. 2,948. De la Haye. Gilding copper wires of telegraphs.

,, December 23. No. 3,323. E. Clifton. Electro-bronzing.

,, December 26. No. 3,339. W. F., Deane. Coating the bottoms of ships with copper.

1866. February 14. No. 469. M. Henry. Electro-deposition with photography.

,, April 27. No. 1,186. M. Nelson. Making moulds for electrotype plates.

,, April 27. No. 1,195. J. B. Thompson. Protecting iron ships from corrosion.

1866. May 8. No. 1,315. W. B. Woodbury. Producing designs upon wood and other substances.

 ,, July 25. No. 1,934. C. E. Brooman. Coating armour-plates with copper.

 ,, August 15. No. 2,095. J. Webster. Coating metals, and recovering metals from solutions.

 ,, September 28. No. 2,513. W. Clark. Re-producing telegraphic signs and characters.

 ,, November 20. No. 3,047. C. E. Brooman. Coating iron and steel, with copper and its alloys.

 ,, November 26. No. 3,113. R. H. Courtenay. Preparing printing surfaces.

 ,, December 11. No. 3,517. A. M. Clark. Reduction of tin.

1867. No. 810. G. Bischof. Coating metals.

 ,, April 1. No. 968. C. E. Brooman. Producing surfaces in relief.

1868. May 29. No. 1,777. G. T. Bousfield. Plating spoons, &c.

 ,, August 14. No. 2,545. J. B. Thompson. Preparing surfaces for gilding, &c.

 ,, October 10. No. 3,117. W. R. Lake. Deposition of nickel.

 ,, October 15. No. 3,155. H. A. Bonneville. Elastic moulds.

 ,, December 15. No. 3,801. A. Watt. Making printing rollers.

 ,, December 24. No. 3,930. W. H. Walenn. Depositing copper and brass.

1869. May 12. No. 1,458. P. W. Flower and H. Nash. Coating sheets of metal.

 ,, July 26. No. 2,268. W. E. Tilley. Coating metals with tin.

 ,, August 17. No. 2,456. M. H. Jacobi. Depositing iron. Forming engraved surfaces, &c.

 ,, October 6. No. 2,961. B. Hunt. Ornamenting metals in relief.

 ,, October 28. No. 3,125. W. Brookes. Deposition of nickel.

 ,, October 30. No. 3,159. A. Minton. Coating iron and other metals.

 ,, November 23. No. 3,377. H. A. Bonneville. Apparatus for keeping electrolytes in motion.

 ,, December 16. No. 3,643. A. Buirat. Producing engraved plates.

1870. February 2. No. 303. I. Adams, jun. Deposition of nickel.

 ,, February 24. No. 554. J. B. Elkington and C. E. Ryder. Making copper tubes and rollers.

1870. April 12. No. 1,068. I. Adams, jun. Preparing surfaces for receiving nickel coating.

,, August 27. No. 2,359. W. R. Lake. Coating tin-tacks with copper.

,, September 28. No. 2,580. J. E. Bingham. Deposition of tin.

,, November 16. No. 3,005. G. Haseltine. Coating iron with gold and silver.

,, December 30. No. 3,396. E. D. Nagel. Coating iron and steel with nickel and cobalt.

1871. June 8. No. 1,511. H. Wilde. Coating iron boilers with copper.

,, June 21. No. 1,626. J. Unwin. Deposition of nickel by magneto-electricity. Nickel solution.

,, July 7. No. 1,777. J. Brough and G. Fletcher. Coating vacuum pans.

,, August 29. No. 2,266. T. Fearn. Depositing alloys of nickel and iron. Solutions for ditto.

,, September 16. No. 2,450. W. H. Maitland. Deposition of copper.

,, October 4. No. 2,623. De Lobstein. Electro-plating.

,, December 21. No. 3,459. J. Unwin. Coating with nickel by immersion. Solution for ditto.

1872. May 6. No. 1,376. Fitzgerald and Molloy. Decomposing substances with electrodes of carbon.

,, June 10. No. 1,742. C. A. Faure. Manufacture of alkalies by electrolysis.

,, November 1. J. A Jeancon. Deposition of aluminium. (N.B. American patent.)

,, December 5. No. 3,680. T. Petitjean. Making and ornamenting articles. Coating glass, &c.

,, December 18. No. 3,839 J. Noad. Making moulds of sulphide of lead, &c., for electrotype.

,, December 31. No. 3,970. J. H. Johnson. Coating iron with copper and its alloys.

,, No. 95,593. Mr. Unwin. Deposition of nickel. (N.B. French patent.)

1873. February 10. No. 474. R. Werdermann. Reducing metals from their ores.

,, No. 476. R. Werdermann. Reducing metals from their ores

,, April 7. T. Fearn. Deposition of tin.

,, April 29. J. T. Sprague. Galvanometer for use in electrolysis.

,, July 2. W. R. Lake. Coating iron with nickel.

1873. December 27. No. 148,459. W. C. Holman. Apparatus to show weight of metal deposited. (N.B. American patent.)

1874. No. 1,492. Brook, Draper, and Unwin. Preparing articles for coatings of nickel and other metals.

,, No. 1,493. W. Baker and J. Unwin. Deposition of nickel.

,, No. 3,033. J. B. Thompson. Coating iron with gold, silver, and alloys.

,, No. 3,148. W. Morgan Brown. Preparing china and glass for being coated.

,, No. 3,432. T. S. Johnson. Producing electro-type plates.

,, April 19. E. Casselbury. Electrolytic apparatus. (N.B. American patent.)

1875. No. 58. Wollaston. Thermo-electric apparatus for coating metals.

,, No. 175. Vera. Decomposing water.

,, No. 473. Clark. Obtaining metals from their salts.

,, No. 519. Terrell. Electro-typing iron plates.

,, No. 714. Brown. Producing copper plates and printing surfaces.

,, No. 1,746. Bartlett and Murray. Facing type with nickel.

,, No. 2,996. Kilner. Magneto-apparatus for electro-coating.

,, No. 3,243. Alexander. Electro-typing.

,, No. 3,440. Jewitt. Making gas-burners by deposition.

,, No. 3,904. Mori. Thermo-regulators for electro-gilding.

,, No. 4,302. Blewitt. Electro-deposition of tin.

,, No. 4,326. Ellerbeck and Syers. Making hydrogen and oxygen.

,, No. 4,515. H. Wilde. Refining copper, and coppering calico-printers' rollers.

1876. No. 1,445. Werdermann. Converting metallic salts.

,, No. 1,704. Fixsen. Compound for galvano-plastic uses.

,, No. 2,500. Lake. Making wax moulds.

,, No. 2,554. Prior. Electro-plating with nickel.

,, No. 2,821. Zanni. Magneto-machine for plating.

,, No. 2,938. Lake. Galvanic battery for plating.

,, No. 3,181. Pitt. Making copper tubes, wire, &c.

,, No. 3,515. Gardner. Reducing and purifying metals.

,, No. 3,569. H. Wilde. Electro-coppering rollers.

,, No. 3,670. Faure. Thermo-battery for coating metals.

,, No. 4,280. Haddan. Magneto-machine for plating.

,, No. 4,302. R. J. Blewitt. Coating iron with tin.

,, No. 4,515. H. Wilde. Making metal rollers. Refining copper.

1877. No. 329. Drummond. Producing printing surfaces.
,, No. 828. Dodd. Electro-plating iron, copper, and nickel.
,, No. 853. Parkes. Separating nickel from copper in alloys.
,, No. 1,023. Hughes. Electro-plating coils of wire.
,, No. 1,259. Wiley. Nickel-plating.
,, No. 1,548. Unwin. Electro-solution of nickel.
,, No. 1,572. Dupuis and Schultz. Gilding non-metallic frames.
,, No. 2,996. Kagenbusch and Kerr. Extracting metals from slags.
,, No. 3,476. Lake. Electro-tinning iron plates.
,, No. 3,743. Johnson. Magneto-machine for electro-metallurgy.
,, No. 4,053. Lake. Magneto-machine for electro-plating.
,, No. 4,708. Lake. Magneto-machine for electro-typing.
,, No. 4,748. Conradi. Electro metallurgy.
1878. No. 288. Johnson. Nickel-plating iron wire.
,, No. 380. Van Winkle. Nickel-plating iron wire.
,, No. 1,054. Parry. Solution for electro-tinning.
,, No. 1,228. Wilde. Generating electric currents for use in deposition.
,, No. 1,979. Michaud. Electro-plating with copper.
,, No. 2,003. Haddan. Machine for electro-plating.
,, No. 2,017. Keith. Refining lead and separating gold and silver.
,, No. 2,407. Lake. Making combs by electro-deposition.
,, No. 3,392. Maxwell Lyte. Coating iron with copper and nickel.
,, No. 3,425. Maxwell Lyte. Coating iron with copper and nickel.
,, No. 3,606. Alexander. Etching plates.
,, No. 3,976. Ward. Magneto-machine for plating.
,, No. 4,074. Arnaud. Dividing electric currents in plating.
,, No. 4,075. Johnson. Voltaic battery for electro-plating.
,, No. 4,206. Higgs. Magneto-machine for depositing metals.
,, No. 4,313. Cochrane. Generating electricity for electro-plating.
,, No. 4,573. Zanni. Machine for regulating electric currents.
,, No. 4,611. Edwards and Normandy. Generating electricity for electro-plating.
,, No. 4,755. Cobley. Precipitating copper.
,, No. 4,921. Lake. Solution for depositing nickel.
,, No. 5,127. Glaser. Plating with nickel and cobalt.
,, No. 5,250. Scott. Thermo-pile for depositing metals.
1879. No. 307. Elphinstone and Vincent. Dynamo-machine for plating.
,, No. 359. Brittain. Compound deposit for electro-metallurgy.
,, No. 529. Blake. Electro-depositing white metal.

1879. No. 696. Desmurs. Producing designs on metals.

„ No. 1,203. Clowes and Batey. Machine for black-leading moulds.

„ No. 1,387. Lake. Dynamo-machines for plating.

„ No. 1,481. Muller and Geisenberger. Making saltpetre.

„ No. 1,592. Muller and Geisenberger. Obtaining ammonia.

„ No. 1,692. Sellon and Edmunds. Regulating currents from dynamo-machines for plating.

„ No. 2,821. Zanni. Closing and opening circuits during deposition.

„ No. 3,565. Elmore. Dynamo-machine for plating, &c.

„ No. 3,586. Lambotte-Doucet. Obtaining metals from ores.

„ No. 4,087. Johnson. Obtaining aluminium and magnesium.

„ No. 4,100. Lake. Dynamo and batteries for plating.

„ No. 4,295. Desmurs. Depositing metals for ornament.

„ No. 4,821. Elmore. Nickel alloys for electro-coating.

„ No. 4,862. Pitt. Coating insides of provision cans with nickel.

„ No. 4,879. Gutensohn. Separating tin from waste tinned iron.

„ No. 5,030. Morgan. Producing alkalies and salts.

„ No. 5,085. Wise. Dynamo-machine for plating.

„ No. 5,175. Joel. Magneto-machine for plating.

1880. No. 458. Lake. Extracting metals from ores.

„ No. 830. Von Buch. Depositing crystalline carbon.

„ No. 1,120. Parry and Cobley. Coating iron and zinc.

„ No. 1,178. Perry. Dynamo-machine for plating.

„ No. 1,392. Lake. Dynamo-machine for plating, with regulator.

„ No. 1,556. Wirth. Electro-plating wood carvings.

„ No. 1,570. Fischer. Electro-typing.

„ No. 1,700. Young. Producing ammonia.

„ No. 1,705. Davies. Solution for depositing aluminium.

„ No. 1,909. Sachs. Etching rollers for printing.

„ No. 2,020. Abel. Separating substances by electrolysis.

„ No. 2,465. Wetter. Benzoic acid used in nickel-plating solution.

„ No. 2,519. Barlow. Electro-typing inlaid metal articles.

„ No. 2,631. Hodge. Decomposing substances by electrolysis.

„ No. 2,966. Sachs. Producing designs on printing rollers.

„ No. 3,043. Glaser. Electro-plating with copper, nickel, and their alloys.

„ No. 4,005. Brewer. Dynamo-machine for electro-deposition.

„ No. 4,094. Elmore. Extracting copper and zinc from liquors.

1880. No. 4,541. Barlow. Producing hydrogen in alcohol by electrolysis.
,, No. 4,985. Morgan. Making alkalies by electrolysis.
1882. No. 6. Chaster. Nickel-plating.
,, No. 358. Moss. Electro-typing.
,, No. 1,639. Walenn. Deposition of copper, brass, and bronze.
,, No. 1,884. Lake. Separating metals from their ores.
,, No. 2,875. Gülcher. Producing hydrogen and oxygen by gas batteries.
,, No. 3,046. Barker. Extracting gold and silver from their ores.
,, No. 4,580. Lake. Decomposing alloys by electrolysis.
,, No. 5,300. Boult. Electro-plating with nickel and cobalt.
,, No. 5,719. Appleton and Horsfield. Nickel-plating engraved rollers.
1883. No. 88. Appleton. Electro-plating metal printing rollers.
,, No. 543. Appleton. Electro-plating metal printing plates.
,, No. 2,281. Clark. Depolarising electrolytes.
,, No. 2,577. Hammersley. Electro-gilding vulcanite.

Table of useful Numerical Data.

1 centimetre =	·3937 inches.
1 decimetre =	3·937 ,,
1 metre =	39·37 ,,
1 gramme =	15·432 grains.
1 kilogramme =	15432· ,,
1 ,, =	35·274 ounces avoirdupois.
1 ,, =	2·2046 pounds ,,
1 ounce avoirdupois =	437·5 grains.
1 pound ,, =	7000· ,,
1 pennyweight troy =	24· ,,
1 ounce troy =	480· ,,
1 pound ,, =	5760· ,,
1 litre of water =	15432· ,,
1 ,, ,, =	1000· grammes.
1 ,, ,, =	35·275 ounces by measure.
1 gallon of water =	4·536 litres.
1 ,, ,, =	70000· grains.
1 cubic inch of water =	252·5 ,,
1 ounce measure =	1·733 cubic inches.
1 pint (or 20 ounces) =	34·659 ,, ,,
1 gallon (or 160 ounces) =	277·276 ,, ,,
1 litre =	61·024 ,, ,,

At the ordinary temperature and pressure of the atmosphere, 100 cubic inches of—

		Grains				Grains
Hydrogen .	. weigh	2·11	Oxygen .	. . weigh	33·80
Ammonia	. . ,,	18·00	Carbonic anhydride.	,,	46·50
Hydrocyanic acid			Sulphurous anhydride	,,	67·78
vapour	. . ,,	28·57	Chlorine.	. . ,,	76·40
Nitrogen	. . ,,	29·70	Sulphuretted hydro-		
Atmospheric air	. ,,	31·00	gen .	. . ,,	80·50

Table of Corresponding Temperatures on the Scales of Centigrade and Fahrenheit Thermometers.

Deg. Cent.	Deg. Fahr.	Deg. Cent.	Deg. Fahr.	Deg. Cent.	Deg. Fahr.
100	212	66	150·8	32	89·6
99	210·2	65	149	31	87·8
98	208·4	64	147·2	30	86
97	206·6	63	145·4	29	84·2
96	204·8	62	143·6	28	82·4
95	203	61	141·8	27	80·6
94	201·2	60	140	26	78·8
93	199·4	59	138·2	25	77
92	197·6	58	136·4	24	75·2
91	195·8	57	134·6	23	73·4
90	194	56	132·8	22	71·6
89	192·2	55	131	21	69·8
88	190·4	54	129·2	20	68
87	188·6	53	127·4	19	66·2
86	186·8	52	125·6	18	64·4
85	185	51	123·8	17	62·6
84	183·2	50	122	16	60 8
83	181·4	49	120·2	15	59
82	179·6	48	118·4	14	57·2
81	177·8	47	116·6	13	55·4
80	176	46	114·8	12	53·6
79	174·2	45	113	11	51·8
78	172·4	44	111·2	10	50
77	170·6	43	109·4	9	48·2
76	168·8	42	107·6	8	46·4
75	167	41	105·8	7	44·6
74	165·2	40	104	6	42·8
73	163·4	39	102·2	5	41
72	161·6	38	100·4	4	39·2
71	159·8	37	98·6	3	37·4
70	158	36	96·8	2	35·6
69	156·2	35	95	1	33·8
68	154·4	34	93·2	0	32
67	152·6	33	91·4		

NOMENCLATURE OF ELECTRICAL UNITS.

Since the publication of this book, a revision has been made of the Nomenclature of Electrical Units.

The Unit of Resistance is still termed an Ohm, and is equal to that offered by 1·0486 mètre length of mercury of one square millimètre section at 0° C. The amount of resistance in a wire, A, is conveniently measured by dividing a current from a very small Daniells' cell, so that one portion passes through A and one wire, B, of a differential galvano-meter, and the other portion through another wire of known resistance, C, and the other wire, D, of the galvanometer in the opposite direction to that through B, and then the length of A (or of C) altered until the needles of the instrument stay at zero; the resistances of A and C are then equal. The measurement of the degree of resistance of an electrolyte is much more difficult, on account of the varying polarisation of the electrodes. It may be effected in a similar manner, by making two measurements by means of a very feeble current after the polarisa-tion has become steady : one when the electrodes are near together, and the other when they are far asunder, using in each case electrodes as large as the transverse section of the electrolyte, and usually of the same metal as that of the salt of the liquid. The difference of resistance in the two measurements is the amount of resistance of the difference of length of the liquid in the two cases.

The Unit of Electro-motive Force still retains its name of a Volt. That of a Daniells' cell is equal to 1·070 volt ; and that of a Clarks' standard cell is equal to 1·457 volt. For measuring feeble electro-motive forces, I have devised a convenient form of thermo-electric pile, consisting of about 300 pairs of iron and German-silver wires, and have employed it in making a large number of measurements not much exceeding that of one Daniell. It is capable of measuring differences of $\frac{1}{38000}$ of a volt (see 'Proceedings of the Birmingham Philosophical Society,' vol. iv. Part 1).

The Unit of Strength of current is now termed an Ampère, and is the strength produced by an electro-motive force of one volt in a circuit, having a resistance of one ohm ; it was formerly termed ' one Weber per second.' It is that current which will deposit in a solution of argento-potassic cyanide, containing the least practical amount of free cyanide of potassium, ·017343 grain of silver per second ; or will liberate from dilute sulphuric acid with platinum electrodes, ·000162 grain of hydrogen per second; or will deposit ·0051035 grain of copper per second from the usual sulphate solution.

The Unit of Quantity of current is now termed a Coulomb (formerly a Weber). It is but little used; and is the amount which one Ampère gives in one second.

An Unit of Density of current would be of great value in electrolysis, but one has not yet been adopted. I have suggested that of one Ampère leaving or entering one square centimètre of surface of electrode per second (see 'Proceedings of the Birmingham Philosophical Society,' vol. iii. p. 277).

INDEX.

LONDON: PRINTED BY
SPOTTISWOODE AND CO., NEW-STREET SQUARE
AND PARLIAMENT STREET